Bio-Inspired Soft Robotics: Design, Fabrication and Applications

Bio-Inspired Soft Robotics: Design, Fabrication and Applications

Guest Editors

Yong Zhong
Pei Jiang
Sun Yi

MDPI

Basel • Beijing • Wuhan • Barcelona • Belgrade • Novi Sad • Cluj • Manchester

Guest Editors

Yong Zhong
Shien-Ming Wu School of Intelligent Engineering
South China University of Technology
Guangzhou
China

Pei Jiang
State Key Laboratory of Mechanical Transmission
University of Chongqing
Chongqing
China

Sun Yi
Graduate School of Information Science and Technology
University of Tokyo
Tokyo
Japan

Editorial Office
MDPI AG
Grosspeteranlage 5
4052 Basel, Switzerland

This is a reprint of the Special Issue, published open access by the journal *Biomimetics* (ISSN 2313-7673), freely accessible at: https://www.mdpi.com/journal/biomimetics/special_issues/NHHSIT717D.

For citation purposes, cite each article independently as indicated on the article page online and as indicated below:

Lastname, A.A.; Lastname, B.B. Article Title. *Journal Name* **Year**, *Volume Number*, Page Range.

ISBN 978-3-7258-4725-9 (Hbk)
ISBN 978-3-7258-4726-6 (PDF)
https://doi.org/10.3390/books978-3-7258-4726-6

Contents

About the Editors

Yong Zhong

Yong Zhong obtained his PhD from The Chinese University of Hong Kong. Subsequently, he served as a Research Fellow for nearly two years in the Department of Biomedical Engineering at the National University of Singapore. He is currently an Associate Professor and Assistant to the Dean at the Shien-Ming Wu School of Intelligent Engineering, South China University of Technology. Dr. Zhong has long been engaged in research on bionic robotics, soft robotics, Intelligent diagnosis and control. He has published over 100 papers in journals such as *IEEE TRO, TII, TEMCH, TIE, Soft Robotics, Additive Manufacturing, Advanced Functional Materials,* etc. He has presided over or participated in more than 20 national, provincial/ministerial-level, and industrial projects.

Pei Jiang

Pei Jiang received his BS and PhD in control science and engineering from Zhejiang University, Hangzhou, China, in 2008 and 2015, respectively. He is currently an Associate Professor with the College of Mechanical and Vehicle Engineering, Chongqing University, Chongqing, China. His research interests include soft robotics, manufacturing, and mechatronic systems. He has also participated in more than ten competitive research projects at the national level, including the National Natural Science Foundation of China's Key Research and Development Program project. Throughout his research career, he has published over twenty articles as a first author or corresponding author in JCR-indexed journals, such as *IEEE T-RO, TIE,* and *T-MECH.* He is also the co-author of more than ten papers that have been presented at international conferences.

Sun Yi

Sun Yi holds a PhD in mechanical engineering (National University of Singapore (NUS)) and is a co-founder of the Swiss startup company –Dexterous Endoscopes. In his academic career, his research has been closely related to soft robotics. Over the years, he has studied soft pneumatic actuators, soft morphing technologies, variable stiffness mechanisms, soft medical robots, soft underwater robots, etc. He has contributed more than 40 papers to high-impact journals and top conferences. Currently, he is focused on transferring variable stiffness technology to medical endoscopes, with multiple patents filed. He is closely working with doctors to develop a new endoscope that can facilititate the procedure with ease and speed.

Preface

We welcome you to read the collection of articles in *Bio-Inspired Soft Robotics: Design, Fabrication and Applications* a Reprint that captures the essence of innovation and progress in the field of bio-inspired soft robotics. As we delve into the intricacies of nature's designs and translate them into robotic systems, we are continually amazed by the potential these technologies hold for transforming various aspects of our lives.

Soft robotics, with its focus on flexibility, adaptability, and safe interaction with humans and the environment, has emerged as a dynamic and promising field within robotics. The inspiration drawn from the natural world, where organisms exhibit remarkable capabilities to adapt and thrive in diverse conditions, has been instrumental in shaping the development of soft robotic systems. This Reprint brings together a selection of ten articles that highlight the latest advancements, challenges, and opportunities in the design, fabrication, and application of bio-inspired soft robots. The articles in this collection showcase a wide range of innovative approaches and applications. From the development of a pneumatic soft glove for hand rehabilitation to the creation of an amphibious robot inspired by cuttlefish, these contributions demonstrate the versatility and potential of bio-inspired soft robotics. Each article provides valuable insights into the design principles, fabrication techniques, and real-world applications that are pushing the boundaries of what soft robots can achieve. As we look to the future, the field of bio-inspired soft robotics holds immense promise. The continued exploration of nature-inspired principles, combined with advancements in materials science, fabrication techniques, and control systems, will undoubtedly lead to even more remarkable innovations. This Reprint aims to provide a snapshot of the current state of the field, while also inspiring researchers, engineers, and scientists to push the boundaries of what is possible.

We would like to express our sincere gratitude to all the authors who have contributed their valuable research to this Reprint. Their dedication and innovative work have enriched the field of bio-inspired soft robotics and laid the foundation for future advancements. We also extend our thanks to the reviewers who provided their expertise and constructive feedback, ensuring the high quality of the articles published in this Special Issue.

This Reprint is a testament to the progress and potential of bio-inspired soft robotics. We hope that it will serve as a valuable resource for researchers, practitioners, and students interested in this exciting field, inspiring further exploration and innovation. As we continue to draw inspiration from the natural world, we are confident that bio-inspired soft robotics will play a crucial role in shaping a more adaptable, resilient, and interconnected future.

Thank you for joining us on this journey of discovery and innovation.

Yong Zhong, Pei Jiang, and Sun Yi

Guest Editors

biomimetics

MDPI

Editorial

Bio-Inspired Soft Robotics: Design, Fabrication and Applications

Yong Zhong [1,*], Pei Jiang [2] and Yi Sun [3]

1 Shien-Ming Wu School of Intelligent Engineering, South China University of Technology, Guangzhou 510640, China
2 The State Key Laboratory of Mechanical Transmission, Chongqing University, Chongqing 400044, China; peijiang@cqu.edu.cn
3 Department of Mechano-Informatics, University of Tokyo, Tokyo 113-8656, Japan; y-sun@isi.imi.i.u-tokyo.ac.jp
* Correspondence: zhongyong@scut.edu.cn

The field of soft robotics has witnessed a remarkable surge in interest and innovation over the past decade, with bio-inspiration playing a pivotal role in driving forward the design, fabrication, and applications of these adaptable and resilient robotic systems. This book, titled "Bio-Inspired Soft Robotics: Design, Fabrication and Applications," compiles ten cutting-edge articles that showcase the latest advancements and diverse applications in this burgeoning field. These contributions highlight the multifaceted ways in which nature's ingenuity can be harnessed to create soft robots that are not only highly functional but also capable of operating in complex and dynamic environments.

One of the key themes explored in this book is the design of soft robots inspired by the unique locomotion strategies of various animals. For instance, Yuqiao Dai et al. [Contribution 1] present a novel approach for controlling flexible continuum robots, leveraging neural networks to model the relationship between motor inputs and shape outputs. This work underscores the potential of bio-inspired design principles in enhancing the precision and adaptability of soft robotic systems. Similarly, Emmanouil Papadakis et al. [Contribution 2] introduce a soft robotic arm that mimics the muscular structure and movement capabilities of an octopus arm. The authors' detailed design process and experimental evaluation demonstrate the arm's ability to perform complex tasks such as bending, elongation, and twisting, thereby expanding the scope of potential applications for soft robotic arms.

The fabrication techniques for bio-inspired soft robots are another focal point of this book. Xiaohui Wang et al. [Contribution 3] describe the development of a soft robotic glove for hand rehabilitation, featuring bidirectional bending actuators and a programmable pneumatic control platform. This work exemplifies how innovative fabrication methods can be employed to create soft robotic devices that are both functional and user-friendly. Additionally, Qian Yang et al. [Contribution 4] explore the design and testing of a high-frequency wire-driven robotic fish, highlighting the importance of optimizing the mechanical design and control strategies to achieve efficient and agile underwater locomotion.

The applications of bio-inspired soft robots span a wide range of fields, ranging from medical rehabilitation to environmental exploration. Yihan Wang et al. [Contribution 5] investigate the hydrodynamic performance of a quadrupedal paddling model, providing valuable insights into the design of amphibious robots that can navigate both land and water environments. This research not only advances our understanding of bio-inspired locomotion but also paves the way for the development of versatile robots capable of performing diverse tasks in challenging terrain. Furthermore, Deli Xia et al. [Contribution 6] present a 3D-printed soft robot that mimics the movement of an inchworm, showcasing the potential of combining advanced fabrication technologies with bio-inspired design to create robots that can adapt to complex environments and perform tasks such as obstacle climbing and surface transition. Huibin Liu et al. [Contribution 7] present a quadruped soft robot driven by magnetic fields. The robot is designed to mimic the movement patterns of quadrupeds and can perform crawling and tumbling motions. This multimodal motion capability enables the robot to navigate complex environments, cross obstacles, and deliver cargo in a targeted manner, demonstrating the potential of soft robots in biomedical applications such as targeted drug delivery.

In addition to these notable contributions, the book features several other articles that delve into various aspects of bio-inspired soft robotics. These include studies on the development of novel actuators, sensors, and control systems [Contribution 8], [Contribution 9], as well as explorations of the potential applications of soft robots in fields such as healthcare, environmental monitoring, and search and rescue operations [Contribution 10]. Collectively, these articles provide a comprehensive overview of the current state of the art in bio-inspired soft robotics and highlight the vast potential of this field for future research and development.

As we compile these ten articles into a book, it is evident that the field of bio-inspired soft robotics is at an exciting juncture, with researchers continually pushing the boundaries of what is possible. The innovative designs, fabrication techniques, and applications showcased in this book serve as a testament to the creativity and ingenuity of the scientific community working in this area. We hope that this collection of articles will inspire further research and development, fostering the growth of bio-inspired soft robotics and its applications in various domains. The potential for these robots to revolutionize industries, improve quality of life, and contribute to our understanding of the natural world is immense, and we look forward to witnessing the continued evolution and impact of this fascinating field.

We would like to express our sincere gratitude to all the authors who contributed their valuable research to this book, as well as to the reviewers who provided their expertise and constructive feedback. Their efforts have made this book a rich and informative resource for the soft robotics community. We also extend our appreciation to the editorial team at MDPI for their support and dedication in bringing this Special Issue to fruition.

Conflicts of Interest: The authors declare no conflicts of interest.

List of Contributions:

1. Dai, Y.; Zhang, S.; Cheng, W.; Li, P. Neural Network-Based Shape Analysis and Control of Continuum Objects. *Biomimetics* **2024**, *9*, 772. https://doi.org/10.3390/biomimetics9120772.
2. Papadakis, E.; Tsakiris, D.P.; Sfakiotakis, M. An Octopus-Inspired Soft Pneumatic Robotic Arm. *Biomimetics* **2024**, *9*, 773. https://doi.org/10.3390/biomimetics9120773.
3. Wang, X.; Cheng, Q.; Wang, Z.; Lu, Y.; Zhang, Z.; Zhao, X. A Pneumatic Soft Glove System Based on Bidirectional Bending Functionality for Rehabilitation. *Biomimetics* **2025**, *10*, 129. https://doi.org/10.3390/biomimetics10030129.
4. Yang, Q.; Wang, Q.; Cao, Z.; Zhao, Z.; Chen, Y.; Zhong, Y. Development of a Wire-Driven Robotic Fish Based on Double Sine Mechanism. *Biomimetics* **2025**, *10*, 136. https://doi.org/10.3390/biomimetics10030136.
5. Wang, Y.; Cai, Y.; Xie, B.; Zhu, C.; Li, Y.; Chen, Y. A Computational Study on the Hydrodynamics of Bio-Inspired Quadrupedal Paddling. *Biomimetics* **2025**, *10*, 148. https://doi.org/10.3390/biomimetics10030148.
6. Xia, D.; Zhang, L.; Nong, W.; Duan, Q.; Ding, J. 3D-Printed Soft Bionic Inchworm Robot Powered by Magnetic Force. *Biomimetics* **2025**, *10*, 202. https://doi.org/10.3390/biomimetics10040202.
7. Liu, H.; Teng, X.; Qiao, Z.; Yang, W.; Zou, B. Magnetically Driven Quadruped Soft Robot with Multimodal Motion for Targeted Drug Delivery. *Biomimetics* **2024**, *9*, 559. https://doi.org/10.3390/biomimetics9090559.

8. Duan, J.; Lei, Y.; Fang, J.; Qi, Q.; Zhan, Z.; Wu, Y. Learning from Octopuses: Cutting-Edge Developments and Future Directions. *Biomimetics* **2025**, *10*, 224. https://doi.org/10.3390/biomimetics10040224.
9. Mo, Y.; Su, W.; Hong, Z.; Li, Y.; Zhong, Y. Finite-Time Line-of-Sight Guidance-Based Path-Following Control for a Wire-Driven Robot Fish. *Biomimetics* **2024**, *9*, 556. https://doi.org/10.3390/biomimetics9090556.
10. Gao, Y.; Pancheri, F.; Lueth, T.C.; Sun, Y. Development of a Cuttlefish-Inspired Amphibious Robot with Wave-Motion Propulsion and Rigid–Flexible Coupling. *Biomimetics* **2025**, *10*, 396. https://doi.org/10.3390/biomimetics10060396.

biomimetics

MDPI

Article

Development of a Cuttlefish-Inspired Amphibious Robot with Wave-Motion Propulsion and Rigid–Flexible Coupling

Yichao Gao , Felix Pancheri, Tim C. Lueth and Yilun Sun *

Institute of Micro Technology and Medical Device Technology, Technical University of Munich, 85748 Garching, Germany; yichao.gao@tum.de (Y.G.); felix.pancheri@tum.de (F.P.); tim.lueth@tum.de (T.C.L.)
* Correspondence: yilun.sun@tum.de

Abstract: Amphibious robots require efficient locomotion strategies to enable smooth transitions between terrestrial and aquatic environments. Drawing inspiration from the undulatory movements of aquatic organisms such as cuttlefish and knifefish, this study introduces a bio-inspired propulsion system that emulates natural wave-based locomotion to improve adaptability and propulsion efficiency. A novel mechanism combining crank–rocker and sliding components is proposed to generate wave-like motions in robotic legs and fins, supporting both land crawling and aquatic paddling. By adopting a rigid–flexible coupling design, the system achieves a balance between structural integrity and motion flexibility. The effectiveness of the mechanism is systematically investigated through kinematic modeling, animation-based simulation, and experimental validation. The developed kinematic model captures the principles of wave propagation via the Crank–Slider–Rocker structure, offering insights into motion efficiency and thrust generation. Animation simulations are employed to visually validate the locomotion patterns and assess coordination across the mechanism. A functional prototype is fabricated and tested in both terrestrial and aquatic settings, demonstrating successful amphibious locomotion. The findings confirm the feasibility of the proposed design and underscore its potential in biomimetic robotics and amphibious exploration.

Keywords: amphibious robot; rigid–flexible coupling; wave-motion propulsion; biomimetic locomotion; cuttlefish-inspired robotics

1. Introduction

Amphibious robots have attracted significant attention due to their capability to operate effectively across diverse terrains, offering great potential for environmental monitoring, search and rescue, and underwater exploration. Natural species such as cuttlefish, salamanders, and mudskippers exhibit highly adaptive locomotion mechanisms that allow efficient movement in both aquatic and terrestrial environments. Inspired by these biological examples, considerable research has been dedicated to developing amphibious robots capable of seamless transitions between land and water [1–3].

As shown in Figure 1, current amphibious robots typically utilize one of four locomotion methods: (1) undulatory motion, (2) legged locomotion, (3) fin-based propulsion, and (4) hybrid wheel–leg systems [1,2,4–6]. Representative examples include snake-inspired undulatory robots employing lateral oscillation [7], salamander-inspired legged robots capable of switching between swimming and walking gaits [2,8], fin-based robots such as the AQUA series that utilize oscillatory fins [4], and hybrid wheel–leg robots like the Whegs series designed for versatility across various terrains [9–11]. Recent advances in soft and biohybrid propulsion have also enabled skeletal-skin fin models for

 https://doi.org/10.3390/biomimetics10060396

low-speed aquatic maneuvering, exemplified by biorobotic cormorant-inspired flippers with anisotropic compliant structures [12]. In parallel, emerging satellite-scale amphibious robots—such as light-driven soft microrobots that fuse inchworm-like terrestrial motion and water-strider-inspired aquatic gait—have demonstrated rapid multimodal transition and remote wireless control in milligram-scale platforms [13]. These miniaturized designs highlight promising directions in untethered amphibious actuation and terrain adaptation.

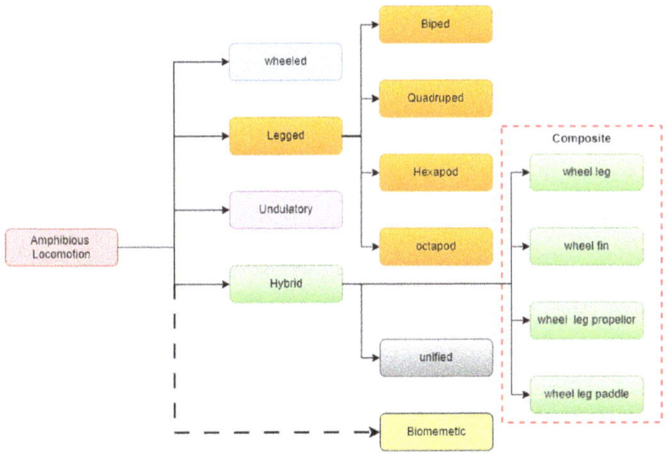

Figure 1. Classification of amphibious robots [3].

Although these designs exhibit promising multimodal capabilities, they often suffer from significant challenges. These include limited adaptability to irregular terrain, high actuator complexity, fragile mechanical designs, and poor energy efficiency during transitions between mediums [14,15]. As highlighted in recent reviews, especially in unstructured environments such as surf zones or rocky transitional areas, robots face obstacles like uneven surfaces, shoals, and slopes that demand not only robust propulsion but also mechanical compliance and real-time adaptability [16,17].

To further complicate the design, amphibious robots employing separate propulsion mechanisms for land and water modes often become bulky and complex to control [18]. Hybrid mechanisms with unified structures can simplify the system but often require intricate transformations or passive compliance to function effectively in both environments [19]. Achieving a balance between adaptability, mobility, and control simplicity remains an ongoing design challenge in the field.

To address the limitations of conventional amphibious locomotion strategies, this study focuses on the principle and realization of a novel bio-inspired wave propulsion mechanism. Drawing inspiration from the undulatory fin motion of cuttlefish (Figure 2), we introduce a Crank–Slider–Rocker transmission system coupled with a rigid–flexible structural design to generate continuous traveling wave-like motion across both aquatic fins and terrestrial legs (Figure 3). While this paper presents the complete development process—including mechanical design, embedded control implementation on an ESP32 platform, and the modulation of key wave parameters such as amplitude, frequency, and direction—the central emphasis is placed on uncovering the underlying locomotion mechanism and validating its effectiveness.

Figure 2. Oscillatory lateral fin movement of a cuttlefish.

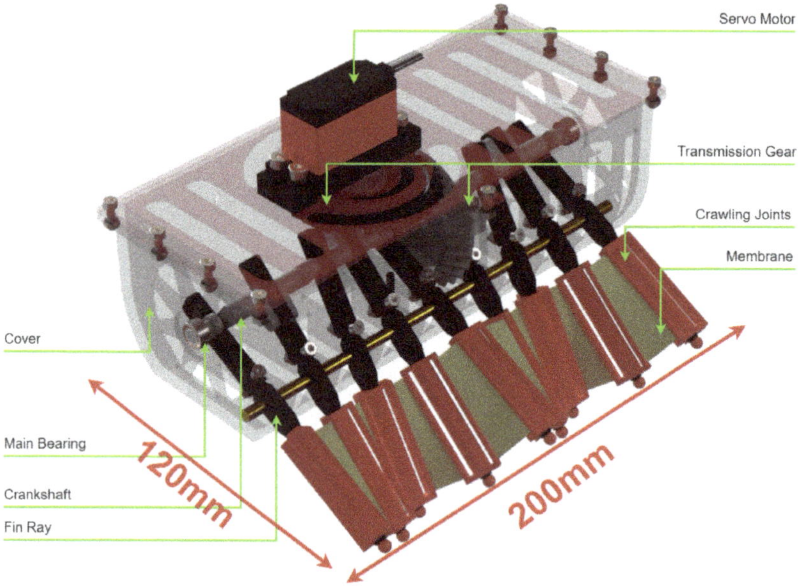

Figure 3. An overview of the proposed wave-like propulsion mechanism.

Unlike previous designs such as the crank–rocker-driven robot in [20], which requires switching between separate swimming and crawling modes, our robot employs a unified Crank–Slider–Rocker mechanism and a rigid–flexible fin structure that enables continuous undulatory motion across both land and water without transformation. This simplifies the mechanical design and improves locomotion adaptability.

Extensive dynamic simulations and real-world experiments were conducted in both water and land environments to evaluate the proposed locomotion approach. The results highlight the system's capacity for smooth and efficient movement using minimal actuation input. The rigid–flexible coupling plays a crucial role in terrain adaptability, while the unified propulsion strategy enhances control simplicity and energy performance. By systematically exploring the kinematics, implementation, and validation of this wave-based locomotion principle, this study contributes a comprehensive experimental framework that advances the development of highly adaptable amphibious robots.

By exploiting passive compliance, our design effectively mimics biomechanical principles found in nature, achieving smooth, efficient, and continuous locomotion across both aquatic and terrestrial environments.

This paper is structured as follows: Section 2 explains the locomotion principles and kinematic models for swimming and crawling. Section 3 details the materials and methods used in mechanical construction and electronic integration. Section 4 presents simulation and experimental results validating the proposed mechanism. Finally, Section 5 concludes the study and outlines directions for future work.

2. Locomotion Principle and Kinematic Modeling

In this section, we investigate the fundamental principles of undulatory crawling and swimming locomotion, followed by a kinematic and dynamic modeling approach to characterize the proposed system. Inspired by biological wave-like motion, we develop a Crank–Slider–Rocker mechanism that generates the required mechanical oscillations to achieve both terrestrial crawling and aquatic swimming. This mechanism leverages a rigid–flexible coupling structure, allowing the smooth propagation of wave motion while maintaining structural stability.

2.1. Swimming Locomotion

Undulatory swimming, characterized by traveling wave propagation along fins, is a fundamental aquatic locomotion strategy extensively observed in gymnotiform, rajiform, and amiiform swimmers [21,22]. Aquatic species adopting such Median or Paired Fin (MPF) propulsion modes typically exhibit superior maneuverability, hovering stability, and energy efficiency, particularly in low-speed and precision maneuvering scenarios [3,23–25].

Fish propulsion mechanisms are generally classified into two distinct modes:

- Body and Caudal Fin (BCF) Propulsion: Utilized by fast swimmers such as tunas and sharks, this mode involves lateral undulation of the body combined with rapid oscillation of the caudal fin, optimized for high-speed swimming but generally sacrificing maneuverability [23,24].
- Median and Paired Fin (MPF) Propulsion: Observed in species like the knifefish, stingrays, and cuttlefish, MPF propulsion involves wave propagation along fins positioned dorsally, ventrally, or laterally, enabling precise control, minimal hydrodynamic drag, and high efficiency at lower speeds [22,25–27].

The MPF propulsion mode is particularly advantageous for biomimetic robotics due to its inherent capabilities for multi-directional maneuverability and efficient thrust generation. Species such as the knifefish generate thrust by propagating undulations along their elongated anal fins, maintaining stationary body positions and reducing hydrodynamic drag [22,26]. Similarly, stingrays utilize broad pectoral fins to produce controlled undulatory waves, providing remarkable agility and efficiency [25,27].

Based on the previously discussed principles of undulatory propulsion, we designed a novel Crank–Slider–Rocker mechanism to transform circular motion into periodic oscillations of the fin rays, as illustrated in Figure 4. By introducing a phase shift between multiple fin rays, this mechanism successfully generates harmonic wave-like motion along the fin array. The angular motion of each fin ray can be described by the following mathematical model:

In our proposed Crank–Slider–Rocker mechanism, the circular motion of the crank is converted into oscillatory motion of the fin rays. The key geometric transformation occurs due to the linkages, which amplify the crank displacement into larger oscillations of the fin ray tip.

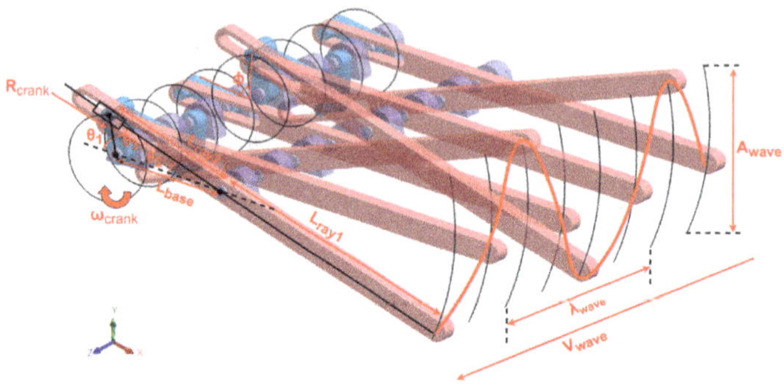

Figure 4. A kinematic model of the Crank–Slider–Rocker mechanism. Key parameters: R_{crank}: crank radius; L_{base}: fixed linkage; $\{L_{\text{ray}i}\}_{i=1,2}$: fin ray segments; $\{\theta_i\}_{i=1,2,3}$: joint angles.

For a crank rotating at angular velocity ω_{crank}, the instantaneous length L_{ray2} follows

$$L_{\text{ray2}}(\theta_1) = \sqrt{R_{\text{crank}}^2 + L_{\text{base}}^2 - 2R_{\text{crank}}L_{\text{base}}\cos\theta_1}, \quad \theta_1(t) = \omega_{\text{crank}}t \tag{1}$$

The oscillation angles θ_2 (fin ray) and θ_3 (closure angle) satisfy

$$\begin{aligned} \theta_2 &= \arcsin\left(\frac{R_{\text{crank}}\sin\theta_1}{L_{\text{ray2}}}\right) \\ \theta_3 &= \pi - (\theta_1 + \theta_2) \end{aligned} \tag{2}$$

The time-dependent ratio between fin ray segments is defined as

$$K(t) \triangleq \frac{L_{\text{ray1}}}{L_{\text{ray2}}(t)} = \frac{L_{\text{ray1}}}{\sqrt{R_{\text{crank}}^2 + L_{\text{base}}^2 - 2R_{\text{crank}}L_{\text{base}}\cos\theta_1}} \tag{3}$$

The maximum oscillation amplitude occurs when the closure angle satisfies $\theta_3 = \frac{\pi}{2} + n\pi$ $(n \in \mathbb{Z})$, leading to

$$A_{\text{wave}} = 2K_{\max}R_{\text{crank}}, \quad \text{where } K_{\max} = \left.\frac{L_{\text{ray1}}}{L_{\text{ray2}}}\right|_{\theta_3=\pi/2} \tag{4}$$

This condition corresponds to the slider reaching its extreme positions, maximizing mechanical advantage.

Sequential fin ray oscillations with phase shift $\Delta\phi_n$ between adjacent rays generate a traveling wave:

$$y(x,t) = A_{\text{wave}}\sin\left(\underbrace{\frac{2\pi}{\lambda_{\text{wave}}}}_{k}x - \underbrace{2\pi f_{\text{wave}}}_{\omega_{\text{wave}}}t + \phi_n\right) \tag{5}$$

The wave characteristics are constrained by

$$\begin{aligned} \lambda_{\text{wave}} &= \frac{2\pi d}{\Delta\phi_n} \quad \textit{(Structural)} \\ f_{\text{wave}} &= \frac{\omega_{\text{crank}}}{2\pi} \quad \textit{(Kinematic)} \\ V_{\text{wave}} &= \lambda_{\text{wave}}f_{\text{wave}} = \frac{d\omega_{\text{crank}}}{\Delta\phi_n} \end{aligned} \tag{6}$$

where d is the fixed spacing between fin rays.

The actual swimming velocity incorporates hydrodynamic efficiency η:

$$V_{\text{sw}} = \eta V_{\text{wave}} = \eta \frac{d\omega_{\text{crank}}}{\Delta\phi_n} \tag{7}$$

The mechanism's performance is governed by three adjustable parameters:

- Wave Amplitude: Controlled via the crank radius R_{crank} through Equation (4).
- Wave Frequency: Determined by ω_{crank} (Equation (6)).
- Wave Direction: Reversed by setting $\Delta\phi_n < 0$.

In this study, η is treated as an empirical coefficient, obtained via the linear regression of experimental velocity measurements under varied crank speeds. It incorporates unmodeled hydrodynamic drag, fin–body interaction effects, and viscous losses. While this linear approximation is valid in the tested low-to-moderate Reynolds number regime, we acknowledge that η may become nonlinear and dependent on ω_{crank} or fin geometry under higher flow speeds or altered configurations.

Based on the above analyses, a comprehensive theoretical framework is established, elucidating the inherent relationships from rotational motion to fin-ray undulations and thrust generation. Specifically, the kinematic foundations (Equations (1) and (3)) reveal how the Crank–Slider–Rocker mechanism geometrically converts rotary motion into fin-ray oscillations. The wave amplitude generation model (Equation (4)) explicitly relates the crank radius to the maximum oscillation amplitude. Wave propagation dynamics (Equations (5) and (6)) characterize the relationships among wavelength, frequency, and structural parameters. Lastly, the thrust conversion model (Equation (7)) quantitatively accounts for hydrodynamic drag and energy losses.

Overall, this theoretical framework provides a robust foundation for optimizing amphibious robot locomotion, enabling the independent tuning of wave amplitude, propagation frequency, and phase differences to effectively balance swimming velocity and maneuverability.

2.2. Crawling Locomotion

Undulatory crawling is a widely adopted locomotion strategy observed in nature, notably among serpentine and limbless organisms such as snakes, worms, and caterpillars, which propagate traveling waves along their bodies to achieve forward thrust [28–30]. Inspired by these biological organisms, robotics research has successfully adapted similar wave-based propulsion methods for terrestrial locomotion. A prominent example is the Single Actuator Wave-like (SAW) robot developed by Zarrouk et al. [28], which utilizes a single rotating actuator to create continuous wave propagation across its rigid body segments. Despite its efficiency, the SAW robot encounters mechanical constraints and reduced adaptability on uneven surfaces due to rigid linkages.

Rigidly linked wave robots often exhibit challenges such as limited adaptability, increased mechanical complexity, and structural fragility on irregular terrain conditions [29,30]. To overcome these limitations, this study introduces a novel Crank–Slider–Rocker mechanism featuring rigid–flexible coupling. Unlike traditional rigid designs, our approach employs elastic membranes to connect rigid segments, significantly enhancing the robot's ability to conform adaptively to terrain irregularities and smoothly propagate waves throughout its structure.

Figure 5 illustrates the Crank–Slider–Rocker mechanism and wave propagation dynamics. The flexible membranes (characterized by flexibility parameters and stiffness) enable segments to rotate passively in response to ground contact forces. At each time step, a sinusoidal wave profile characterized by amplitude, wavelength, and link spacing propagates along the robot's body from head to tail. Adjacent segments oscillate with alternating phases, forming a coordinated traveling wave.

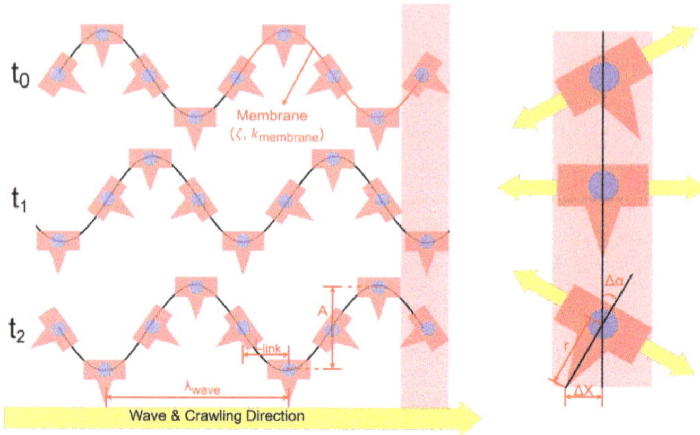

Figure 5. Kinematic model and wave propagation in the crawling mechanism. Key parameters: A: wave amplitude; λ_{wave}: wavelength; L_{link}: link length; ζ: flexibility factor; k_{membrane}: membrane stiffness; $\Delta\alpha$: joint rotation angle; r: tip radius; ΔX: horizontal displacement.

The traveling wave motion can be mathematically described by

$$y(x,t) = A\sin\left(\frac{2\pi}{\lambda_{\text{wave}}}x - 2\pi f t\right) \tag{8}$$

where the amplitude directly corresponds to the vertical displacement magnitude, and the wavelength denotes the distance between wave peaks. The wave propagation velocity is given by the product of wavelength and frequency:

$$V_{\text{wave}} = \lambda_{\text{wave}} \cdot f \tag{9}$$

The actual forward crawling speed depends on how efficiently this wave motion translates into linear displacement. Following the advance ratio (AR) model in [28], we have:

$$V_{\text{robot}} = \text{AR} \cdot V_{\text{wave}} = \text{AR} \cdot \lambda_{\text{wave}} f \tag{10}$$

The AR is influenced by the curvature radius at the segment tip and wave amplitude, expressed as

$$\text{AR} = \frac{2\pi r A}{\lambda_{\text{wave}}^2} \tag{11}$$

As further detailed in Figure 5, each segment's rotational displacement is passively driven by the elastic deformation of the membrane, characterized by membrane stiffness:

$$\Delta\alpha \approx \frac{k_{\text{membrane}} A}{\lambda_{\text{wave}}} \tag{12}$$

This rotational displacement, combined with the tip curvature radius, translates into horizontal displacement:

$$\Delta X \approx r\Delta\alpha \tag{13}$$

In summary, the proposed crawling mechanism integrates mechanical wave generation, flexible joint deformation, and thrust conversion into a unified theoretical framework. The wave dynamics (Equations (8) and (9)) govern the propagation of traveling waves along the robot body, defining the amplitude, frequency, and wavelength parameters critical for effective locomotion. Meanwhile, the efficiency of converting these waves into forward displacement is quantified by the advance ratio (AR) model (Equations (10) and (11)), which incorporates the compliance characteristics of the elastic membranes. Additionally, the kinematic relationships (Equations (12) and (13)) highlight how passive elastic deformation directly contributes to rotational and translational motion.

3. Materials and Methods

To comprehensively evaluate the effectiveness of the proposed wave-motion propulsion mechanism, we conducted both dynamic simulations and physical experiments under swimming and crawling conditions. The validation process was divided into two stages: first, a simulation was performed based on the developed kinematic model to verify the feasibility and coherence of wave propagation across the robot's limbs; second, a series of controlled experiments were carried out using a physical prototype to assess real-world locomotion performance.

In the simulation phase, time-resolved wave propagation and leg displacement trajectories were analyzed to ensure that the Crank–Slider–Rocker mechanism with rigid–flexible coupling could generate stable, continuous traveling waves. Subsequently, physical experiments were conducted to quantify the robot's locomotion performance, including its straight-line displacement speed and in-place rotational capability. These experiments relied on a monocular camera system for visual tracking, an onboard IMU for angular motion measurement, and photointerrupters for event-based timing.

The combined simulation and experimental results confirm the feasibility, controllability, and versatility of the proposed propulsion mechanism in both aquatic and terrestrial environments.

3.1. Simulation of Wave-like Propulsion

To verify the effectiveness of the kinematic model and assess the feasibility of the proposed Crank–Slider–Rocker mechanism, a dynamic simulation was carried out to visualize the time-dependent wave propagation along the robot's limbs. The model parameters were based on the mechanical design and equations detailed in Section 2, incorporating flexible link deformation, phase shift, and crank-driven oscillations.

The simulation focused on both qualitative wave propagation and quantitative leg displacement analysis over a complete motion cycle. Figure 6 illustrates the spatiotemporal evolution of the leg trajectories from $t = 0\,\text{s}$ to $t = 2.0\,\text{s}$, showing a continuous traveling wave propagating from left to right. The red dotted line indicates the vertical position of each limb tip at successive time intervals, and clearly demonstrates the desired phase-shifted oscillatory pattern.

Figure 7 presents the vertical displacement of a representative leg tip over time, extracted from the simulation output. The waveform confirms harmonic motion consistent with the sinusoidal input defined in the model:

$$y(t) = A\sin(2\pi ft + \phi_n)$$

where A is the amplitude, f is the frequency, and ϕ_n is the phase shift. The periodicity and smoothness of the trajectory validate the correctness of the mechanical transmission and phase control strategy.

These simulation results confirm that the proposed rigid–flexible coupling design successfully generates continuous traveling waves with tunable parameters. The consistency between model predictions and simulated leg trajectories provides strong evidence for the mechanical plausibility of the wave-like propulsion mechanism prior to physical prototyping.

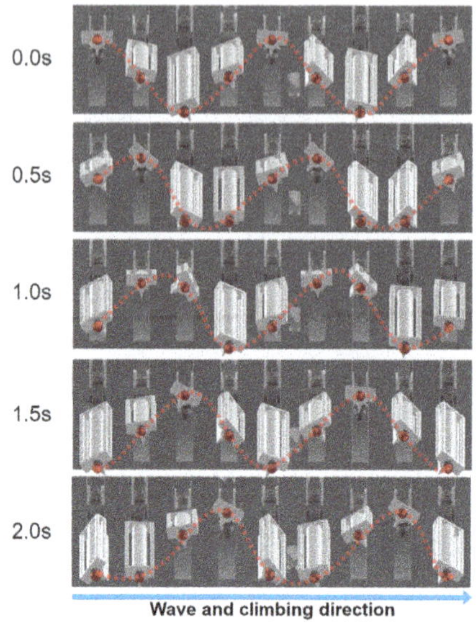

Figure 6. A timeline of the motion of the wave-like propulsion.

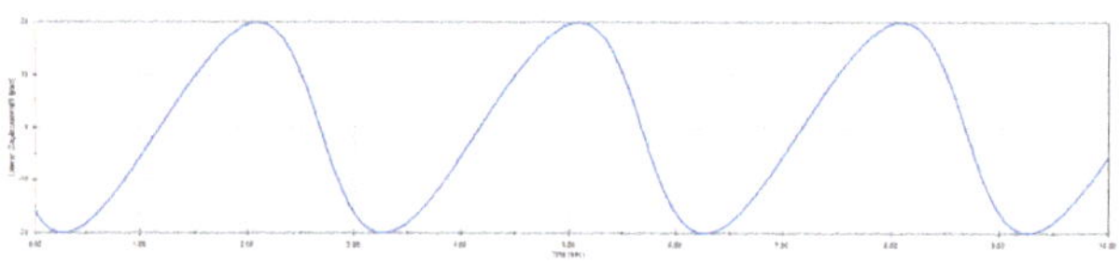

Figure 7. Displacement of the end of legs.

3.2. *Experimental Setup*

The prototype was primarily constructed using polylactic acid (PLA) material through fused deposition modeling (FDM) 3D printing, chosen for its suitability in rapid prototyping and adequate strength-to-weight ratio (Figure 8). PLA's ease of processing and biodegradability also align with sustainable development goals. To reduce friction and improve wear resistance in the high-cycle transmission system, the crank components were manufactured via zinc alloy die casting, leveraging the material's low surface friction coefficient and durability in repetitive mechanical contact, as shown in Figure 9.

Figure 8. The crawling prototype on grass.

Figure 9. Zinc-casted crankshaft.

The flexible membrane joints between rigid link segments were fabricated using latex rubber, selected for its high elasticity and tear resistance. Custom 3D-printed molds were designed and printed to form the membranes, with a uniform thickness of 1 mm, ensuring repeatability and mechanical consistency. This molding process allowed tight control over geometry while preserving compliance essential for wave propagation. The membranes were bonded to the rigid body using flexible adhesive to allow passive joint deformation under dynamic loading, as illustrated in Figure 10.

Figure 10. Latex-casted membrane.

For the electronic system, all sensors and actuators were connected via conventional jumper wires to a general-purpose ESP32 expansion board, eliminating the need for custom PCB fabrication. The board integrated power distribution and GPIO breakout functionalities, facilitating straightforward development and debugging. To ensure safe operation in aquatic environments, the entire electronic assembly was enclosed within a waterproof sealed box. This protective housing enabled stable underwater performance while maintaining accessibility for maintenance and updates, as shown in Figure 11.

To quantitatively assess steady-state straight-line locomotion, controlled experiments were conducted in both terrestrial and aquatic environments using calibrated measurement tracks. For crawling trials, the robot traversed a flat surface bounded by two parallel guide rails. The test path was marked at 500 mm intervals over a total length of 1000 mm. A top-mounted monocular camera operating at 30 fps continuously recorded the motion. To eliminate transient effects, the robot was allowed to reach a constant speed before entering the measurement region.

Figure 11. Prototype electronics system with waterproof housing.

For aquatic trials, the robot was deployed in a narrow test channel measuring 250 mm in length. The same camera system was positioned laterally to capture the robot's progression across the waterway. Frame-by-frame position tracking was performed to determine the effective swimming velocity once the robot achieved consistent forward movement.

For in-place rotation, the angular velocity was recorded using the onboard IMU (MPU6050, TDK InvenSense, Boston, MA, USA). The yaw angle over time was used to compute the average rotation speed in degrees per second.

4. Results and Discussion

4.1. Swimming Performance

To evaluate the real-world aquatic locomotion capability of the prototype, the fin modules were actuated at a wave frequency of 1.25 Hz (i.e., period of 0.8 s) with a phase shift of $\Delta\phi = 90^\circ$ between adjacent rays. This configuration corresponds to a wave propagation speed of

$$V_{\text{wave}} = \frac{d \cdot \omega_{\text{crank}}}{\Delta\phi_n} = \frac{50 \cdot 2\pi \cdot 1.25}{\pi/2} = 160\,\text{mm/s}$$

Experimental evaluation was performed in a test tank with transparent sides for video analysis. Using calibrated ground markers, the swimming velocity was measured to be approximately 12.5 mm/s. This result agrees with the theoretical estimation using the hydrodynamic efficiency factor:

$$V_{\text{sw}} = \eta V_{\text{wave}} = 0.078125 \times 160 = 12.5\,\text{mm/s}$$

The close match between the model and experimental measurements supports the validity of the wave propulsion framework derived in Equation (7). The robot maintained neutral buoyancy and demonstrated stable forward propulsion with low pitch oscillations (Figures 12–14).

The coefficient η was obtained through linear regression of experimental data collected at various crank speeds. No computational hydrodynamic model was employed. The resulting η represents an averaged empirical fit under the tested conditions.

In-place turning was achieved by reversing the wave propagation direction on the left fin while maintaining the forward wave on the right fin. The IMU recorded an average yaw rate of 5°/s during this maneuver, confirming effective directional control.

Figure 12. Snapshot sequence of swimming test at 20 s intervals under water.

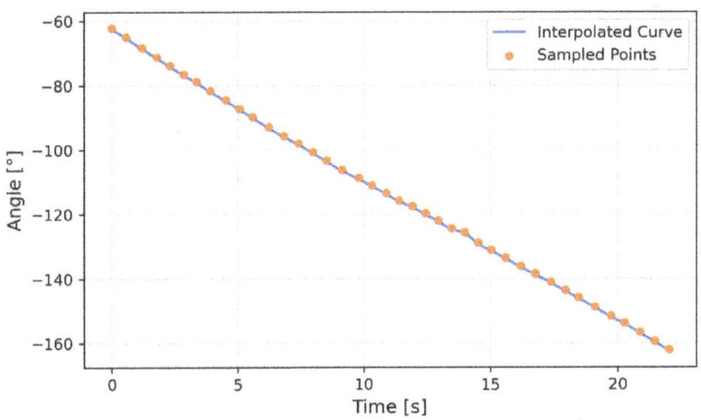

Figure 13. Angular position on water over time.

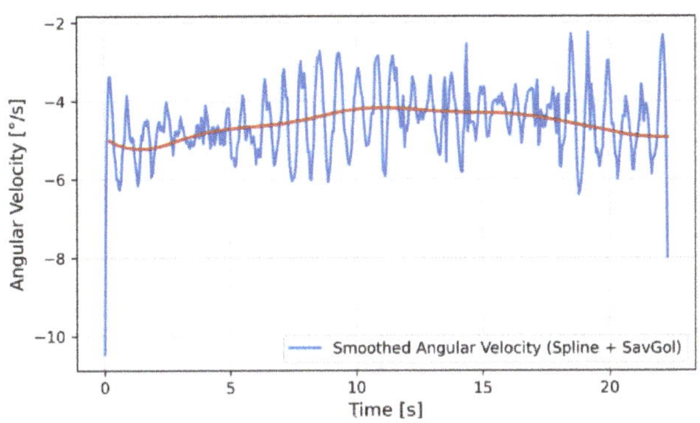

Figure 14. Angular velocity on water over time. Here, the red line indicates the smoothed angular velocity.

4.2. Crawling Performance

For terrestrial locomotion, identical wave parameters were applied to the leg segments to achieve crawling motion. Based on the derived kinematic model using a wave amplitude of $A = 40$ mm, radius $r = 5$ mm, and wavelength $\lambda = 100$ mm, the theoretical advance ratio was calculated to be $AR \approx 0.126$. Given a wave propagation speed of $V_{\text{wave}} = 160\,\text{mm/s}$ (based on actuation frequency), the estimated forward speed becomes

$$AR = \frac{2\pi r A}{\lambda^2} = \frac{2\pi \cdot 5\,\text{mm} \cdot 40\,\text{mm}}{(100\,\text{mm})^2} \approx 0.126$$
$$V_{\text{robot}} = AR \cdot V_{\text{wave}} = 0.126 \times 160 = 20.2\,\text{mm/s}$$

In practical crawling experiments (shown in Figure 15), the robot's forward velocity was measured using side-mounted camera tracking and calibrated ground markers. The observed straight-line speed was approximately 20 mm/s (0.1 BL/s), which is in strong agreement with the predicted value, thereby validating the accuracy of the kinematic model for land-based crawling. This result demonstrates that the proposed wave-based leg actuation mechanism can produce consistent and predictable locomotion performance on solid ground (Figures 16 and 17). The fin–ground interaction exhibits anisotropic friction behavior, where forward slippage is minimized during the power stroke. Sliding is allowed to a limited extent but does not dominate the propulsion mechanism.

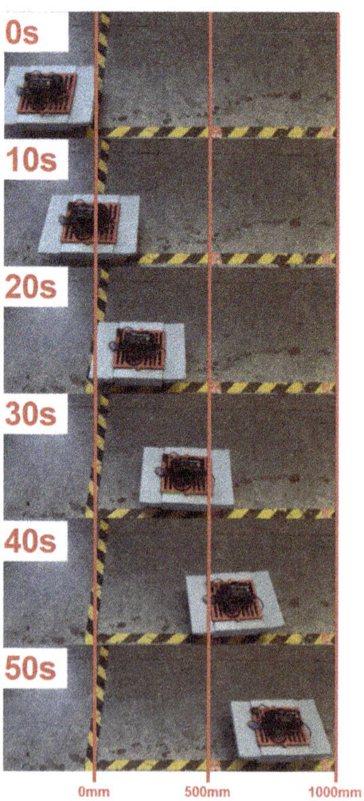

Figure 15. Snapshot sequence of crawling test over surface at 50 s intervals.

Additionally, in-place rotation was achieved by introducing phase asymmetry between the left and right leg arrays. The onboard IMU recorded an average rotational velocity of 10°/s, confirming the robot's ability to maneuver directionally in place.

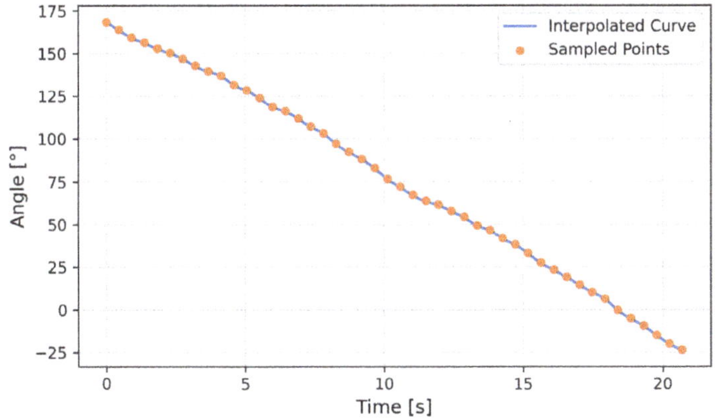

Figure 16. Angular position over time.

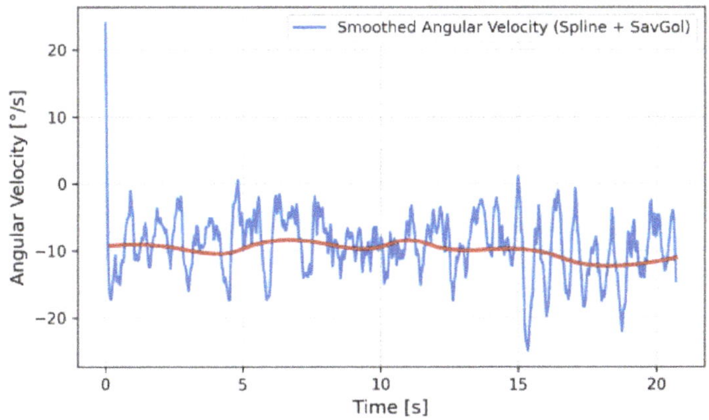

Figure 17. Angular velocity over time. Here, the red line indicates the smoothed angular velocity.

4.3. Discussion

The experimental results affirm the feasibility and functional robustness of the proposed wave-based propulsion system across both water and land environments. The observed locomotion behavior is in close agreement with the theoretical model derived from the Crank–Slider–Rocker mechanism and the corresponding kinematic equations.

In aquatic settings, the measured swimming velocity of 12.5 mm/s corresponds closely to the predicted value based on the applied wave frequency and calculated hydrodynamic efficiency. The robot exhibited consistent directional motion with minimal pitch oscillation, highlighting the advantages of using biologically inspired undulatory waveforms for stable aquatic locomotion. The in-place turning capability, validated by an average yaw rate of $5^\circ/s$, further demonstrated precise maneuverability using phase-modulated wave control.

However, the relatively low swimming speed suggests opportunities for further optimization. One strategy is to increase the actuation frequency of the crank input to raise the overall wave propagation speed. Additionally, extending the length of the fin rays could amplify the effective amplitude of oscillation, resulting in greater thrust generation. Increasing the surface area of the fin membrane may also improve hydrodynamic interaction and propulsion efficiency. Alternative material selections for the fin structure could reduce drag or enhance stiffness in key directions, allowing for more effective energy transfer from mechanical input to fluid displacement. These improvements could collectively contribute to a substantial boost in underwater performance.

On land, the crawling performance also aligned with theoretical predictions. The measured forward velocity of 20 mm/s closely matched the calculated value using the advance ratio and wave propagation speed. This agreement validates the effectiveness of the kinematic model even under friction-dominant conditions. The robot successfully demonstrated in-place turning on flat terrain with an angular velocity of $10°/s$, confirming the applicability of wave-phase control in terrestrial locomotion as well.

Overall, these results support the conclusion that the rigid–flexible coupled transmission system enables predictable and controllable locomotion through tunable waveform parameters. The successful integration of simulation and experimental outcomes reinforces the value of the proposed design framework for future amphibious robotic applications. Further improvements could focus on optimizing hydrodynamic efficiency, improving surface interaction during crawling, extending sensor feedback integration, and exploring multi-degree-of-freedom actuation schemes for enhanced agility and adaptability.

5. Conclusions

This study presents a novel mechanical architecture for amphibious robotic propulsion, featuring a Crank–Slider–Rocker mechanism integrated with a rigid–flexible coupling configuration. The proposed system enables biomimetic undulatory locomotion through a single-degree-of-freedom rotary input, achieving continuous wave propagation along articulated limbs without requiring complex multi-actuator coordination.

A comprehensive kinematic model was derived to capture the geometric and temporal transformation from crank motion to fin-ray oscillation, enabling the analytical tuning of critical wave parameters such as amplitude, wavelength, frequency, and propagation direction. This framework not only guided the physical design but also allowed predictive analysis of propulsion performance across aquatic and terrestrial environments.

Time-resolved dynamic simulations verified that the mechanical structure supports coherent phase propagation and stable traveling wave generation. The simulated trajectories confirmed the harmonic nature of oscillation and validated the phase modulation strategy embedded in the Crank–Slider–Rocker mechanism.

Physical experiments were conducted in both water and land testbeds under controlled conditions. The measured swimming speed of 12.5 mm/s and crawling speed of 20 mm/s closely matched theoretical predictions derived from the model, confirming its accuracy and robustness. Angular velocity tracking during turning maneuvers further demonstrated the feasibility of phase-based direction control. The close agreement across modeling, simulation, and empirical validation reinforces the mechanical and kinematic viability of the proposed approach.

Compared to conventional robotic propulsion systems, this architecture achieves functionally rich wave-based motion with significantly reduced actuator complexity. The system offers promising scalability and modularity, laying the groundwork for high-efficiency, low-cost amphibious locomotion platforms. Future work will focus on enhancing control autonomy, integrating soft sensor feedback for adaptive gait modulation, and extending the approach to three-dimensional fin actuation for improved hydrodynamic maneuverability.

Supplementary Materials: The following supporting information can be downloaded at: https://www.mdpi.com/article/10.3390/biomimetics10060396/s1, Video S1: Crawling and swimming motions of the cuttlefish-inspired robot.

Author Contributions: Conceptualization, Y.G. and Y.S.; methodology, Y.G., F.P. and Y.S.; validation, Y.G.; formal analysis, Y.G.; resources, F.P. and Y.S.; writing—original draft preparation, Y.G.;

writing—review and editing, Y.S.; supervision, Y.S.; project administration, Y.S.; funding acquisition, T.C.L. and Y.S. All authors have read and agreed to the published version of the manuscript.

Funding: This research was funded by TUM School of Engineering and Design (grant number: 1512/1107010300).

Institutional Review Board Statement: Not applicable.

Informed Consent Statement: Not applicable.

Data Availability Statement: The original contributions presented in this study are included in the article/Supplementary Materials. Further inquiries can be directed to the corresponding author.

Conflicts of Interest: The authors declare no conflicts of interest.

References

1. Wang, Y.; Pancheri, F.; Lueth, T.C.; Sun, Y. DuckyDog: An Erect Amphibious Robot with Variable-Stiffness Legs and Passive Fins. *Adv. Intell. Syst.* **2025**, 2500267.
2. Ijspeert, A.J.; Crespi, A.; Ryczko, D.; Cabelguen, J.M. From Swimming to Walking with a Salamander Robot Driven by a Spinal Cord Model. *Science* **2007**, *315*, 1416–1420. [CrossRef] [PubMed]
3. Rafeeq, M.; Toha, S. F.; Ahmad, S.; Razib, M. A. Locomotion strategies for amphibious robots-a review. *IEEE Access* **2021**, *9*, 26323–26342. [CrossRef]
4. Georgiades, C.; German, A.; Hogue, A.; Liu, H.; Prahacs, C.; Ripsman, A.; Sim, R.; Torres, L.A.; Zhang, P.; Buehler, M.; et al. AQUA: an aquatic walking robot. In Proceedings of the Proceedings of the 2004 IEEE/RSJ International Conference on Intelligent Robots and Systems (IROS), Sendai, Japan, 28 September–2 October 2004; pp. 3525–3531.
5. Qu, J.; Cai, Q.; Fish, F.E.; Li, Y.; Chen, Y.; Zhong, Y.; Xia, J.; Fu, S.; Xie, W.; Luo, H.; et al. Amphibious robotic dog: Design, paddling gait planning, and experimental characterization. *Bioinspiration Biomim.* **2025**, *20*, 036012. [CrossRef]
6. Sun, Y.; Pancheri, F.; Rehekampff, C.; Lueth, T.C. TurBot: A turtle-inspired quadruped robot using topology optimized soft-rigid hybrid legs. *IEEE/ASME Trans. Mechatron.* **2024**, *29*, 3193–3202. [CrossRef]
7. Hirose, S.; Mori, M. Biologically Inspired Snake-like Robots. In Proceedings of the 2004 IEEE International Conference on Robotics and Biomimetics, Shenyang, China, 22–26 August 2004; pp. 1–7. [CrossRef]
8. Wang, J.; Zheng, J.; Zhao, Y.; Yang, K. Structure design and coordinated motion analysis of bionic crocodile robot. *Biomim. Intell. Robot.* **2024**, *4*, 100157. [CrossRef]
9. Boxerbaum, A.S.; Werk, P.; Quinn, R.D.; Vaidyanathan, R. Design of an Autonomous Amphibious Robot for Surf Zone Operation: Part I Mechanical Design for Multi-mode Mobility. In Proceedings of the 2005 IEEE/ASME International Conference on Advanced Intelligent Mechatronics, Monterey, CA, USA, 24–28 July 2005, pp. 1459–1464.
10. Tanaka, M.; Nakajima, M.; Suzuki, Y.; Tanaka K. Development and Control of Articulated Mobile Robot for Climbing Steep Stairs. *IEEE/ASME Trans. Mechatron.* **2018**, *23*, 531–541. [CrossRef]
11. Wang, Y.; Pancheri, F.; Lueth, T.C.; Sun, Y. Design of a spider-inspired wheeled compliant leg for search mobile robots. *Biomim. Intell. Robot.* **2024**, *4*, 100182. [CrossRef]
12. Huang, J.; Sun, Y.; Wang, T.; Lueth, T.C.; Liang, J.; Yang, X. Fluid-structure interaction hydrodynamics analysis on a deformed bionic flipper with non-uniformly distributed stiffness. *IEEE Robot. Autom. Lett.* **2020**, *5*, 4657–4662. [CrossRef]
13. Zhu, P.; Shang, K.; Huang, Y.; Jiang, Z.; Zhou, J.; Lu, X.; Yang, T. Light-Driven Amphibious Mini Soft Robot Mimicking the Locomotion Gait of Inchworm and Water Strider. *Adv. Intell. Syst.* **2024**, *6*, 2300466. [CrossRef]
14. Yu, J.; Ding, R.; Yang, Q.; Tan, M.; Wang, W.; Zhang, J. On a bio-inspired amphibious robot capable of multi-modal motion. *IEEE/ASME Trans. Mechatron.* **2012**, *17*, 847–856. [CrossRef]
15. Kim, H.; Lee, D.; Jeong, K.; Seo, T. Water and Ground-Running Robotic Platform by Repeated Motion of Six Spherical Footpads. *IEEE/ASME Trans. Mechatron.* **2016**, *21*, 175–183. [CrossRef]
16. Rafeeq, M.; Toha, S.F.; Ahmad, S.; Razib, M.A.; Idris, A.S.; Tokhi, M.O. Amphibious Robots Locomotion Strategies in Unstructured Complex Environments: A Review. *Platform A J. Eng.* **2024**, *8*, 12–22. [CrossRef]
17. Baines, R.; Fish, F.; Kramer-Bottiglio, R. Amphibious Robotic Propulsive Mechanisms: Current Technologies and Open Challenges. In *Bioinspired Sensing, Actuation, and Control in Underwater Soft Robotic Systems*; Springer International Publishing: Cham, Switzerland, 2021; pp. 41–69. [CrossRef]
18. Li, X.; Lei, J.; Li, X.; Ruina, D.; Li, F.; Wang, W.; Tong, Z.; Guoshun, W. Modeling and implementation of a novel amphibious robot with multimode motion. *Ind. Robot. Int. J. Robot. Res. Appl.* **2022**, *49*, 947–961. [CrossRef]
19. Cohen, A.; Zarrouk, D. Design, Analysis and Experiments of a High-Speed Water Hovering Amphibious Robot: AmphiSTAR. *IEEE Access* **2023**, *11*, 80874–80885. [CrossRef]

20. Arslan, E.; Akça, K. A Design Methodology for Cuttlefish Shaped Amphibious Robot. *Eur. J. Sci. Technol.* **2019**, 214–224. [CrossRef]
21. Low, K.H. Modelling and parametric study of modular undulating fin rays for fish robots. *Mech. Mach. Theory* **2009**, *44*, 615–632.
22. Li, Y.; Chen, L.; Wang, Y.; Ren, C. Design and experimental evaluation of the novel undulatory propulsors for biomimetic underwater robots. *Bioinspiration Biomim.* **2021**, *16*, 056005. [CrossRef]
23. Sfakiotakis, M.; Lane, D.M.; Davies, J.B.C. Review of Fish Swimming Modes for Aquatic Locomotion. *IEEE J. Ocean. Eng.* **1999**, *24*, 237–252. [CrossRef]
24. Hu, T.; Shen, L.; Lin, L.; Xu, H. Biological inspirations, kinematics modeling, mechanism design and experiments on an undulating robotic fin inspired by Gymnarchus niloticus. *Mech. Mach. Theory* **2009**, *44*, 633–645. [CrossRef]
25. Uddin, M.I.; Garcia, G.A.; Curet, O.M. Force scaling and efficiency of elongated median fin propulsion. *Bioinspiration Biomim.* **2022**, *17*, 046004. [CrossRef] [PubMed]
26. Curet, O.M.; Patankar, N.A.; Lauder, G.V.; MacIver, M.A. Mechanical properties of a bio-inspired robotic knifefish with an undulatory propulsor. *Bioinspiration Biomim.* **2011**, *6*, 026004. [CrossRef] [PubMed]
27. Shi, X.; Chen, Z.; Zhang, T.; Li, S.; Zeng, Y.; Chen, L.; Hu, Q. Hydrodynamic analysis and performance evaluation of a biomimetic undulating-fin underwater robot. *Ocean Eng.* **2023**, *288*, 116068. [CrossRef]
28. Zarrouk, D.; Mann, M.; Degani, N.; Yehuda, T.; Jarbi, N.; Hess, A. Single actuator wave-like robot (SAW): Design, modeling, and experiments. *Bioinspiration Biomim.* **2016**, *11*, 046004. [CrossRef]
29. Shachaf, E.; Inbar, O.; Zarrouk, D. RSAW, A Highly Reconfigurable Wave Robot: Analysis, Design, and Experiments. *IEEE Robot. Autom. Lett.* **2019**, *4*, 4475–4482. [CrossRef]
30. Hirose, S.; Yamada, H. Snake-like robots [Tutorial]. *IEEE Robot. Autom. Mag.* **2009**, *16*, 88–98. [CrossRef]

biomimetics

MDPI

Article

3D-Printed Soft Bionic Inchworm Robot Powered by Magnetic Force

Deli Xia [1], Luying Zhang [1], Weihang Nong [1], Qingshan Duan [2] and Jiang Ding [1,3,*]

[1] Guangxi Key Laboratory of Manufacturing System and Advanced Manufacturing Technology, School of Mechanical Engineering, Guangxi University, Nanning 530004, China; 2211301072@st.gxu.edu.cn (D.X.); 2311301081@st.gxu.edu.cn (L.Z.); 2411401173@st.gxu.edu.cn (W.N.)

[2] School of Light Industry and Food Engineering, Guangxi University, Nanning 530004, China; qs_duan@gxu.edu.cn

[3] State Key Laboratory of Featured Metal Materials and Life-Cycle Safety for Composite Structures, Guangxi University, Nanning 530004, China

* Correspondence: jding@gxu.edu.cn

Abstract: Based on soft body structure and unique gait of bending and stretching, Soft Bionic Inchworm Robots (SBIRs) are used in pipeline inspection and terrain exploration. Many existing SBIRs rely on complex production mechanisms and are cable-driven, which hinders rapid production and smooth movement through complex environments, respectively. To address these challenges, this paper introduces a 3D-printed SBIR, featuring a 3D-printed body actuated by magnetic forces. We introduce the design and production process of the 3D-SBIR and analyze its motion gait. Subsequently, the material composition model and bending deformation model of the robot are developed based on the theory of hyper-elastic materials. The accuracy of the model is validated using simulation analysis and experimental testing of the robot. Meanwhile, we carry out a magnetic simulation analysis and discuss the factors influencing the size of the magnetic force. Finally, a series of experiments are conducted to prove the excellent locomotion capability of the robot. The 3D-SBIR demonstrates remarkable flexibility and multimodal movement capabilities. It can navigate through narrow curved passages with ease, passively overcome obstacles, climb steps up to 0.8 times its body height, and perform a seamless transition while moving across a horizontal plane onto a vertical plane. The 3D-SBIR proposed in this paper is characterized by rapid production, cable-free actuation, and multimodal motion capabilities, making it well suited for moving in unstructured environments.

Keywords: bionic inchworm; soft robot; magnetic force; 3D-printed

1. Introduction

Soft bionic robots inspired by soft-bodied animals in nature have become a popular research topic because of their potential to perform various specialized tasks in complex environments, such as space exploration, biomedical exploration, and more [1–3]. The inchworm, a common mollusk, relies on the coordination between a flexible body and alternately moving legs to achieve a special gait of bending and stretching. Inspired by the soft body structure and distinct gait of the inchworm, the soft bionic inchworm robot (SBIR) adapts well to environmental changes and shows excellent adaptability to unstructured environments [4–6].

However, most existing SBIRs are driven by voltage actuation [7–9], fluid actuation [10–12], or shape memory alloy actuation [13]. These actuation methods require external electric wires or pneumatic pipes, which will hinder the movement of the robot in special environments. Meanwhile, most of the existing SBIRs are tedious to produce

 https://doi.org/10.3390/biomimetics10040202

and require more assembly processes. Conversely, cable-free actuation, exemplified by magnetic actuation, can liberate the robot from the bondage of cables. In certain special scenarios, such as conducting inspections within semi-enclosed pipes or delivering medications inside animal bodies, this actuation allows the robot to execute tasks more effectively [14–18]. Rapid production technologies, exemplified by 3D-printed technology, are capable of rapidly translating researchers' design concepts into physical models. This allows for immediate testing and analysis to identify potential design flaws. Consequently, it effectively shortens the research and development cycle, reduces research and development costs, and expeditiously facilitates rapid design iterations [19–21]. Some researchers also combine the advantages of magnetic actuation and 3D-printed technology to develop new SBIRs. As an example, Erina et al. combine 3D-printed technology and magnetic actuation to present a fully 3D-printed, inchworm-like soft robot that is magnetically actuated to move and crawl on the horizontal plane [22]. While the robot achieves cable-free actuation and rapid production, it loses capability for multimodal movement and only moves on a plane surface. It is noteworthy that current SBIRs primarily focus on pipe-climbing or horizontal surface locomotion. Significant challenges remain when these robots attempt to climb over obstacles or execute transitional movements from horizontal to vertical surfaces.

In this study, a 3D-printed soft bionic inchworm robot (3D-SBIR) powered by magnetic force is designed and fabricated, achieving rapid production, cable-free actuation, and multimodal movement capability. The flexible body of the robot is produced in one single step by 3D-printed, eliminating the necessity for a complex production process. By embedding magnets in the flexible body, a non-contact, cable-free power input is provided to the robot. Furthermore, the robot exhibits multimodal movement capabilities, including climbing obstacles and the ability to climb from horizontal to vertical surfaces. A systematic comparison, as shown in Table 1, of the 3D-SBIR with previous inchworm-like robots demonstrates the spatial mobility of the proposed 3D-SBIR design.

Table 1. Comparison of the 3D-SBIR with previous inchworm-like robots.

Ref	Actuation	Cable-Free	Product Process	Movement Capability
[4]	Voltage actuation	No	Complex	Climbing pipes
[5]	Voltage actuation	No	Complex	Moving on a horizontal surface
[6]	Pneumatic actuation	No	Complex	Climbing pipes and slopes
[11]	Pneumatic actuation	No	Simple	Moving in two-dimensional space
[15]	Magnetic actuation	Yes	Simple	Moving on a horizontal surface
[16]	Magnetic actuation	Yes	Simple	Moving inside the pipeline
[17]	Magnetic actuation	Yes	Simple	Moving inside the pipeline
[22]	Magnetic actuation	Yes	Simple	Moving on a horizontal surface
This work	Magnetic actuation	Yes	Simple	Moving on a horizontal surface, climbing obstacles, steps and transitions from horizontal to vertical surfaces

Similarly to the natural inchworm, the robot moves by coordination between the bending deformation of its belly and the alternate motion of its legs. This research focuses on analyzing the robot's gait and force conditions as it climbs an obstacle. Based on the theory of hyper-elastic materials [23–25], the constitutive model of the material and a bending deformation model of the robot are established. Finite element analysis and experiments of the bending deformation process of the robot are conducted to validate the material constitutive model and the bending deformation model. Finally, a series of experiments are performed to evaluate the movement performance of the 3D-SBIR.

2. The Design and Production Process of the 3D-SBIR

As shown in Figure 1a, the 3D-SBIR consists of a flexible body capable of bending in multiple directions and four neodymium-iron-boron (NdFeB) magnets. The flexible body is designed to imitate the body structure of the inchworm, with a head, belly, anterior legs, and posterior legs. The belly undergoes bending or stretching deformation under the combined action of the anterior and the posterior legs. Both the anterior and posterior legs feature gaps to embed NdFeB magnets serving as the power source of the 3D-SBIR. As shown in Figure 1b, the flexible body is made of anhydrous eutectic gel, providing excellent flexibility. The body is directly formed by printing with light-curing printing technology. Its adhesion after printing enables four NdFeB magnets to be embedded directly into the gap of the four legs without using additional glue. After inlaying the NdFeB magnets into the gaps, the flexible body undergoes drying to minimize surface adhesion. The 3D-SBIR measures 60 mm in total length, 12 mm in width, and 20 mm in height. Four NdFeB magnets are cylindrical, each measuring 15 mm in length and 2.0 mm in diameter.

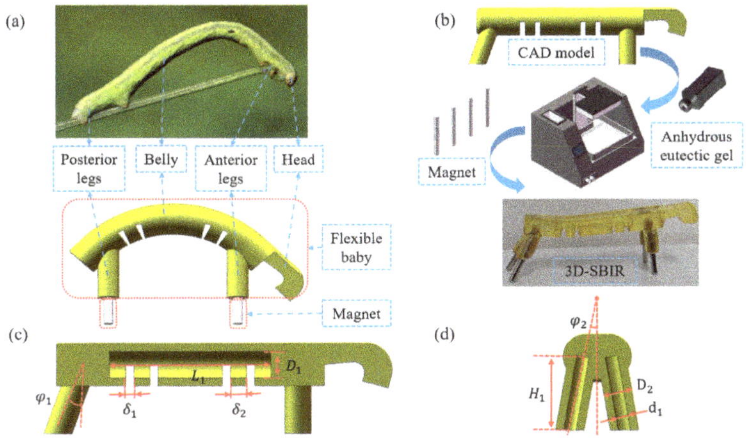

Figure 1. 3D-SBIR design and production process: (**a**) Structure comparison between the robot and an inchworm. (**b**) The production process of the flexible body. (**c**) Cutaway view of the robot's flexible body. (**d**) Cutaway view of robot's anterior legs.

Figure 1c provides a cutaway view of the flexible body of the 3D-SBIR, sliced along the axial midsection of the belly. The robot relies on the coordination of the alternating movement of the legs and the bending deformation of the belly to achieve a gait of bending and stretching while moving. The belly is a hollow structure with four gaps at the bottom to allow for better bending deformation. Figure 1d shows a cutaway view of the anterior legs of the 3D-SBIR, with the anterior legs tilted outward relative to the centerline of the belly. The posterior legs are similar to the anterior legs, tilted outward and backward relative to the centerline of the belly and the anterior legs, respectively, enhancing the stability of the robot during movement. The design parameters of the flexible body are outlined in Table 2.

Table 2. Design parameters of the flexible body.

Part	Parameter Name	Symbol	Value
Belly	Cavity diameter	D_1	5.0 mm
	Cavity length	L_1	30.0 mm
	Gap size	δ_1	1.5 mm
	Gap distance	δ_2	3.0 mm

Table 2. *Cont.*

Part	Parameter Name	Symbol	Value
Legs	Backward tilt angle	φ_1	20°
	Outward tilt angle	φ_2	10°
	Outside diameter	D_2	5.0 mm
	Inside diameter	d_1	2.2 mm
	Height	H_1	10.0 mm

3. The Motion Gait Analysis of the 3D-SBIR

3.1. Gait Analysis of Motion on a Horizontal Plane

When the 3D-SBIR moves on a horizontal plane, it achieves a bending and stretching motion gait by utilizing the alternating movement of the legs and the bending deformation of the belly. As shown in Figure 2I, Similarly to its biological counterpart in nature, the robot replicates the concertina-style gait characterized by alternating bending and stretching motions. However, unlike the natural inchworm, which relies on synchronized muscle contractions, the 3D-SBIR accomplishes a full locomotion cycle through a four-step actuation sequence. As shown in Figure 2a, in the initial state of the 3D-SBIR moving on the horizontal plane, the robot has vertical anterior legs and posterior legs tilted backward, which ensures the stability of the robot. As shown in Figure 2b, the magnetic field of the anterior legs remains stationary, whereas the magnetic field of the posterior legs moves to the right, forcing the belly to undergo a bending deformation. The posterior legs gradually change from a backward-tilted state to a vertical state, while the anterior legs are tilted forward under the influence of the bending deformation of the belly. As shown in Figure 2c, the magnetic field of the anterior legs moves to the right, whereas the magnetic field of the posterior legs remains stationary, causing the belly to undergo a stretching deformation. The anterior legs gradually change from a forward-tilted state to a backward-tilted state, while the posterior legs are tilted forward under the influence of the stretching deformation of the belly. As shown in Figure 2d, the magnetic field of the anterior legs remains stationary, and the magnetic field of the posterior legs moves to the right. The belly returns to the natural state, and the robot advances a distance of s to the right.

3.2. Gait Analysis of Climbing Obstacles

Research on rigid robots when climbing obstacles often prevents direct contact or collision between robots and obstacles to avoid damage to environments or injury to robots. However, the 3D-SBIR, with its soft body structure, is highly adapted to different environments and minimizes environmental damages. This study makes use of this feature to assist the robot in climbing obstacles quickly, leveraging direct contact between the robot and the obstacle as well as contact deformation of the robot while climbing the obstacle.

To better explain the advantages of the 3D-SBIR in climbing obstacles, we analyze the force conditions when the robot climbs obstacles. As shown in Figure 2e, in the initial state of the 3D-SBIR climbing obstacles, there are obstacles on both anterior and posterior legs of the robot. As shown in Figure 2f, in the state of potential energy accumulation for the 3D-SBIR climbing obstacles, when the magnetic field of the anterior legs remains stationary and the magnetic field of the posterior legs moves forward, forcing the belly to undergo bending deformation and accumulate elastic potential energy. In this moment, the anterior legs are affected by a combination of five forces: G_1 (the force of gravity on the anterior legs), F_{t1} (the force of the attraction of magnetic fields to the anterior legs, which can be decomposed into the force F_{X1} in the horizontal direction and the force F_{Z1} in the vertical direction), F_{N1} (the force of the environmental support for the anterior legs), F_1 (the force of the obstacle on the anterior legs), and F_2 (the force of the belly on the anterior legs). The forces on the anterior legs satisfy the following force balance equation:

$$F_{X1} + F_2\text{Cos}(\theta_2) = F_1\text{Cos}(\theta_1) \tag{1}$$

$$F_{N1} + F_1\text{Sin}(\theta_1) + F_2\text{Sin}(\theta_2) = G_1 + F_{Z1} \tag{2}$$

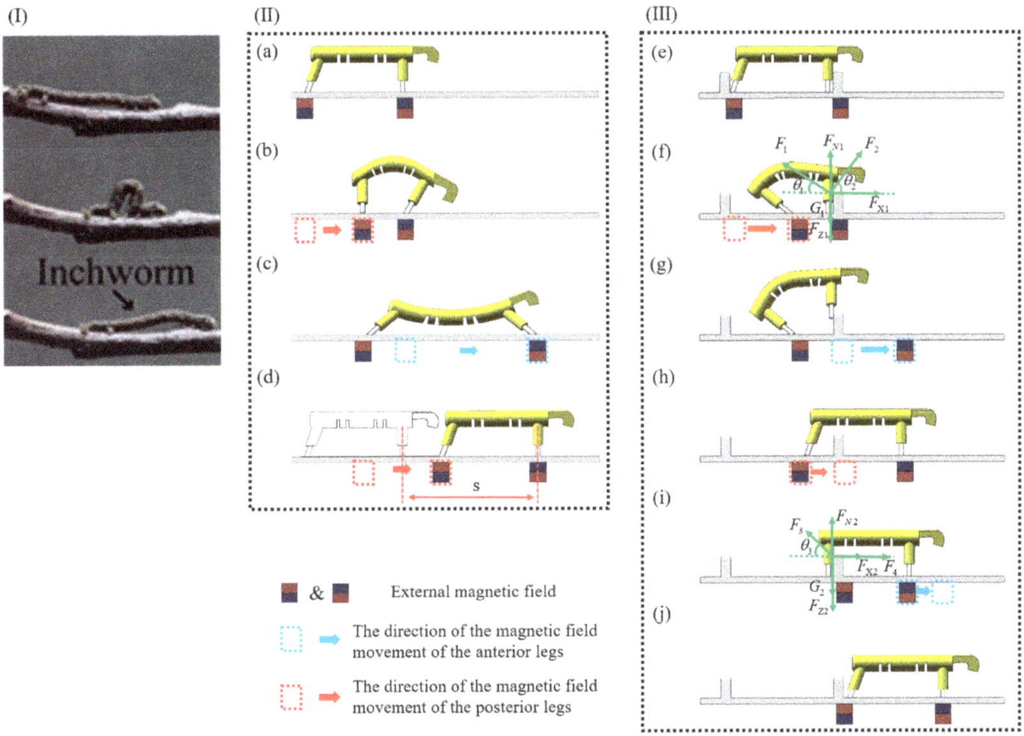

Figure 2. Motion gait diagram of the 3D-SBIR: (**I**) The motion gait of the inchworm [16]. (**II**) Gait diagram of motion on a horizontal plane: (**a**) Initial state of the robot moving on the horizontal plane. (**b**) Bending state of the robot moving on the horizontal plane. (**c**) Stretching state of the robot moving on the horizontal plane. (**d**) Natural state of the robot moving on the horizontal plane. (**III**) Gait diagram of climbing obstacles: (**e**) Initial state of the robot climbing obstacles. (**f**) State of potential energy accumulation for the robot climbing obstacles. (**g**) State of potential energy release for the robot climbing obstacles. (**h**) Moving state of the robot climbing obstacles. (**i**) Passive deformation state of the robot climbing obstacles. (**j**) Natural state of the robot climbing obstacles.

Figure 2g shows the state of potential energy release for 3D-SBIR during obstacle climbing. The magnetic field of the anterior legs moves forward while that of the posterior legs remains stationary, resulting in a gradual decrement in the vertical magnetic attraction force on the anterior legs (F_{Z1}) and a concurrent increase in the horizontal attraction force (F_{X1}). This consequently leads to an enhanced reaction force from the obstacle on the anterior legs (F_1), whereas the belly force acting on the anterior legs remains constant. These combined effects ultimately cause the environmental support force on the anterior legs (F_{N1}) to decrease until it reaches zero. At this critical juncture, the anterior legs are raised upward, the belly releases its stored elastic potential energy, and the robot successfully climbs over the obstacle with its anterior legs. Table 3 lists the trends of each force change during anterior legs obstacle climbing.

Table 3. Changes in forces on the legs.

Figure	Eternal Magnetic Field Movement	Changing Trend of the Forces					
Figure 2g	Forward movement of the magnetic field of the anterior legs	F_{N1} ↓	F_{X1} ↑	F_{Z1} ↓	G_1 —	F_1 —	F_2 —
Figure 2i	The magnetic field of the anterior legs continues to move forward	F_{N2} ↓	F_{X2} —	F_{Z2} —	G_2 —	F_3 —	F_4 ↑

As shown in Figure 2h, in the moving state of the 3D-SBIR climbing obstacles, the magnetic field of the anterior legs remains stationary, and the magnetic field of the posterior legs moves until the posterior legs reach the front of the obstacle. Figure 2i shows the passive deformation state of the 3D-SBIR climbing obstacles. When the magnetic field of the anterior legs continues to move forward, the motion of the posterior legs is obstructed, and the belly is forced to undergo stretching deformation. In this moment, the posterior legs are affected by a combination of five forces: G_2 (the force of gravity on the posterior legs), F_{t2} (the force of the attraction of magnetic fields to the posterior legs, which can be decomposed into the force F_{X2} in the horizontal direction and the force F_{Z2} in the vertical direction), F_{N2} (the force of the environmental support for the posterior legs), F_3 (the force of the obstacle on the posterior legs), and F_4 (the force of the belly on the posterior legs). The forces on the posterior legs satisfy the following force balance equation:

$$F_4 + F_{X2} = F_3\mathrm{Cos}(\theta_3) \tag{3}$$

$$F_{N2} + F_3\mathrm{Sin}(\theta_3) = G_2 + F_{Z2} \tag{4}$$

As the magnetic field of the anterior legs continues to move forward, the force of the belly on the posterior legs (F_4) gradually increases while the force of the attraction of magnetic fields to the posterior legs (F_{t2}) remains constant, resulting in a gradual increase in the force of the obstacle on the posterior legs (F_3). As the force of the obstacle on the posterior legs (F_3) gradually increases, the force of gravity on the posterior legs (G_2) remains unchanged, resulting in a gradual decrease in the force of the environmental support on the posterior legs (F_{N2}), until this force reaches zero. Figure 2j shows the natural state of the 3D-SBIR while climbing obstacles. The belly releases elastic potential energy, and the posterior legs of the robot climb over the obstacle. The trend of each force change when climbing obstacles with the posterior legs is shown in Table 3.

4. Model Construction and Simulation Analysis

4.1. Constitutive Model of the Materials

The body of the 3D-SBIR is an anhydrous eutectic gel, which provides the same hyperelasticity as silicone rubber. The constitutive model of the material and the bending deformation model of the robot are developed based on the theory of hyperelastic materials. The mechanical behavior of silicone rubber-based hyperelastic materials is generally described using two approaches. One approach establishes stress–strain relationships based on microscopic statistical mechanics, such as the Neo-Hookean model. The other approach establishes stress–strain relationships based on continuum phenomenological theory, including models such as the Ogden model, the Mooney-Rivlin model, and the Yeoh model. The Yeoh model, primarily applied to the large deformation behavior of carbon black filling, requires only simple data from uniaxial tensile experiments to simulate the deformation results well and is suitable for large deformation conditions. The relationship between main stress σ_1 and main tensile ratio λ_1 is established based on the Yeoh model as follows (The detailed procedure of derivation is given in S1):

$$\frac{\sigma_1}{2\left(\lambda_1 - \frac{1}{\lambda_1^2}\right)} = 2C_{20}\left(\lambda_1^2 + \frac{2}{\lambda_1}\right) + C_{10} - 6C_{20} \quad (5)$$

where C_{10} and C_{20} are the material constants of the Yeoh model, measured from uniaxial tensile tests of the material. As shown in Figure 3a, we conduct uniaxial tensile tests of the material, the thickness of the specimen is 3 mm.

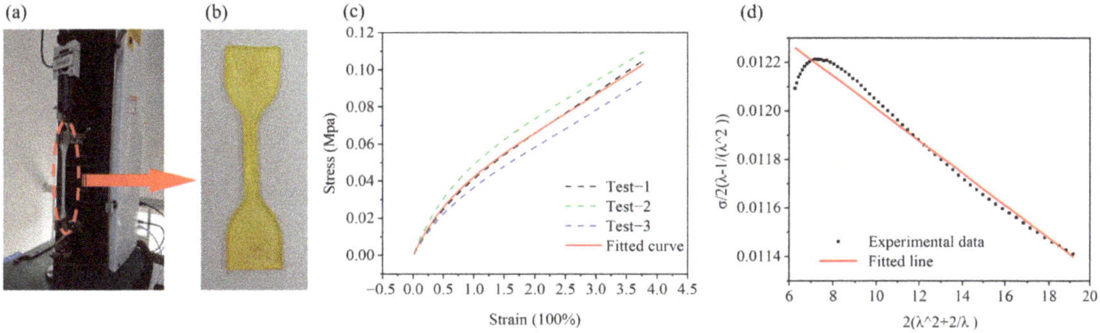

Figure 3. Testing of the material in uniaxial tension: (**a**) Material is stretched on a universal testing machine. (**b**) Standard sample of the material. (**c**) Stress–strain curve of the material. (**d**) Linear fitting of the material constants to the Yeoh model.

The displacement and load data are obtained from uniaxial tensile experiments on the material. Corresponding data within the displacement range of 0~1 mm is averaged. Similarly, data with displacements of 1~2 mm, 2~3 mm, 3~4 mm and so on are processed in the same manner to calculate stress and strain values. Three sets of tests are conducted, and the stress–strain curve of the material was fitted, as shown in Figure 3c. To determine the material constants C_{10} and C_{20}, let $2(\lambda_1^2 + 2/\lambda_1)$ and $\sigma_1/\left[2\left(\lambda_1 - 1/\lambda_1^2\right)\right]$ represent the horizontal and vertical axes of the coordinate system, respectively. Considering that the uniaxial tensile test exhibits a significant error in the initial loading of several sets of data, the data within 20% of the strain are removed. Considering that the 3D-SBIR is working with a strain of 200% or less, the data sets where the strain is larger than 200% are excluded. As shown in Figure 3d, the fitted straight line of material constants for the Yeoh model is obtained, $C_{10} = 0.0123$, $C_{20} = -6.68\text{e} - 5$.

4.2. The Bending Deformation Model of the 3D-SBIR

When the 3D-SBIR is bent and deformed by an external magnetic field, it is divided into a deformed part and a non-deformed part (no stretching and compressive deformation of the m and n parts). The belly of the 3D-SBIR undergoes bending deformation under the action of external forces. Assuming that the middle layer of the belly does not undergo tensile deformation and serves as a neutral layer, the uppermost layer of the belly undergoes tensile deformation, while the lowermost layer of the belly undergoes compressive deformation. Assuming that the belly undergoes bending by equal curvature, as shown in Figure 4, let r is the radius of curvature of the neutral layer, s is the curvature length of the neutral layer, θ is the center angle of the circle of the neutral layer, the distance of OC is $L_{OC} = y$, the height of the belly lifting is H, and the distance between the anterior and posterior legs is $L_{AB} = x$. If the neutral layer does not undergo tensile deformation, $s = 40$, then there is,

$$\frac{2\pi r\theta}{360} = s, \quad (6)$$

$$H = r - 4 - y. \quad (7)$$

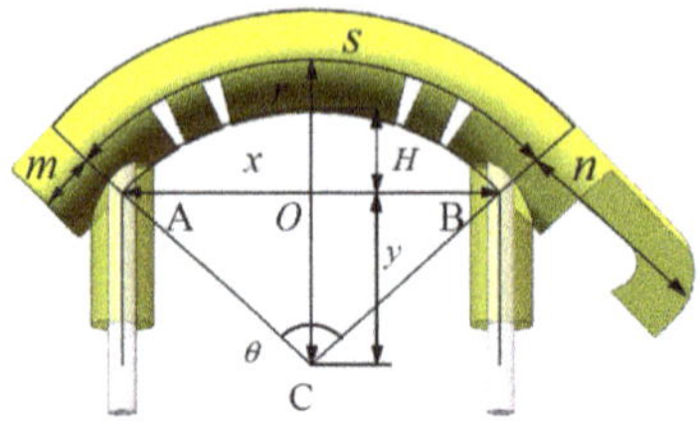

Figure 4. The diagram of the bending deformation of the 3D-SBIR.

According to the law of sines for ΔABC and ΔAOC, it can be obtained,

$$\frac{x}{\mathrm{Sin}\theta} = \frac{r-4}{\mathrm{Sin}(90-\theta/2)}, \tag{8}$$

$$\frac{r-4}{\mathrm{Sin}(90)} = \frac{y}{\mathrm{Sin}(90-\theta/2)}. \tag{9}$$

From Equations (6) and (8), it can be obtained,

$$\frac{x}{\mathrm{Sin}\left(\frac{7200}{\pi r}\right)} = \frac{r-4}{\mathrm{Sin}\left(90-\frac{3600}{\pi r}\right)}. \tag{10}$$

Therefore, we can conclude that r is a function of x, denoted as,

$$r = r(x). \tag{11}$$

From Equations (6) and (9), it can be obtained,

$$y = (r-4)\mathrm{Sin}\left(90-\frac{3600}{\pi r}\right). \tag{12}$$

Therefore, we can conclude that y is a function of r, denoted as,

$$y = y(x). \tag{13}$$

From Equations (7), (11) and (13), the relationship between the height of the belly lift (H) and the distance between the anterior and posterior legs (x) is obtained,

$$H = r(x) - 4 - y(x). \tag{14}$$

4.3. Simulation and Experimentation of the Bending Deformation

To verify the correctness of the material constitutive model and the bending deformation model of the 3D-SBIR, and to more intuitively demonstrate the bending deformation process of the robot under the action of external force, finite element analysis of the robot is carried out based on Abaqus. The 3D model is directly established in Abaqus, with a hyperelastic material assigned to the body and the Yeoh model selected for the strain energy density function. Steel material is assigned to the NdFeB magnet. The NdFeB magnet interacts with the flexible body, and its deformation is negligible compared to the deformation of the flexible body. Rigid body constraints are applied to the NdFeB magnet to simplify the simulation process. With the anterior legs' magnet completely fixed, a body force is applied to the posterior legs' magnet, moving it 20 mm forward.

We conducted an experiment on the bending deformation of the 3D-SBIR to verify the accuracy of the above results of bending deformation simulation. By applying external force directly to the robot, with the magnets of the anterior legs remaining stationary and the magnets of the posterior legs moving forward, we observed the relationship between the height of the belly lift (H) and the distance between the anterior legs and the posterior legs (x). Figure 5a shows the theoretical, simulated, and experimental values of the bending deformation process of the robot. Simulation values closely match the experimental data, demonstrating the accuracy of simulation results, and confirming the validity of the material constitutive model developed using the Yeoh model. Figure 5b shows the difference between the theoretical and simulated values. When the distance x is between 23 mm and 34 mm, the difference is within 1 mm and the bending deformation model is applicable to the robot. When $x > 34$ mm, the abdomen primarily undergoes compressive de-formation, making the bending deformation model inapplicable. When $x < 23$ mm, there is not only a bending deformation of the abdomen, but also an upward slip, so the bending deformation model is not applicable in this range either.

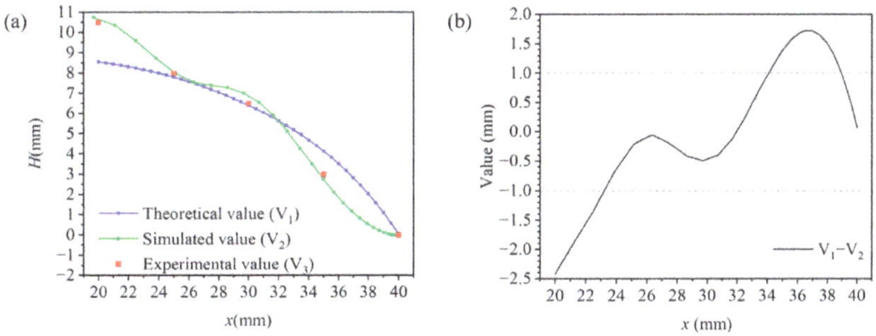

Figure 5. Simulation and experiment of the 3D-SBIR bending deformation: (**a**) Theoretical, simulated and experimental values during bending deformation of the robot (**b**) Difference between theoretical and experimental values.

4.4. Magnetic Simulation and Analysis

The motion of the robot is dependent on the magnetic force provided by external magnets, so we discuss the main factors affecting the size of magnetic force. We perform magnetic simulation of the robot based on Comsol. The 3D model is built directly in Comsol, as shown in Figure 6a. A magnetically insulated simulation space is set up, along with a divider having magnetic permeability of one. The two magnets on the divider represent the two magnets on the anterior or posterior legs, while the two magnets beneath the divider represent the magnets that provide the external magnetic field to the anterior or posterior legs. We analyze the influence of different numbers of external magnets and varying divider thicknesses on the magnetic force exerted by the magnets.

The top two magnets are maintained in a stationary position while the bottom magnets are permitted to translate along the X-axis to investigate the relationships between the X-direction magnetic forces (F_X) and displacement (X), as well as between the Z-direction magnetic forces (F_Z) and displacement (X). The corresponding results are illustrated in Figure 6b,c. The results show that the F_X reaches its maximum when the moving distance is between 4 and 5 mm, while the F_Z keeps decreasing. Additionally, both F_X and F_Z increase with the number of external magnets, and using two magnets is the most suitable solution. Next, we keep magnets stationary and vary the thickness of the divider to obtain the relationship between the F_Z and the thickness of the divider, as shown in Figure 6d. The results reveal that the F_Z decreases gradually as the divider thickness increases and increases with the number of external magnets. Furthermore, we offset the top and bottom magnets by 5 mm and vary the thickness of the divider to examine the relationship between the F_X and the divider thickness, as shown in Figure 6e. The results are similar to those of Figure 6d, where the F_X decreases with increasing thickness of the divider and increases with the number of magnets. Unlike, as shown in Figure 6f, the F_Z is increasing and then decreasing with the thickness of the divider. Combining the above magnetic simulation results, we conclude that the best performance is achieved with two external magnets, providing a magnetic force of at least 0.16 N in the horizontal direction and 0.29 N in the vertical direction when the divider thickness is 5 mm or less. (The robot weighs only 5.9 g).

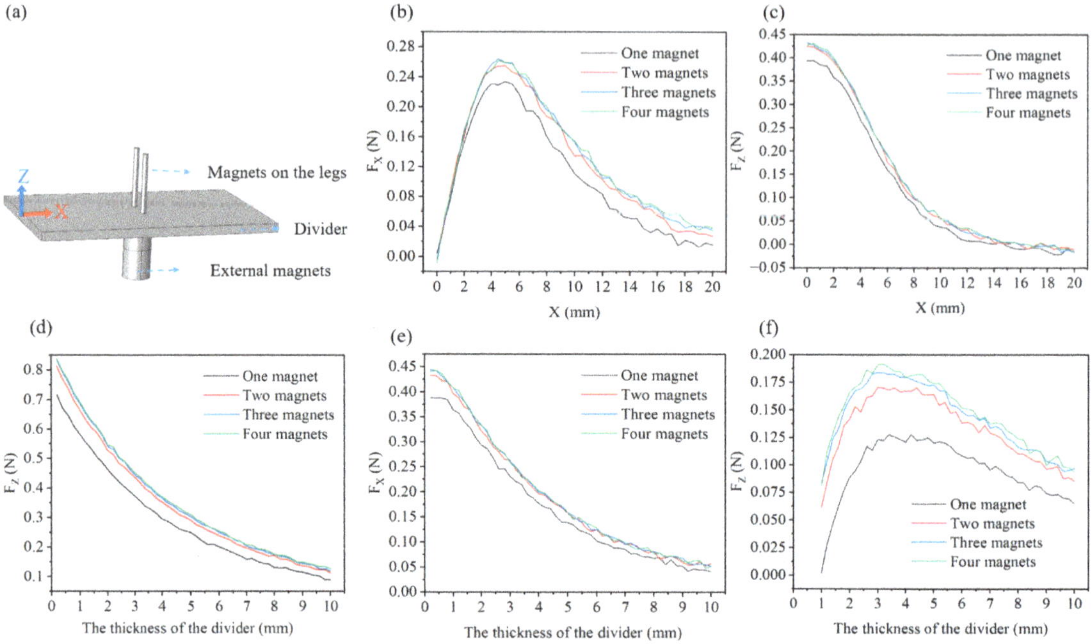

Figure 6. Simulation analysis of the magnetic forces: (**a**) Model for the magnetic forces simulation analysis. (**b**) Relationship between the magnetic forces in the horizontal direction (F_X) and forward displacement of the bottom magnets (X). (**c**) Relationship between the magnetic forces in the vertical direction (F_Z) and forward displacement of the bottom magnets (X). (**d**) Relationship between the magnetic forces in the vertical direction (F_Z) and the thickness of the divider. (**e**) Relationship between the magnetic forces in the horizontal direction (F_X) and the thickness of the divider when the bottom magnets are horizontally offset by 5 mm. (**f**) Relationship between the magnetic forces in the vertical direction (F_Z) and the thickness of the divider when the bottom magnets are horizontally offset by 5 mm.

Meanwhile, we investigate the variation in the magnetic induction intensity produced by the magnets as it moves. The magnetic induction intensity simulation is modeled as shown in Figure 7a. Initially, the external front and rear magnets are spaced 40 mm away from each other. We keep the external magnets in front stationary so that the external magnets at the rear moves forward (D mm). Create a moving axis (Z), which is always centered between the front and rear external magnets. The zero scale of the Z-axis is in the middle of the divider, oriented vertically upward. We obtain the variation in magnetic induction intensity at each point on the Z-axis for every D mm of forward movement of the rear magnet, as shown in Figure 7b. As D increases and the two external magnets continue to move closer together, the magnetic induction intensity at points on the Z-axis will gradually increase. We observe that the magnetic induction intensity appears to peak at $Z = -21.5$ and $Z = -1.5$, and the value of $Z = -21.5$ is greater than the value of $Z = -21.5$. This is due to the fact that at $Z = -1.5$, the magnetic induction intensity is subjected to the combined action of the external magnets and the leg magnets. The simulation results show that when the distance between the external front and rear magnets is greater than 20 mm, the magnetic induction intensity between the two magnets is less than 0.17 T, and the interferences between the two magnets are small.

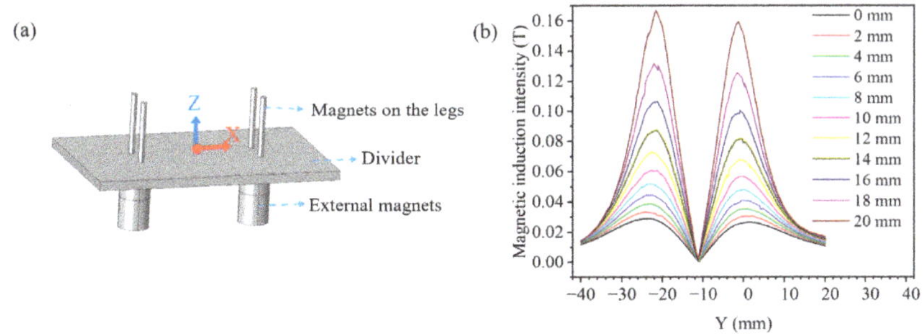

Figure 7. Simulation analysis of magnetic induction intensity: (**a**) Model of magnetic induction intensity simulation. (**b**) Variation in magnetic inductive on the Z-axis.

5. Test Experiments on Movement Ability of the 3D-SBIR

5.1. Test of the 3D-SBIR Movement on a Horizontal Plane

To elaborate on the movement principle of the D-SBIR, the robot's forward movement is observed on a horizontal plane. The process of the robot moving one step on a horizontal acrylic plate is shown in Figure 8. As illustrated in Figure 8i, to clearly describe the movement process of the robot, the upper endpoint of the anterior legs is labeled as O_1, the lower endpoint of the anterior legs is labeled as O_2, the upper endpoint of the posterior legs is labeled as O_3, and the lower endpoint of the posterior legs is labeled as O_4. Based on experimental results, the movement can be divided into four distinct states: initial state, bending state, stretching state, and natural state. In the initial state, shown in Figure 8i, the anterior legs of the robot are vertical and the posterior legs are tilted backward. In the bending state, shown in Figure 8ii, the posterior legs move forward, with the lower endpoints of the posterior legs shifting by about 33 mm, while the lower endpoints of the anterior legs remain motionless. This causes the belly of the robot to bend upward. The stress distribution of the robot during bending deformation is shown in Figure 8a, and the maximum pressure stress is mainly concentrated in the legs, with a maximum value of 0.67 MPa. In the stretching state, shown in Figure 8iii, the anterior legs move forward, and the lower endpoints of the anterior legs move forward by about 60 mm, whereas the lower endpoints of the posterior legs stay in place. The belly forms a downward bend, and the robot undergoes stretching deformation. The stress distribution of the robot during stretching deformation is shown in Figure 8b, and the maximum tensile stress is mainly concentrated in the legs and belly, with a maximum value of 0.53 MPa. In the natural state of the robot, shown in Figure 8iv, the posterior legs move forward, with the lower endpoints of the posterior legs shifting by about 35 mm, and the lower endpoints of the anterior legs remain in place, causing the robot to return to the natural state. This completes a full motion cycle, with the robot moving about 70 mm forward. Additionally, the robot is tested on a grass-like plastic pad as shown in Figure 9a and on a plastic track with a width 0.3 times the robot's body length, as shown in Figure 9b. Videos of the experiment are available in the Supplementary Information, and the experimental results prove that the robot exhibits excellent mobility.

5.2. Test of Climbing Obstacles for the 3D-SBIR

The 3D-SBIR has a soft body structure, allowing it to adapt well to its environment. In this study, this advantage is leveraged to assist the robot quickly in climbing over obstacles by utilizing contact deformation between the robot and the obstacles during the climbing process. As shown in Figure 10i, in the initial state of climbing obstacles, there are obstacles on both the anterior and posterior legs of the robot. In Figure 10ii, during the potential energy accumulation phase, the magnetic field of the anterior legs remains stationary, while the magnetic field of the posterior legs moves forward, causing the belly of the robot to bend and accumulate elastic potential energy. In Figure 10iii, during the potential energy release phase, the magnetic field of the anterior legs moves forward and the magnetic field of the posterior legs remains stationary. As the magnetic field of the anterior legs moves, the attraction force of the magnetic field of the anterior legs decreases, and the robot successfully jumps up to climb over the obstacle with the help of the elastic potential energy accumulated in the belly. In Figure 10iv, in the mobile state of climbing obstacles, after the anterior legs cross the obstacle, the posterior legs move to the front of the obstacle. In Figure 10v, during the state of passive deformation while climbing obstacles, the magnetic field of the posterior legs remains stationary, the magnetic field of the anterior legs moves forward, and the posterior legs are hindered by the obstacle, causing the belly to undergo stretching deformation. Finally, in Figure 10vi, in the natural state of climbing obstacles, the magnetic field of the anterior legs continues to move forward, and the robot climbs the obstacle successfully with the help of the pulling force of the belly.

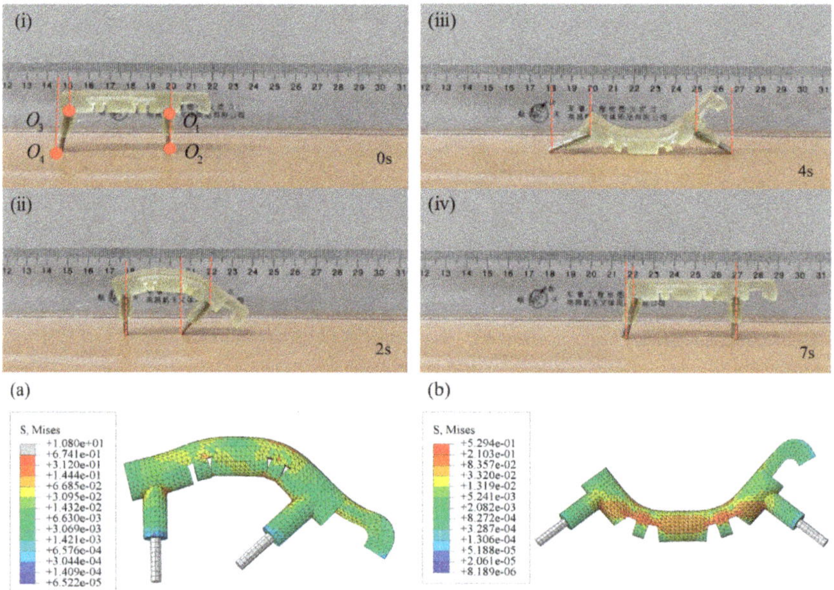

Figure 8. Process of the 3D-SBIR moving one step on a horizontal acrylic plate: (**i**) Initial state: Anterior legs are vertical and posterior legs are tilted backward. (**ii**) Bending deformation: Posterior legs move rightward, and belly is bending. (**iii**)Stretching deformation: Anterior legs move rightward and the belly is stretching. (**iv**) Natural state: Robot moves by a distance of 70 mm. (**a**) Stress distribution under bending deformation. (**b**) Stress distribution under stretching deformation.

Experiments on climbing steps and climbing vertical planes are conducted to further validate the climbing ability of the 3D-SBIR. As shown in Figure 11a, the process of the robot climbing steps is similar to that of crossing obstacles, with the deformation of the belly assisting in the accumulation of elastic potential energy for climbing. Using a bending deformation of the belly to make the anterior legs climb steps and a stretching deformation of the belly to make the posterior legs climb steps. The 3D-SBIR is assisted by the bending deformation of its head when climbing obstacles that are significantly larger than its own height (e.g., vertical walls). The head of the robot contacts the vertical surface as shown in Figure 11b. The magnetic field of the anterior legs moves forward, causing the head to bend and deform, fitting tightly against the vertical plane. The magnetic field of anterior legs moves upward, and the robot is guided upward by the head. The magnetic field of the anterior legs moves upward, followed by the magnetic field of the posterior legs moving forward, enabling the robot to successfully climb up the vertical surface.

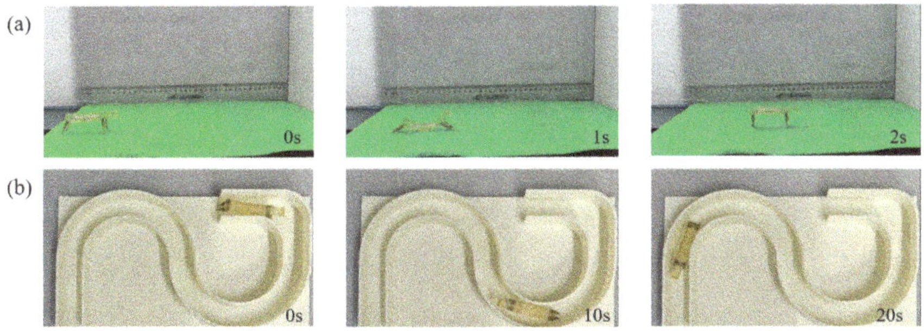

Figure 9. Test of the 3D-SBIR movement: (**a**) Robot moves on a grass-like plastic pad. (**b**) Robot moves through a narrow plastic runway.

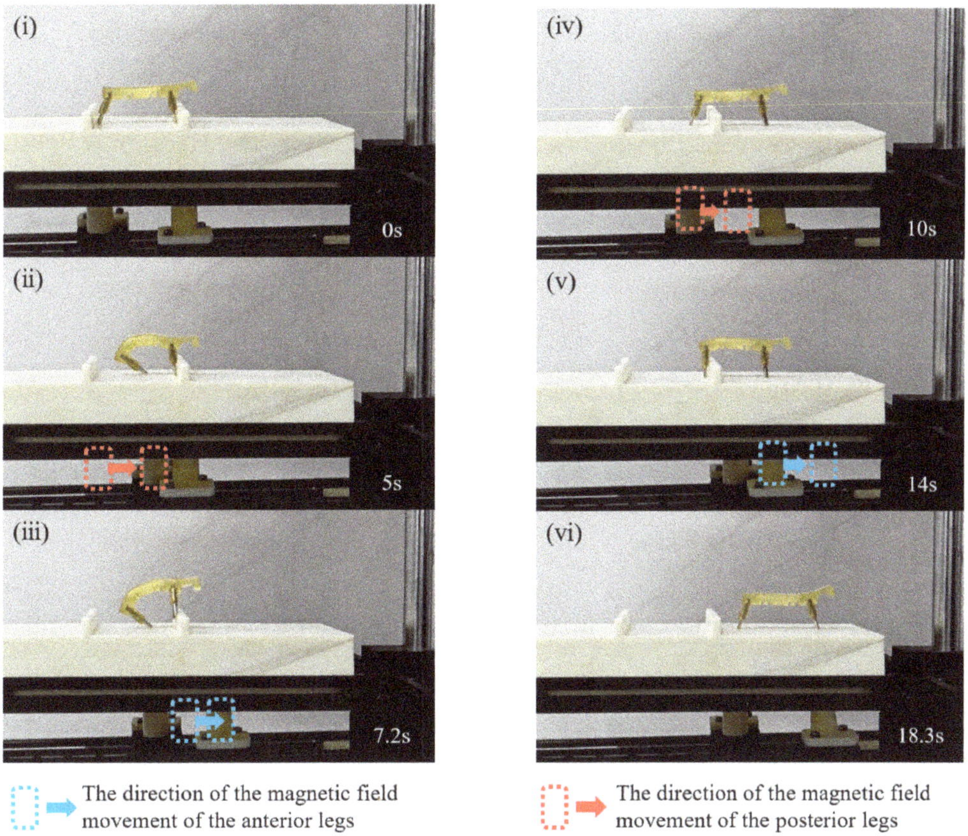

Figure 10. Experiment of the 3D-SBIR climbing obstacles: (**i**) Initial state of climbing obstacles. (**ii**) State of potential energy accumulation for climbing obstacles. (**iii**) State of potential energy release for climbing obstacles. (**iv**) Mobile state of climbing obstacles. (**v**) State of passive deformation of climbing obstacles. (**vi**) Natural state of climbing obstacles.

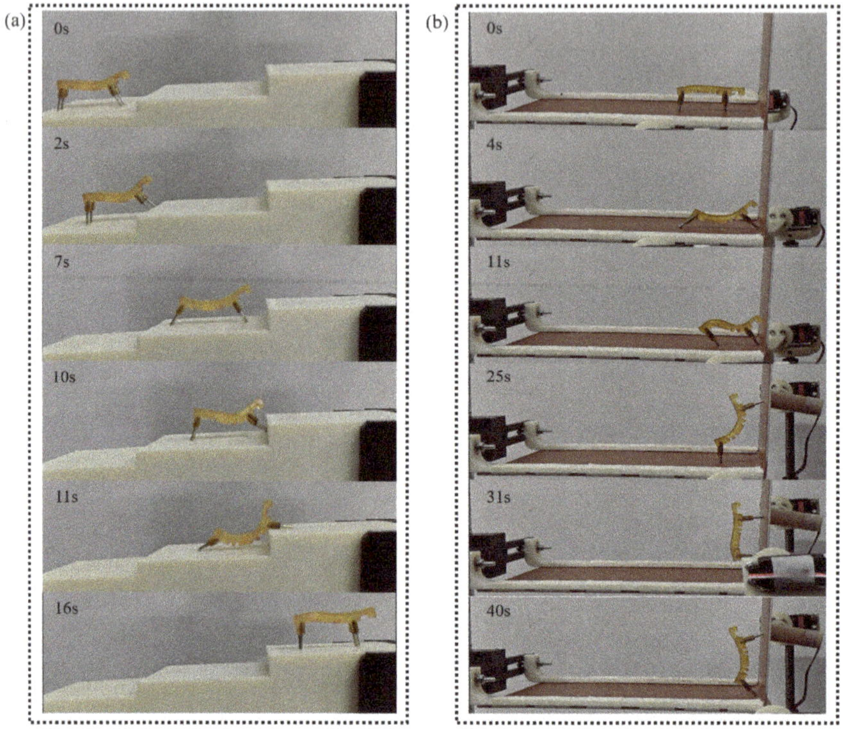

Figure 11. Test of motion ability in complex environments: (**a**) Climbing steps of 16mm in height. (**b**) Climbing vertical surfaces.

6. Conclusions

To address the problems of existing soft bionic inchworm robots, such as the complex production process and the influence of cables on the movement, this study proposes a 3D-printed soft bionic inchworm robot powered by magnetic force. First, the robot body is made directly by light-curing printing technology, and the NdFeB magnet is directly embedded in the gaps of the body as the power source. This eliminates the necessity for intricate production steps and is powered by magnetic drive, allowing cable-free operation. Additionally, the robot is capable of multimodal movement. Next, the material composition model and the bending deformation model of the robot are developed based on the Yeoh model. To validate the accuracy of the above models, simulation analysis and bending deformation experiments are conducted. The simulations and experimental results confirm the suitability of the models for the 3D-SBIR. At the same time, we discuss the main factors that influence the size of the magnetic force. With a divider thickness of 5 mm or thinner, two external magnets can provide a magnetic force of at least 0.16 N in the horizontal direction and 0.29 N in the vertical direction. Finally, unlike traditional SBIR, the 3D-SBIR utilizes its soft body structure to passively climb obstacles by making direct contact with them. As a result, the robot does not require a complex control and sensing system to have excellent obstacle-climbing capabilities. It is capable of climbing obstacles that are 0.5 times its body height with ease. The robot demonstrates multimodal movement capabilities, including the ability to traverse rough mats and narrow curved runways that are only 0.3 times its own length in width, climb steps that are 0.8 times its height, and transition from horizontal to vertical surfaces. In summary, the 3D-SBIR features a simple production process, cable-free actuation, and multimodal motion capabilities.

Supplementary Materials: The following supporting information can be downloaded at: https://www.mdpi.com/article/10.3390/biomimetics10040202/s1, Video S1: The motion gait analysis of the 3D-SBIR. Video S2: Test experiments on movement ability. Pdf S1: Supplementary.

Author Contributions: Conceptualization, D.X.; Funding acquisition, J.D.; Investigation, Q.D.; Methodology, D.X.; Project administration, L.Z.; Software, W.N.; Validation, D.X.; Writing—original draft, D.X.; Writing—review and editing, D.X. and J.D. All authors have read and agreed to the published version of the manuscript.

Funding: This research was funded by the National Natural Science Foundation of China [52475096]; Natural Science Foundation of Guangxi, China [2025GXNSFFA069009]; and National Science and Technology Innovation Development Doubling Plan Project of Guangxi University [2024BZRC010].

Institutional Review Board Statement: Not applicable.

Data Availability Statement: Data are contained within the article and Supplementary Materials.

Conflicts of Interest: The authors declare no conflicts of interest.

References

1. Niu, H.; Feng, R.; Xie, Y.; Jiang, B.; Sheng, Y.; Yu, Y.; Baoyin, H.; Zeng, X. MagWorm: A Biomimetic Magnet Embedded Worm-Like Soft Robot. *Soft Robot.* **2021**, *8*, 507–518. [CrossRef] [PubMed]
2. Romano, D.; Giovanni, A.; Stefanini, C. Assessing black solider fly pupation and survival in lunar regolith simulant: Implications for sustainable controlled habitats on the Moon. *Acta Astronaucica* **2024**, *223*, 505–511.
3. Kandhari, A.; Wang, Y.; Chiel, H.J.; Quinn, R.D.; Daltorio, K.A. An analysis of peristaltic locomotion for maximizing velocity or minimizing cost of transport of earchworm-like robots. *Soft Robot.* **2021**, *8*, 458–505.
4. Li, R.; Liu, Y.; Guo, A.; Shou, M.; Zhao, M.; Zhu, D.; Yang, P.-A.; Lee, C.-H. An Inchworm-Like Climbing Robot Based on Cable-Driven Grippers. *IEEE/ASME Trans. Mechatron.* **2024**, *29*, 1591–1600.
5. Liu, J.R.; Liu, L.W.; Liu, Y.J.; Leng, J.S. Dielectric Elastomer Spring-Roll Bending Actuators: Applications in Soft Robotics and Design. *Soft Robot.* **2019**, *6*, 69–81.
6. Xu, Z.F.; Hu, L.K.; Xiao, L.Y.; Jiang, H.J.; Zhou, Y.T. Modular Soft Robotic Crawlers Based on Fluidic Prestressed Composite Actuators. *J. Bionic Eng.* **2024**, *21*, 694–706.
7. Jiao, P.C.; Hao, Z.; Hong, L.Q.; Yang, Y.; Li, W.T. Piezo-Wormbots for Continuous Crawling. *Soft Robot.* **2024**, *11*, 260–269.
8. Cao, J.; Qin, L.; Liu, J.; Ren, Q.; Foo, C.C.; Wang, H.; Lee, H.P.; Zhu, J. Untethered soft robot capable of stable locomotion using soft electrostatic actuators. *Extrem. Mech. Lett.* **2018**, *21*, 9–16.
9. Gu, G.; Zou, J.; Zhao, R.; Zhao, X.; Zhu, X. Soft wall-climbing robots. *Sci. Robot.* **2018**, *3*, 2874.
10. Ding, J.; Su, H.; Nong, W.; Huang, C. An inchworm-inspired soft robot with combined functions of omni-directional steering and obstacle surmounting. *Ind. Robot. Int. J. Robot. Res. Appl.* **2023**, *50*, 456–466.
11. Zhang, Y.; Huang, P.; Li, D.; Zhou, J.; Li, Y.; You, B.; Zhu, Y. Multi-modal Bionic Motion Analysis of A Cpg-controlled Pneumatic Soft Robot. *J. Bionic Eng.* **2024**, *21*, 2247–2257. [CrossRef]
12. Huang, H.; Xie, J.; Huang, C.; Wu, D.; Ouyang, J.; Pang, C.; Zheng, L. Soft robot based on dual-mode deformation driving variable configuration. *J. Shenzheng Univ. Sci. Eng.* **2024**, *41*, 602–611. [CrossRef]
13. Lara-Ortiz, V.; Guarneros, A.; Llorente-Vidrio, D.; Cruz-Ortiz, D.; Salgado, I.; Chairez, I. Active Disturbance Rejection Controller for a Flexible Walking Bioinspired Inchworm Mobile Robot Actuated with Shape Memory Alloy Devices. *IEEE Trans. Control. Syst. Technol.* **2022**, *30*, 1790–1797. [CrossRef]
14. Hu, W.Q.; Lum, G.Z.; Mastrangeli, M.; Sitti, M. Small-scale soft-bodied robot with multimodal locomotion. *Nature* **2018**, *554*, 81–85. [CrossRef]
15. Shen, H.; Cai, S.; Wang, Z.; Yuan, Z.; Yu, H.; Yang, W. A Programmable Inchworm-Inspired Soft Robot Powered by a Rotating Magnetic Field. *J. Bionic Eng.* **2023**, *20*, 506–514. [CrossRef]
16. Steiner, J.A.; Pham, L.N.; Abbott, J.J.; Leang, K.K. Modeling and Analysis of a Soft Endoluminal Inchworm Robot Propelled by a Rotating Magnetic Dipole Field. *J. Mech. Robot.* **2022**, *14*, 051002. [CrossRef]
17. Steiner, J.A.; Nagel, W.S.; Leang, K.K. Magnetically-Actuated Endoluminal Soft Robot with Electroactive Polymer Actuation for Enhanced Gait Performance. *J. Mech. Robot.* **2024**, *16*, 104503. [CrossRef]
18. Xu, R.; Xu, Q. Design of Bio-Inspired Untethered Soft Octopodal Robot Driven by Magnetic Field. *Biomimetics* **2023**, *8*, 269. [CrossRef]
19. Xie, D.S.; Liu, J.B.; Kang, R.J.; Zuo, S.Y. Fully 3D-Printed Modular Pipe-Climbing Robot. *IEEE Robot. Autom. Lett.* **2021**, *6*, 462–469. [CrossRef]

20. Sun, B.; Jia, R.; Yang, H.; Chen, X.; Tan, K.; Deng, Q.; Tang, J. Magnetic Arthropod Millirobots Fabricated by 3D-Printed Hydrogels. *Adv. Intell. Syst.* **2022**, *4*, 2100139. [CrossRef]
21. Zhu, Y.; Liu, N.; Chen, Z.; He, H.; Wang, Z.; Gu, Z.; Chen, Y.; Mao, J.; Luo, Y.; He, Y. 3D-Printed High-Frequency Dielectric Elastomer Actuator toward Insect-Scale Ultrafast Soft Robot. *ACS Mater. Lett.* **2023**, *5*, 704–714.
22. Joyee, E.; Pan, Y. A Fully Three-Dimensional Printed Inchworm-Inspired Soft Robot with Magnetic Actuation. *Soft Robot.* **2019**, *6*, 333–345. [PubMed]
23. Li, X.B.; Wei, Y.T. An Improved Yeoh Constitutive Model for Hyperelastic Material. *Eng. Mech.* **2016**, *33*, 38–43.
24. Fei, Y.Q.; Peng, W.; Yu, W.B. Movement of Air-driven Soft Robot. *J. Mech. Eng.* **2017**, *53*, 14–18.
25. Li, X.; Liu, S.; Xu, Q.Y.; Zhang, Y.; Huo, Q.; Zhang, Y. Research on Prediction Method of Bending of Imitation Manta Soft Body Driver. *Chin. Hydraul. Pneum.* **2023**, *47*, 63–69.

biomimetics

MDPI

Article

A Computational Study on the Hydrodynamics of Bio-Inspired Quadrupedal Paddling

Yihan Wang [1,†], Yumeng Cai [1,†], Bin Xie [1], Chi Zhu [2], Yunquan Li [1,*] and Ye Chen [1,*]

1 Shien-Ming Wu School of Intelligent Engineering, South China University of Technology, Guangzhou 510641, China; wiyihanwang@mail.scut.edu.cn (Y.W.); wicaiyumeng@mail.scut.edu.cn (Y.C.); xieb7160@gmail.com (B.X.)
2 Department of Mechanics and Engineering Science, Peking University, Beijing 100871, China; chi.zhu@pku.edu.cn
* Correspondence: yunquanli@scut.edu.cn (Y.L.); yechen@scut.edu.cn (Y.C.)
† These authors contributed equally to this work.

Abstract: Due to its exceptional terrain mobility, quadrupedal locomotion has been used in the design of many amphibious robots for broad applications including resource exploration, disaster rescue, and reconnaissance. In this work, swimming of a quadrupedal paddling model is considered, and the effects of the legs' initial swing angle, swing amplitude, and power phase duration are numerically investigated through three paddling gaits, namely, the trotting gait, the diagonal, and the lateral sequence gaits. Three different modes for drag-based thrust generation, the "Trotting Mode", the "Hindering Mode", and the "Separate Mode", are identified. In the "Trotting Mode", each pair of diagonal legs contributes equally and alternately to the thrust within the paddling cycle, and its contribution is impaired by the other pair of diagonal legs. In the "Hindering Mode", the thrust contribution of an individual leg is significantly undermined by the drag resulting from the preceding leg leaving its current power phase and entering the following recovery phase. In the "Separate Mode", the four legs independently contribute to the total thrust by forming a compact four-peak waveform equally distributed within one paddling cycle. At a given swing amplitude, the leg configuration at peak thrust moment is identical, regardless of initial swing angle and power phase ratio. Meanwhile, a forward-tilted leg configuration with flatter upper- and lower-limb alignment at peak thrust moment consistently indicates a lower thrust generation. Hydrodynamic moments in the diagonal and lateral sequence gaits are much larger than those in the trotting gait. In addition, enhanced thrust is typically accompanied by larger hydrodynamic moments and a higher energy expenditure.

Keywords: quadrupedal paddling; thrust generation; immersed-boundary method

1. Introduction

During the past two decades, various types of amphibious robot have been developed with the aim of quickly transitioning between land and water for broad applications including resource exploration, disaster rescue, and reconnaissance [1]. Normally, locomotion on land and locomotion in water are achieved by different mechanisms for amphibious robots. For land locomotion, legged robots have long captured the interest of researchers because of their outstanding mobility and versatility on rough and transitional zones such as sandy and muddy terrains [2–7].

A variety of bio-inspired propulsion mechanisms have been employed in water with inspiration drawn from nature. Body undulations, fish fin-like propulsors, and paddling legs are three common methods of bio-inspired waterborne propulsion for amphibious robots. For example, the snake-like robot ACM can propel itself through body undulation or spiral spinning with a "HELIX" mode at a speed of 0.4 m/s [8,9]. A similar snake-like Multi-Link Mobile Robot MLMR II is proposed with forward and lateral turning ability underwater [10]. A salamander robot can reproduce an overall natural-body-like undulation and swim at velocities ranging from 0.07 to 0.12 m/s [11]. In addition to body undulation, bio-inspired fish fins have also been used as propulsors by amphibious robots for efficient underwater locomotion. AmphiRobot-II uses a caudal fin to perform a BCF-type carangiform swimming with a forward speed of up to 45 cm/s [2]. A pair of ribbon fins similar to those of sting rays is manufactured and the maximum MPF forward swimming speed is about 0.31 m/s at an undulating frequency of 2 Hz [12]. Such bio-locomotion using either body undulation or fin flapping relies on the generation of lift-based thrust. In contrast, drag-based thrust represents another mechanism for propulsion, which is usually deployed by terrestrial and semi-aquatic mammals with paddling appendages [13]. Although propulsion through body undulation and fin flapping is known to be more efficient in water, their terrestrial movements capability is very limited and can hardly be compared with legged locomotion [13,14]. Therefore, legged robots are suitable for amphibious applications despite their relatively low swimming efficiency. Recently, learning-based methods have been developed for gait pattern learning and energy efficiency improvement in complex environments [15–20]. However, the interaction between legged robots and fluid medium during the learning process remains unexplored. Other approaches, e.g., a combination of wheels for land movement and fins for swimming, may lead to a much more complicated and heavier amphibious robot design that may not be desirable.

Quite a few legged amphibious robots have been developed in the past. For example, a six-legged crab-like robot ALUV is reported to be capable of walking underwater for mine hunting in the surf zone [4]. A cockroach-inspired robot RHex was first built between 1999 and 2004, based on which the subsequent AQUA platform is further developed with underwater mobility [6,21]. AmphiRobot-I uses a transformable fin-leg module and obtains a moving speed of about 0.5 m/s on land and in water [22]. A bio-inspired quadruped amphibious robot with agile legs based on the five-bar mechanism is also constructed with a successful walking and swimming mobility [23]. With the assistance of the computational fluid dynamics (CFD) technique, one is able to study the hydrodynamics and swimming efficiency of underwater locomotion. In this field, body undulation and fin flapping have been extensively researched [24–29].

Unfortunately, to date the study of hydrodynamics for the quadrupedal paddling using the CFD technique is scarce. Among the limited existing studies, the force analysis of quadrupedal hydrodynamics depends on the drag equation and an assumption of the drag coefficient C_d, and the complex three-dimensional geometrical shape of the legs and the interaction between different limbs are ignored [14,27,30,31]. Such simplified force models are easy to use but may lead to significant errors in practice. Relevant CFD research on leg paddling include computational studies on crawfish, beaver-, and turtle-like robots [31–33]. Besides, the synchronous and alternate paddling of the two beaver-like bendable webbed feet is reconstructed and the hydrodynamic forces are investigated through both theoretical analysis and CFD simulations. Table 1 is summarized for previous relevant studies on leg paddling using CFD techniques. However, the models in these studies are distinct from the quadrupedal paddling of mammals, which deserves a further in-depth study since a detailed discussion on paddling hydrodynamics and many other factors, such as paddling kinematics and gait sequences, has not been explored yet. Although a qualitative understanding of the drag-based thrust generation caused by paddling legs may be straightforward (e.g., faster backswing of a paddling leg leads to stronger reaction force by surrounding fluid), a detailed quantitative study is necessary to better understand a few important questions: (I) How propulsion thrust is affected by paddling kinematic parameters including initial swing angle, swing amplitude and power phase ratio? (II) How is the propulsion thrust influenced by different gait sequences? (III) How are the hydrodynamic moments and energetics modulated by paddling kinematics and gait sequences? Motivated by the above fact, it is our novelty to solve the three-dimensional flow field and investigate these questions by performing a series of numerical simulations with the aid of a previously developed parallel CFD solver based on an immersed boundary method.

Table 1. Previous relevant CFD research on leg paddling.

Authors	Year	Leg Number	Simulation Approach	Application Type
Zhang, Calvin, et al. [32]	2014	4	Immersed Boundary Method	NA
Chen, Gang, et al. [31]	2021	2	Sliding/dynamic mesh	Fluent
Baines, Robert, et al. [33]	2022	4	SIMPLEC solver	Fluent

In this work, we first investigate the paddling-induced thrust of a single leg with initial swing angle $\alpha_0 \in [60^\circ, 70^\circ, 80^\circ]$ and swing amplitude $\beta_0 \in [20^\circ, 30^\circ, 40^\circ]$ for three fixed power phase ratios DR $\in [50\%, 33\%, 25\%]$. Next, we examine the effect of the gait sequence on the quadrupedal paddling thrust generation in the same parameter space of α_0 and β_0. Three different mechanisms for thrust generation are categorized and analyzed, and the corresponding hydrodynamic moments and energetics are also discussed.

2. Model Description and the Numerical Approach

The three-dimensional computational model of quadrupedal paddling used in this work is constructed based on an actual quadrupedal robot and is illustrated in Figure 1a,b. The left-fore leg (LF), left-hind leg (LH), right-fore leg (RF), and right-hind leg (RH) are immersed in an incompressible viscous fluid and perform coordinated paddling motion. The lengths of the upper and lower limbs for each leg, L_1 and L_2, are 10.6 and 12.0 cm, respectively. The width W of the limb is 4.65 cm. The distance L between the fore and hind legs is 32 cm, while the width D between the left and right legs is 12.4 cm. These dimensions are taken from an actual robot. A coordinate system is chosen such that its origin O lies in the middle of the upper limb pivots of the four legs. The influence of three major kinematic parameters, the initial swing angle α_0, the swing amplitude β_0, and the power phase ratio DR are illustrated in the figure. The following values of the parameters are chosen: $\alpha_0 \in [60^\circ, 70^\circ, 80^\circ]$, $\beta_0 \in [20^\circ, 30^\circ, 40^\circ]$, and DR $\in [25\%, 33\%, 50\%]$. Considering the constant stepping speed of the step motor used in the experimental study in which the duration of the power phase is fixed, DR $=$ 50%, 33% and 25% correspond proportionally to a paddling cycle of $T = 0.8$ s, 1.2 s, and 1.6 s, respectively. The prescribed cyclic paddling motion of upper and lower limbs are defined by position angles $\alpha(t)$ and $\gamma(t)$ (relative to its upper limb), and their transient variations are obtained from the video of the swimming dog provided by [34]. The average lower limb tip velocity $\bar{U}$ within the power phase under DR $=$ 50% is 0.64, 0.49 and 0.42 m/s for $\beta_0 = 20^\circ$, 30° and 40°, respectively. This velocity increases further to 0.96, 0.73, 0.60 for DR $=$ 33% and 1.34, 1.03, 0.84 m/s for DR $=$ 25%, respectively, for $\beta_0 = 20^\circ$, 30° and 40°. Figure 1c presents the kinematic sequence of a single paddling leg under $\alpha_0 = 70^\circ$ and $\beta_0 = 20^\circ$. In addition to the three paddling kinematic parameters mentioned above, three distinct coordinated leg movements, trotting, diagonal sequence gait, and lateral sequence gait, are also included in this study (Figure 1d). We also point out that when designing an amphibious robotic dog for experimental validation, several key factors must be considered, including actuation, structural stability, waterproofing, and control. The selection of actuators plays a crucial role in swimming performance, and waterproof servos with sufficient torque, such as GXservo-X50 (50 kg·cm), can provide the necessary power for paddling. Additionally, ensuring a proper relationship between the center of buoyancy (CB) and the center of gravity (CG) is essential for maintaining stability in water. Inspired by biological quadrupeds, the CB should be positioned above the CG, reducing the risk of rolling and improving maneuverability. A robust control system can be used to coordinate leg movements, while a wireless remote control allows for flexible gait adjustments. Furthermore, waterproofing measures—including sealed body joints, waterproof adhesives, and protective covers for electronic ports—are essential to prevent water ingress and ensure long-term operation in aquatic environments.

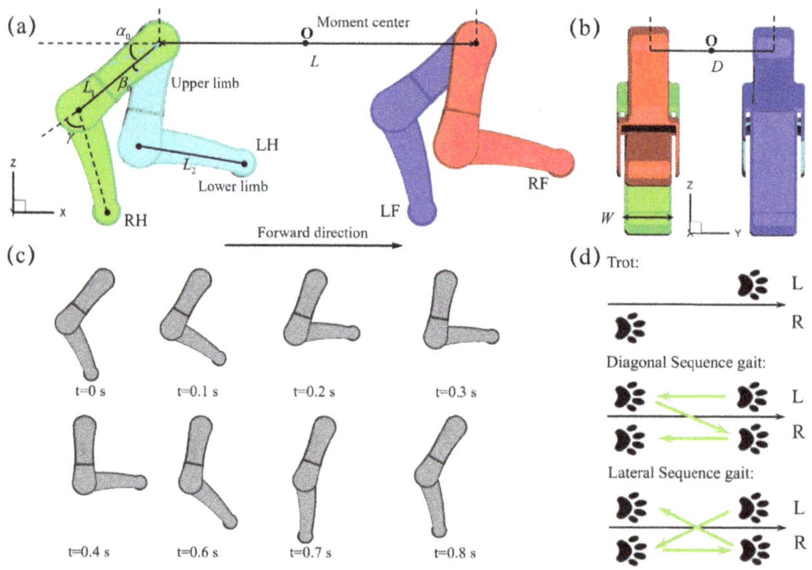

Figure 1. The geometry of the quadrupedal paddling model, (**a**) side view and (**b**) front view. (**c**) Reconstructed paddling sequence of both upper and lower limbs of a single leg under $\alpha_0 = 70°$ and $\beta_0 = 20°$. (**d**) The three distinct quadrupedal paddling gaits of trotting, the diagonal and lateral sequence gaits considered in this work.

The legs are assumed to be rigid without deformation and the quadrupedal paddling system is tethered without displacement. Similar tethered models to ours are often used for the study of self-propulsion of fish swimming [35–38]. Assuming the structure is rigid, one-way coupling is employed to model the fluid-structure interaction, where the motion of the quadrupedal paddling significantly influences the fluid flow. However, structural deformation due to fluid forces is disregarded in this analysis. For spatial discretization, each leg is divided into 29,940 triangular surface elements with a total number of 14,974 nodes. The fluid is assumed to be Newtonian and incompressible. The governing equation of the flow is the unsteady three-dimensional viscous Navier-Stokes equation. Since the peak Reynolds number Re is on order of $O(10^4)$, we do not consider turbulence effect in present work, although it may become necessary when the background environmental flow is chaotic or the robotic system size of quadrupedal paddling becomes larger. The fluid domain is represented by $20L \times 32D \times 12L$ cm^3 rectangular bounding box and is divided by a $290 \times 160 \times 160$ nonuniform Cartesian grid. Fine resolution with

$\Delta x = \Delta y = \Delta z = 0.25$ cm is used in the region around the four paddling legs (Figure A1a in Appendix A). No-slip and no-penetration boundary conditions are imposed on limb surface and zero velocity and zero normal pressure gradient boundary conditions are applied on the six outer faces of the fluid domain. Initially, the flowfield is static, and the flow is driven in motion by coordinated paddling. The density and dynamic viscosity of the fluid are, respectively, $\rho = 1\ \mathrm{g/cm^3}$ and $\mu = 0.001$ Pa·s. To ensure the numerical stability of the flow solver, the time step used in the simulation is $\Delta t = 0.0002$ s, so that each paddling cycle of DR = 50%, 33%, and 25% contains 4000, 6000 and 8000 time steps, respectively. The choice of mesh size $\Delta x = 0.25$ cm and time step $\Delta t = 0.0002$ s is based on the mesh and time step convergence studies described in Appendix A. The propulsion force discussed in this work is normalized by $\rho_f U_0{}^2 W^2$, where a common limb tip velocity $U_0 = 0.42$ m/s under $\beta_0 = 40°$ and DR = 50%, and limb width W are chosen as our velocity and length reference, respectively. The viscous three-dimensional unsteady Navier-Stokes equation is numerically solved using a parallel in-house sharp-interface immersed-boundary method; more information on the computational implementation and mesh convergence study is provided in Appendix A.

3. Results

3.1. Hydrodynamics of a Single Paddling Leg

We first conduct numerical simulations to investigate the hydrodynamics of a single paddling leg with the initial swing angle $\alpha_0 \in [60°, 70°, 80°]$ and swing amplitude $\beta_0 \in [20°, 30°, 40°]$ under three different power phase ratios DR = 50%, 33% and 25%. The time duration of each paddling cycle is $T = 0.8$ s, 1.2 s, and 1.6 s for $\beta_0 = 20°$, 30°, and 40°, respectively. To avoid the influence caused by the initial condition that sets zero fluid velocity everywhere, simulation results from the 5th cycle is shown in Figure 2. Figure 2a presents the transient profile of the thrust $\bar{F}_x(t)$ with varying initial swing angle $\alpha_0 \in [60°, 70°, 80°]$ at a fixed swing amplitude of $\beta_0 = 20°$. We can see that, although the paddling motion gives rise to thrust ($\bar{F}_x(t) > 0$) and drag ($\bar{F}_x(t) < 0$) in both recovery and power phases, most of the thrust is generated during the power phase. Due to the rapid backswing of the lower limb, $\bar{F}_x(t)$ first rises quickly from $t = 3.5$ s ($t = 0.375T$) before reaching a peak value $\bar{F}_{x,pk}$ at $t = 3.7$ s ($t = 0.625T$), it then experiences a fast drop as the backswing motion continues. As the initial swing angle α_0 increases from 60° to 70° then to 80°, the corresponding peak thrust $\bar{F}_{x,pk}$ reduces from 10.53 to 9.90 then to 8.79. However, $\bar{F}_{x,pk}$ occurs almost at the same instant of $t_{pk} = 3.7$ s, which corresponds to $t_{pk} = 0.625T$ within a paddling cycle with period $T = 0.8$ s for $\beta_0 = 20°$. Within the complete paddling cycle, the average thrust $\bar{F}_{x,avg}$ also decreases from 0.75 to 0.59 then to 0.43 with increasing $\alpha_0 = 60°, 70°$, and 80°. When the single leg is accelerated posteriorly, a high pressure region is generated on the back side of the lower limb, while leaving a negative-pressure region simultaneously in its front side (Figure 2b). This pressure difference across the lower limb for $\alpha_0 = 60°, 70°$, and 80° shares a similar pattern and leads to the so-called drag-based thrust. At the peak thrust moment t_{pk}, the orientation of the upper limb is different with $\alpha_{pk} = 70°, 80°$, and 90° for $\alpha_0 = 60°, 70°$ and 80°, respectively; however, the relative position of the lower limb with respect to the upper limb remains the same with $\gamma_{pk} = 55°$ (Figure 2b). Similar varying trends for $\bar{F}_{x,pk}$, t_{pk}, α_{pk}, γ_{pk} and $\bar{F}_{x,avg}$, as well as the pressure contour around the paddling leg at $\bar{F}_{x,pk}$, are also observed for the cases with larger swing angles $\beta_0 = 30°$ and 40°. It can be seen that the average thrust $\bar{F}_{x,avg}$ drops with respect to both increasing initial swing angle α_0 and swing amplitude β_0. The lowest $\bar{F}_{x,avg}$ is generated at the largest $\alpha_0 = 80°$ and $\beta_0 = 40°$. In Figure 2c, we also presented the profiles of transient thrust generated under other combinations of initial swing angle α_0, swing amplitude β_0 and power phase ratio DR, which will be used for the discussion of inter-limb interaction later. In addition, we computed the thrust coefficient via $C_T = 2F_{x,avg}/\rho_f \bar{U}^2 L^2 W$, where $\bar{U}$ is the average lower limb tip velocity of individual cases. Upon fixing β_0 and DR, a consistently higher C_T can be achieved at smaller α_0. Relevant results are summarized in Table 2.

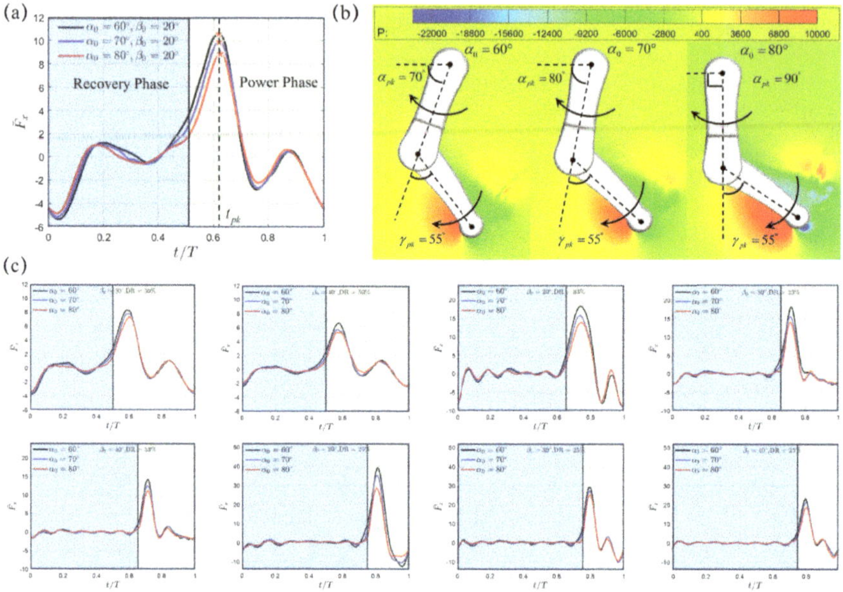

Figure 2. (**a**) The profiles of the transient thrust $\bar{F}_x(t)$ generated by a single paddling leg under $\alpha_0 = 60°$, $70°$, and $80°$ at fixed $\beta_0 = 20°$ and DR $= 50\%$. (**b**) The instantaneous pressure contour (Unit: 0.1 Pa) and the corresponding leg configuration with its upper- and lower-limb alignment at the peak thrust moment t_{pk}. The transient variation of angles (**c**) Transient thrust at different initial swing angle α_0, swing amplitude β_0 and power phase ratio DR. Recovery and power phases are marked using blue and green windows, respectively.

Table 2. Values of $\bar{U}$, α_{pk}, γ_{pk}, $\bar{F}_{x,pk}$, $\bar{F}_{x,avg}$ and C_T at swing amplitude $\beta_0 \in [20°, 30°, 40°]$, initial swing angle $\alpha_0 \in [60°, 70°, 80°]$ and power phase ratio of power phase DR $\in [25\%, 33\%, 50\%]$.

$(\beta_0, \alpha_0, \text{DR})$	$\bar{U}$ (m/s)	T (s)	t_{pk} (T)	$(\alpha_{pk}, \beta_{pk})$	$\bar{F}_{x,pk}$	$\bar{F}_{x,avg}$	C_T
(20, 60, 50)	0.64	0.8	0.625	(70, 55)	10.53	0.75	0.24
(20, 70, 50)	0.64	0.8	0.625	(80, 55)	9.90	0.59	0.19
(20, 80, 50)	0.64	0.8	0.625	(90, 55)	8.79	0.43	0.14
(30, 60, 50)	0.49	1.2	0.583	(80, 45)	8.26	0.62	0.35
(30, 70, 50)	0.49	1.2	0.583	(90, 45)	7.71	0.57	0.32
(30, 80, 50)	0.49	1.2	0.583	(100, 45)	7.15	0.47	0.26
(40, 60, 50)	0.42	1.6	0.581	(90, 35)	6.62	0.53	0.41
(40, 70, 50)	0.42	1.6	0.581	(100, 35)	5.71	0.45	0.34
(40, 80, 50)	0.42	1.6	0.581	(110, 35)	5.37	0.40	0.30
(20, 60, 33)	0.96	0.8	0.740	(70, 55)	18.24	1.27	0.19
(20, 70, 33)	0.96	0.8	0.740	(80, 55)	15.68	0.89	0.13
(20, 80, 33)	0.96	0.8	0.740	(90, 55)	13.75	0.72	0.11
(30, 60, 33)	0.73	1.2	0.710	(80, 45)	17.50	0.82	0.21
(30, 70, 33)	0.73	1.2	0.710	(90, 45)	15.57	0.70	0.18
(30, 80, 33)	0.73	1.2	0.710	(100, 45)	13.95	0.54	0.14
(40, 60, 33)	0.60	1.6	0.715	(90, 35)	14.06	0.73	0.27
(40, 70, 33)	0.60	1.6	0.715	(100, 35)	12.38	0.56	0.21
(40, 80, 33)	0.60	1.6	0.715	(110, 35)	11.11	0.42	0.16

Table 2. *Cont.*

$(\beta_0, \alpha_0, \mathrm{DR})$	$\bar{U}$ (m/s)	T (s)	t_{pk} (T)	$(\alpha_{pk}, \beta_{pk})$	$\bar{F}_{x,pk}$	$\bar{F}_{x,avg}$	C_T
(20, 60, 25)	1.34	0.8	0.810	(70, 55)	39.32	2.35	0.18
(20, 70, 25)	1.34	0.8	0.810	(80, 55)	34.98	1.80	0.14
(20, 80, 25)	1.34	0.8	0.810	(90, 55)	28.54	0.86	0.06
(30, 60, 25)	1.03	1.2	0.800	(80, 45)	31.81	1.31	0.17
(30, 70, 25)	1.03	1.2	0.800	(90, 45)	28.68	1.17	0.15
(30, 80, 25)	1.03	1.2	0.800	(100, 45)	25.86	0.84	0.11
(40, 60, 25)	0.84	1.6	0.800	(90, 35)	22.80	1.22	0.23
(40, 70, 25)	0.84	1.6	0.800	(100, 35)	21.08	1.11	0.21
(40, 80, 25)	0.84	1.6	0.800	(110, 35)	18.95	0.76	0.15

3.2. *Quadrupedal Paddling Using the Trotting Gait*

To investigate the impact of quadrupedal paddling on total thrust, we next consider the trotting gait sequence, which is adopted by many terrestrial mammals. In this sequence, each pair of diagonal legs share the same movement, while the two legs on the same side have opposite movement. The duration of the power phase is equal to that of the recovery phase, which corresponds to a power phase ratio of DR $= 50\%$. Figure 3a,b show the transient profiles of the horizontal and vertical forces, $\bar{F}_x(t)$ and $\bar{F}_z(t)$, within a complete paddling cycle of $T = 0.8$ s under $\beta_0 = 20^\circ$. Distinct from the single paddling leg situation, under the fixed swing amplitude of $\beta_0 = 20^\circ$, both $\bar{F}_x(t)$ and $\bar{F}_z(t)$ now demonstrate two peak values with an equal time interval of $T/2$. $\bar{F}_{x,pk}$ occurs at $t = 3.3$ s ($t = 0.125T$) and 3.7 s ($t = 0.625T$), while $\bar{F}_{z,pk}$ is achieved at $t = 3.25$ s ($t = 0.0625T$) and 3.65 s ($t = 0.5625T$), which corresponds to a slight 6.25% phase lead over $\bar{F}_{x,pk}$. When one pair of diagonal legs generates thrust in the power phase, the remaining two diagonal legs are in the recovery phase and have a counterproductive effect by generating drag (Figure 3a). Figure 3c shows the pressure and velocity fields in the plane that cuts through RF and RH at $t/T = 0, 1/8, 1/4, 3/8, 1/2, 5/8, 3/4$, and $7/8$. Using the length scale of $W = 4.65$ cm and velocity scale of $MDPI\bar{U} = 0.64$ m/s, the corresponding Reynolds number for this case is around $Re = 3.0 \times 10^4$. Vortex shedding from the tip of the lower limb as a result of its rapid paddling motion is conspicuous. During the first half of the trotting cycle $t/T \in [0, 1/2]$, RF is in the power phase and a counterclockwise vortex is produced near the tip of the lower limb when $t/T = 1/8$ and $1/4$. Meanwhile, RH is in the recovery phase and a clockwise vortex can be observed near its lower limb tip. During the latter half of the trotting cycle $t/T \in [1/2, 1]$, the vortex behavior that RF and RH have previously experienced is flipped, and clockwise and counterclockwise vortices are formed near the lower limb tips of RF and RH respectively.

As the initial swing angle α_0 increases from 60° to 70° to 80°, the instants of $\bar{F}_{x,pk}$ and $\bar{F}_{z,pk}$, the time intervals between $\bar{F}_{x,pk}$ and $\bar{F}_{z,pk}$, and the time lead of $\bar{F}_{z,pk}$ over $\bar{F}_{x,pk}$ remain unchanged. However, the changes of $\bar{F}_{x,pk}$ and $\bar{F}_{z,pk}$ are in an opposite trend. $\bar{F}_{x,pk}$ reduces from 20.95 to 20.30 to 18.95 when α_0 increases from 60° to 70° to 80°, respectively, while at the same time $\bar{F}_{z,pk}$ experiences a slight raise from 12.75 to 13.02 to 13.55 (Figure 3a,b). Similarly, $\bar{F}_{x,avg}$ drops from 3.34 to 2.76 to 1.83, while $\bar{F}_{z,avg}$ increases from 3.23 to 3.42 to 3.97. Within the power phase, two diagonal legs contribute equally to $\bar{F}_x(t)$. If the phase difference of $T/2$ in $\bar{F}_x(t)$ between the two pairs of diagonal legs is ignored, the magnitudes of horizontal and vertical forces of the four individual legs are almost identical, indicating that the four legs also contribute equally to the total forces of $\bar{F}_x(t)$ and $\bar{F}_z(t)$. Hereafter, we denote this mechanism of propulsion generation as the "Trotting Mode".

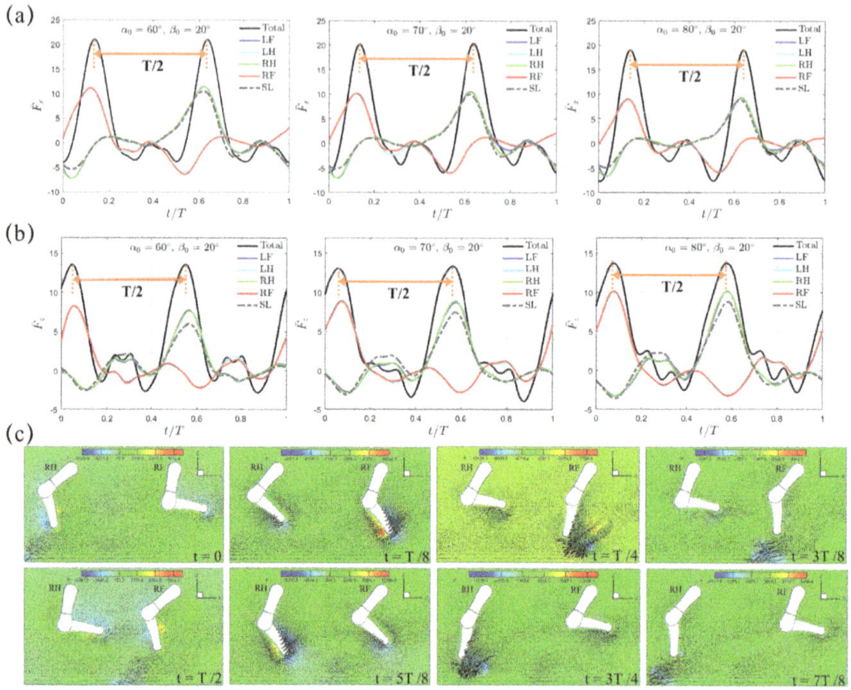

Figure 3. The transient profiles of (**a**) horizontal thrust $\bar{F}_x(t)$ and (**b**) vertical lift $\bar{F}_z(t)$ generated by the quadrupedal paddling using the trotting gait under $\beta_0 = 20°$. Results from the single leg (SL) paddling are also included. (**c**) Velocity vectors along with pressure contour (Unit: 0.1 Pa) in the plane that cuts through RF and RH at $t/T = 0, 1/8, 1/4, 3/8, 1/2, 5/8, 3/4$, and $7/8$. Only every one out of two grid points is shown. An animation of the velocity field is also provided (Supplementary Materials available online).

As shown in Figure 4, similar trends for the "Trotting Mode" mentioned above still remain for the cases with larger swing amplitudes of $\beta_0 = 30°$ and $40°$. For $\beta_0 = 30°$, the average thrust $\bar{F}_{x,avg}$ reduces from 2.80 to 2.39 to 1.90 for $\alpha_0 = 60°$, $70°$ and $80°$; for $\beta_0 = 40°$, it also drops from 2.54 to 2.24 to 1.90. Overall, the combination with smaller magnitudes of $\alpha_0 = 60°$ and $\beta_0 = 20°$ provides the highest average propulsion thrust in the "Trotting Mode".

With the data available for a single paddling leg as listed in Table 2, we are able to quantitatively evaluate the inter-limb interaction and thus trotting gait effectiveness under different combinations of initial swing angle and swing amplitude (α_0, β_0). To do so, we compute the force ratio of the average thrust $\bar{F}_x/4\bar{F}_{SL}$ between the quadrupedal and single-leg paddling motions. In Table 3, we can see that the inter-limb interaction in the trotting gait consistently enhances the thrust with $\bar{F}_x/4\bar{F}_{SL} > 1$, which indicates that the hydrodynamic benefit of enhanced thrust is successfully captured through the cooperation of diagonal legs. For example, $\bar{F}_x/4\bar{F}_{SL}$ can reach 1.26 with $(\alpha_0, \beta_0) = (70°, 40°)$, which means that the average thrust $\bar{F}_x$ of an individual leg in this particular trotting gait is 1.26 times the magnitude of that generated by the single paddling leg with the same combination of (α_0, β_0). From the comparison in Figure 3a between the single-leg and quadrupedal paddling, it can be seen that coordinated paddling slightly increases thrust but significantly reduces drag. From the comparison in Figure 3b, the 4-leg paddling significantly reduce the magnitude of vertical force.

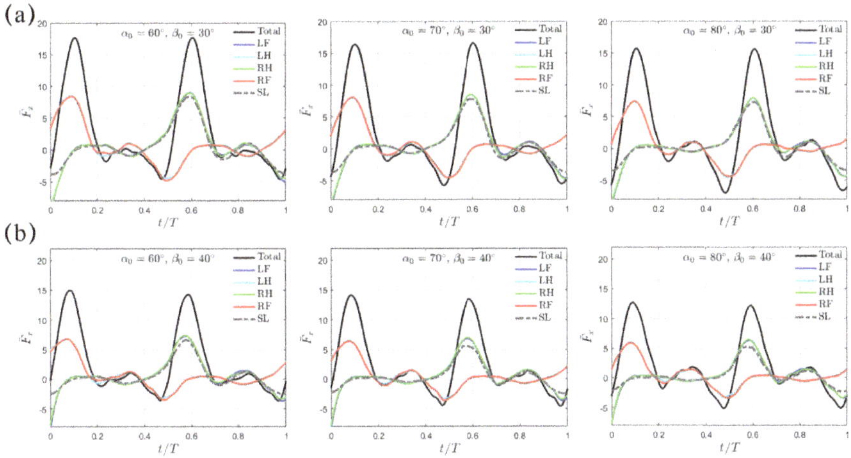

Figure 4. The transient profiles of thrust $\bar{F}_x(t)$ for swing amplitude (**a**) $\beta_0 = 30°$ and (**b**) $\beta_0 = 40°$. Results from the SL paddling are also included.

Table 3. Values of $\bar{F}_x/4\bar{F}_{SL}$ under different combinations of initial swing angle and swing amplitude (α_0, β_0). SL stands for single leg.

$\bar{F}_x/4\bar{F}_{SL}$	Trotting	Diagonal, DR = 33%	Diagonal, DR = 25%
(60, 20)	1.12	1.20	1.00
(70, 20)	1.18	1.42	1.16
(80, 20)	1.09	1.52	1.87
(60, 30)	1.13	1.33	1.26
(70, 30)	1.06	1.37	1.34
(80, 30)	1.02	1.50	1.56
(60, 40)	1.20	1.17	1.04
(70, 40)	1.26	1.44	1.20
(80, 40)	1.19	1.71	1.47

3.3. Quadrupedal Paddling Using the Diagonal Sequence Gait

Next, we consider the diagonal sequence gait that is actually used when swimming by dogs of different breeds [34]. In this sequence (LF→LH→RF→RH), The power stroke of the hindlimb on one side of the body is succeeded by that of the opposite side's forelimb. Unlike the trotting gait with DR = 50%, two shorter power phases with DR = 33% and 25% are considered with the intuition that faster backswing motion should be beneficial for larger thrust generation. In reality, DR = 33% is representative for dogs of different breeds during swimming. For DR = 33%, the average lower limb tip velocity $\bar{U}$ in the power phase is 0.96, 0.73, and 0.60 m/s for $\beta_0 = 20°, 30°$, and $40°$, respectively; for DR = 25%, it becomes 1.34, 1.03, and 0.84 m/s for $\beta_0 = 20°, 30°$, and $40°$, respectively. Figure 5a,b present the transient profiles of horizontal and vertical forces, $\bar{F}_x(t)$ and $\bar{F}_z(t)$, generated when $\beta_0 = 20°$ and DR = 33%. It can be seen that, both $\bar{F}_x(t)$ and $\bar{F}_z(t)$ exhibit a compact four-peak waveform that is equally distributed within the gait cycle with interval $T/4$, which is in contrast to the two-peak waveform observed in the "Trotting Mode". In diagonal sequence gait, the power phase of each leg contributes to one wave of thrust $\bar{F}_x(t)$. Moreover, the thrust $\bar{F}_x(t)$ generated by one individual leg is significantly impaired by the induced drag from the preceding leg leaving the current power phase and entering the following recovery phase, thus leaving an M-shaped profile around $\bar{F}_{x,pk}$. This represents a different hindering mechanism (denoted as the "Hindering Mode" hereafter) of $\bar{F}_x(t)$ in Figure 5a as compared with that generated in the "Trotting Mode" in Figure 3a. Under the same initial swing angle α_0 and swing amplitude β_0, even though the magnitude of peak thrust $\bar{F}_{x,pk}$ in the "Hindering Mode" is lower than that in the "Trotting Mode", the average thrust $\bar{F}_{x,avg}$ in the "Hindering Mode" is much higher due to its shorter power phase and faster leg motion. For example, when $\alpha_0 = 60°$ and $\beta_0 = 20°$, $\bar{F}_{x,pk}$ is 20.95 and 14.77 for the "Trotting Mode" and the "Hindering Mode", respectively, while the corresponding $\bar{F}_{x,avg}$ is 3.34 and 6.12, respectively, which represents a large increment of about 83.23%. A slight phase lead of $t = 0.05$ s ($t = 0.0625T$), which is identical to that of the trotting gait, is observed between $\bar{F}_x(t)$ and $\bar{F}_z(t)$. Due to its accelerated power phase, the average lower limb tip velocity in power phase increases to $\bar{U} = 0.96$ m/s, yielding a Reynolds number $Re = 4.5 \times 10^4$. Within one complete cycle of diagonal sequence gait, counterclockwise vortex is shed from the lower limb tip by following the LF→LH→RF→RH sequence during power phase. For example, a counterclockwise vortex near the lower limb tip of RF and RH is observed at $t/T = 1/4$ and $3/8$, and $t/T = 1/2$ and $5/8$, respectively (Figure 5c).

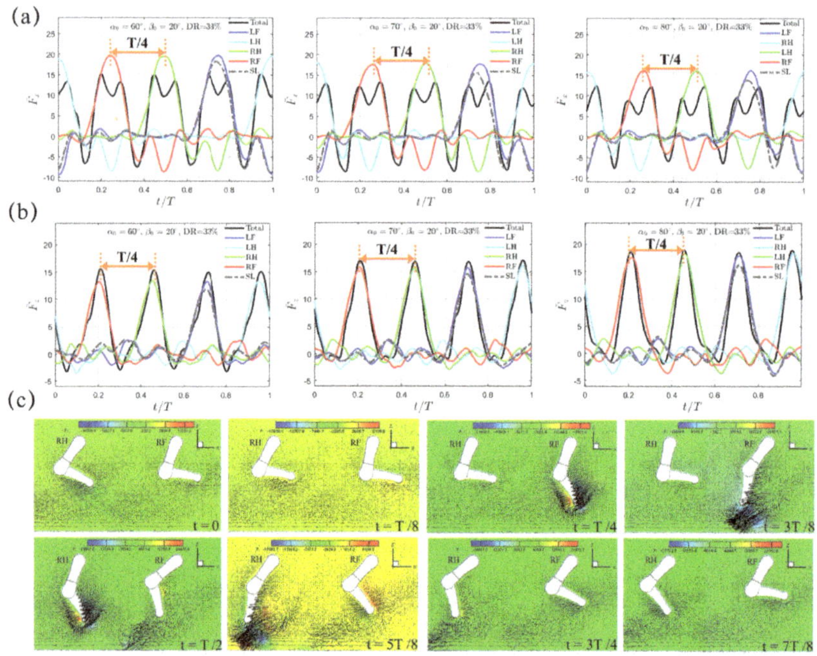

Figure 5. The transient profiles of (**a**) thrust $\bar{F}_x(t)$ and (**b**) vertical force $\bar{F}_z(t)$ generated by the quadrupedal paddling using the diagonal sequence gait when $\beta_0 = 20°$ and DR = 33%. Results from the SL paddling are also included. (**c**) Velocity vectors along with pressure contour (Unit: 0.1 Pa) in the plane that cuts through RF and RH at $t/T = 0, 1/8, 1/4, 3/8, 1/2, 5/8, 3/4$, and $7/8$. Only every one out of two grid points is shown. An animation of the velocity field is also provided (Supplementary Materials available online).

The varying trends for the instants of $\bar{F}_{x,pk}$ and $\bar{F}_{z,pk}$, the time intervals between $\bar{F}_{x,pk}$ and $\bar{F}_{z,pk}$, and the slight time lead of $\bar{F}_{z,pk}$ over $\bar{F}_{x,pk}$ are maintained when we increase the initial swing angle α_0 to 70° and then 80°. $\bar{F}_{x,pk}$ reduces from 14.77 to 13.29 to 12.22 and $\bar{F}_{x,avg}$ also reduces from 6.12 to 5.12 to 4.41; while $\bar{F}_{z,pk}$ increases from 14.93 to 16.95 to 18.65 with increasing $\bar{F}_{z,avg}$ from 4.91 to 5.54 to 6.24. At the moment of peak thrust $\bar{F}_{x,pk}$, the corresponding α_{pk} is 70°, 80°, and 90° for $\alpha_0 = 60°, 70°$, and 80°, respectively, which are identical with that from the trotting gait.

In Figure 6, the transient profiles of $\bar{F}_x(t)$ are presented for the cases with larger swing amplitudes of $\beta_0 = 30°$ and 40°. The varying trend is similar to the situation with $\beta_0 = 20°$ discussed above. For $\beta_0 = 30°$, $\bar{F}_{x,avg}$ decreases from 4.37 to 3.81 to 3.23 for $\alpha_0 = 60°, 70°$ and 80°; for $\beta_0 = 40°$, it further drops from 3.41 to 3.20 to 2.89, respectively. At the same α_0 and β_0, $\bar{F}_{x,avg}$ is evidently higher than that generated with the trotting gait. In addition, different from the M-shaped profile left around $\bar{F}_{x,pk}$ under $\beta_0 = 20°$, the thrust impairment is gentler as β_0 increases to 30° and 40°. Thus, a peak value in total thrust $\bar{F}_{x,pk}$ is still observable. Like the trotting gait, α_{pk} increases with increasing α_0 and β_0, while γ_{pk} maintains the same with respect to α_0 but only decreases as β_0 increases. However, for the same swing amplitude β_0, the values of α_{pk} and γ_{pk} remain identical to those in the trotting gait.

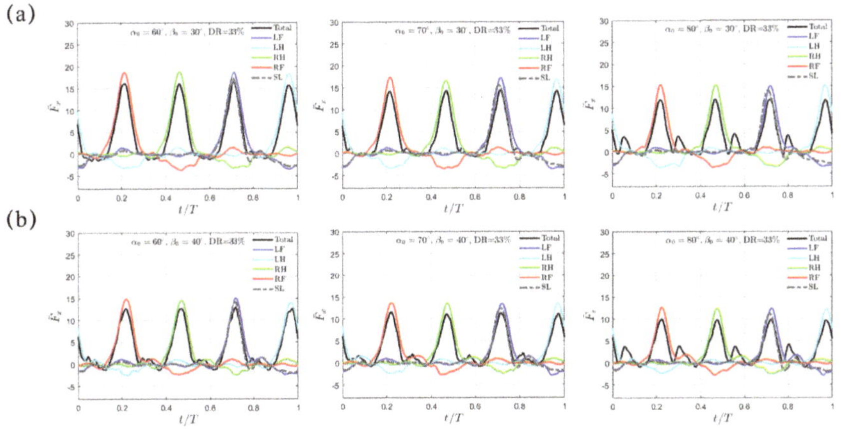

Figure 6. The transient profiles of thrust $\bar{F}_x(t)$ for swing amplitude (**a**) $\beta_0 = 30°$ and (**b**) $\beta_0 = 40°$ when DR = 33%. Results from the SL paddling are also included.

Next, we consider an even shorter power phase using the diagonal sequence gait with DR = 25%. The average lower limb tip velocity $\bar{U}$ in the power phase is about 1.34 m/s, which yields a Reynolds number of $Re = 6.2 \times 10^4$. Compared to that of DR = 33%, $\bar{F}_x(t)$ and $\bar{F}_z(t)$ increase significantly, as shown in Figure 7a,b. As α_0 increases from 60°, 70°, to 80°, $\bar{F}_{x,avg}$ increases to 9.36, 8.35 and 6.42 as compared to 6.12, 5.12, and 4.41 when DR = 33%. Although $\bar{F}_x(t)$ and $\bar{F}_z(t)$ still exhibit a compact four-peak waveform that is equally spaced within one paddling cycle with a uniform interval of $T/4$, the drag caused by the preceding leg leaving the current power phase and entering the following recovery phase in the "Hindering Mode" disappears. Instead, the four legs now contribute independently in LH→RF→RH→LF sequence to the total thrust $\bar{F}_x(t)$. For instance, the paddling of LH generates thrust ($t/T \in [0, 1/8]$) and a drag ($t/T \in [1/8, 1/4]$) during the first quarter of the whole diagonal sequence gait; outside of the time interval $t/T \in [0, 1/4]$, the contribution of LH to $\bar{F}_x(t)$ becomes negligible. Thus, we denote this mechanism of thrust generation as the "Separate Mode". Compared to the case of DR = 33% with $\alpha_0 = 60°$, both α_{pk} and γ_{pk} share the same value, resulting in a configuration where the upper-lower limb alignment is tilted forward. Thus, the reduced power phase ratio from DR = 33% to 25% does not change the arrangement between the upper and lower limbs at t_{pk}. We again present the velocity field in the slice that cuts through RF and RH in Figure 7c. Like the previous situation with DR = 33%, a counterclockwise vortex is shed from the lower limb tip in LF→LH→RF→RH sequence during power phase. At $t/T = 1/4$ and 3/8, and $t/T = 1/2$

and 5/8, such counter-clockwise vortex is clearly observed near the lower limb tips of RF and RH (Figure 7c).

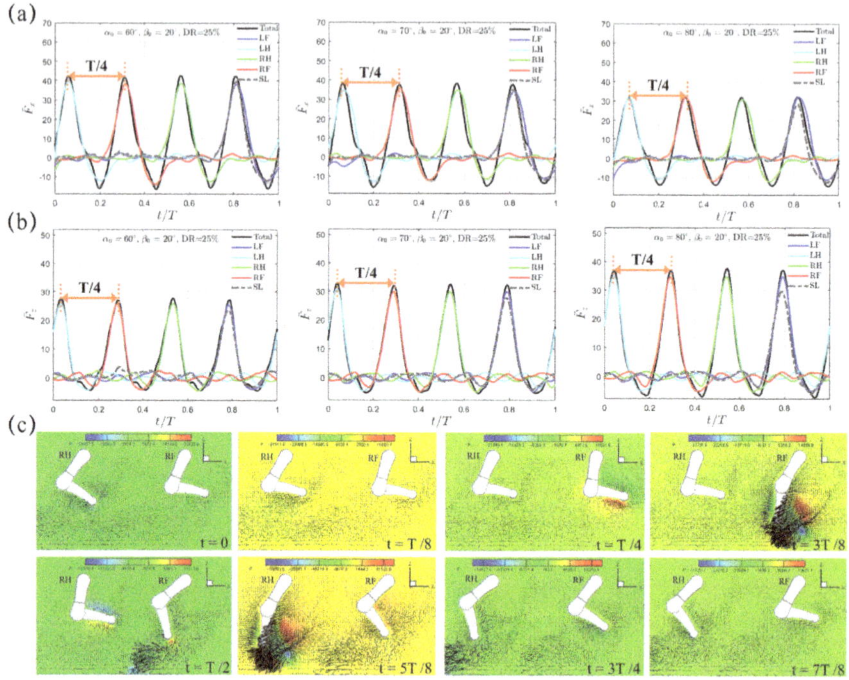

Figure 7. The transient profiles of (**a**) thrust $\bar{F}_x(t)$ and (**b**) vertical force $\bar{F}_z(t)$ generated by the quadrupedal paddling using the diagonal sequence gait when $\beta_0 = 20°$ and DR = 25%. Results from the SL paddling are also included. (**c**) Velocity vectors along with pressure contour (Unit: 0.1 Pa) in the plane that cuts through RF and RH at $t/T = 0, 1/8, 1/4, 3/8, 1/2, 5/8, 3/4$, and $7/8$. Only every one out of two grid points is shown. An animation of the velocity field is also provided (Supplementary Materials available online).

As the swing amplitude β_0 increases to 30° and 40°, the "Separate Mode" discussed above for $\beta_0 = 20°$ still holds (Figure 8). Like the "Trotting Mode" and the "Hindering Mode", the thrust $\bar{F}_x(t)$ decreases as the initial swing angle α_0 increases. For $\beta_0 = 30°$, $\bar{F}_{x,avg}$ is 6.62, 6.28, and 5.23 for $\alpha_0 = 60°$, 70°, and 80°, respectively; for $\beta_0 = 40°$, it decreases further to 5.03, 4.81, and 4.49 respectively. Again, α_{pk} increases with both increasing α_0 and β_0, while γ_{pk} remains unchanged with respect to α_0 but only decreases as β_0 increases. Taking all the cases with DR = 33% and 25% together, we note that a leg configuration with a larger forward-tilted angle and a flatter upper-lower limb alignment at t_{pk} indicates a lower thrust generation.

In Table 3, we also present the force ratio of $\bar{F}_x/4\bar{F}_{SL}$ pertaining to the diagonal sequence gait. In general, inter-limb interaction again enhances the thrust with $\bar{F}_x/4\bar{F}_{SL} > 1$, with the exception of $(\alpha_0, \beta_0, \mathrm{DR}) = (60°, 20°, 25\%)$. When the power phase ratio is reduced to DR = 33% and 25%, even higher value of $\bar{F}_x/4\bar{F}_{SL}$ can be achieved. For example, $\bar{F}_x/4\bar{F}_{SL}$ can achieve 1.87 under $(\alpha_0, \beta_0, \mathrm{DR}) = (80°, 20°, 25\%)$, which means that the average thrust $\bar{F}_x$ of an individual leg from this particular diagonal sequence gait is 1.87 times the magnitude of that generated by a single leg under the same combination of $(\alpha_0, \beta_0, \mathrm{DR})$. Hence, the hydrodynamics of quadrupedal paddling is not necessarily a simple linear superposition with phase shift of those from single-leg paddling, and the effect of inter-limb interaction needs to be put into consideration during the design of paddling gaits. Since only three different values of α_0, β_0 and DR are considered in the present study, which represents the limitation of the present work, a closer investigation of the detailed correlation between the force ratio $\bar{F}_x/4\bar{F}_{SL}$ and the kinematic parameters is not straightforward with limited data and thus is deferred to a future study that can also incorporate geometry optimization and a comparison of energy expenditure with other propulsion mechanisms.

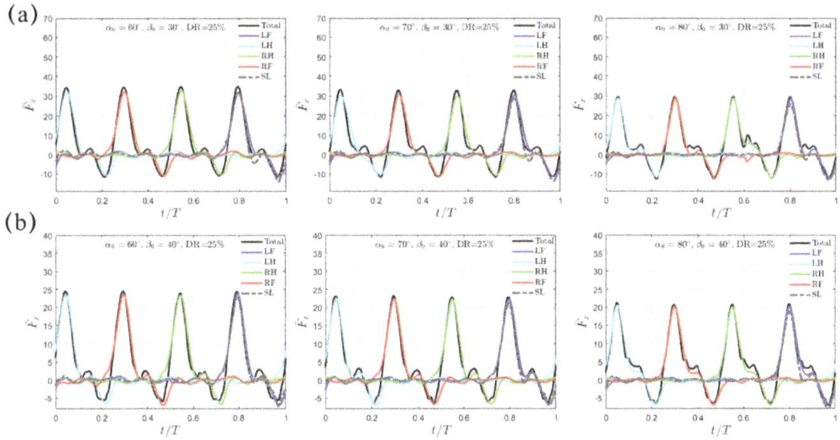

Figure 8. The transient profiles of thrust $\bar{F}_x(t)$ for swing amplitude (**a**) $\beta_0 = 30°$ and (**b**) $\beta_0 = 40°$ when DR = 25%. Results from the SL paddling are also included.

3.4. Quadrupedal Paddling Using the Lateral Sequence Gait

Using the same power phase ratio of DR = 25% and 33%, we continue our study to consider the lateral sequence gait and investigate how the hydrodynamics is affected by the paddling sequence. The lateral gait has the hindleg contact of the ground followed by the contact of the foreleg on the same side, and thus is directly opposite to the diagonal sequence gait. Our simulation results reveal that the transient thrust generated by the lateral sequence gait is almost the same as that generated by the diagonal sequence gait except that the paddling movement now follows LF→RH→RF→LH sequence. Cases with DR = 25% experience the "Separate Mode", in which each leg's contribution to the total thrust $\bar{F}_x(t)$ is limited to its own paddling phase with a time duration of DR $\cdot$ T. On the other hand, cases with DR = 33% experience the "Hindering Mode", in which the thrust $\bar{F}_x(t)$ contributed by an individual leg is significantly reduced by the induced drag of the preceding leg leaving its current power phase and entering the following recovery phase. The thrust generation is closely related to the pressure difference across the lower limb during the backswing in the power phase. In general, the "Hindering Mode" and the "Separate Mode" produce higher thrust efficiency than the "Trotting Mode" through interlimb interaction. However, thrust efficiency may experience degradation under certain combinations of coefficients, such as $(\alpha_0, \beta_0, \mathrm{DR}) = (60°, 20°, 25\%)$, as presented in Table 3.

3.5. Moments and Energetics of Quadrupedal Paddling

After obtaining detailed information about the computed hydrodynamic forces, we continue our study by examining the hydrodynamic moments, which can be expressed by

$$M_x = \int_\Gamma (F_z \cdot y - F_y \cdot z) d\Gamma, \; M_y = \int_\Gamma (F_x \cdot z - F_z \cdot x) d\Gamma, \; M_z = \int_\Gamma (F_y \cdot x - F_x \cdot y) d\Gamma$$

where M_x, M_y, and M_z might instigate rolling, pitching, and yawing motions and significantly impact the stability of the real robotic dog paddling system. The origin O, as shown in Figure 1, is selected as the center, and the integration for moments is performed on the fluid-structure interface Γ. Figure 9 demonstrates the moments M_x, M_y, and M_z with varying initial swing angle ($\alpha_0 = 60°, 70°$, and 80°) under a fixed swing amplitude $\beta_0 = 20°$ and power phase ratio DR $= 33\%$. The results of the trotting gait with DR $= 50\%$ are also included for comparison. We can see that M_x, M_y, and M_z in the diagonal and lateral sequence gaits are very close to each other, which indicates a quite similar stability behavior, even though a phase difference of $t = T/4$ does exist for both M_x and M_z. In contrast, M_x, M_y, and M_z generated in the trotting gait are much smaller, especially for M_x and M_z. Due to the symmetry of the paddling motion with respect to the $y = 0$ plane for diagonal and lateral sequence gaits within a complete gait cycle, the profiles of M_x and M_z in the power and recovery phases are also symmetric about $y = 0$ with average values $M_{x,avg}$ and $M_{z,avg}$ equal to zero, as shown in Figure 9a,c. Thus, the four-leg paddling system would experience sustainable roll and yaw vibrations but is overall stable. In contrast, an overall negative $M_{y,avg}$ is detected throughout the entire paddling cycle in all three distinct gaits. As the initial swing angle α_0 increases, the magnitudes of both M_x and M_y increase while that of M_z reduces. Similar tendencies for M_x, M_y and M_z persist at larger swing amplitudes of $\beta_0 = 30°$ and 40°. In Figure 10, we also present the M_x, M_y, and M_z profiles for cases with even shorter power phase under DR $= 25\%$. Clearly, compared with the case of DR $= 33\%$, the magnitudes of M_x, M_y, and M_z become consistently larger for a fixed combination of α_0 and β_0. Overall, the "Hindering Mode" (DR $= 33\%$) results in larger pitching, rolling, and yawing moments than the "Trotting Mode". Moreover, in the "Separate Mode", the hydrodynamic moments are even higher when DR $= 25\%$. Therefore, achieving larger thrust generation is not without expense. The associated stronger vibrations in hydrodynamic moments will enhance the control difficulty of the robotic quadrupedal paddling system. In real robotic systems of quadrupedal paddling, we can use appropriate sensors to receive the real-time three-dimensional vibration signals and meanwhile feedback control algorithms can be developed to attenuate strong vibrations to a desired level.

Finally, we present the transient profiles of power, which can be expressed by

$$P(t) = -\int_{\Gamma} \boldsymbol{F}(t) \cdot \boldsymbol{u}(t) d\Gamma$$

where $P(t)$ is exerted on the fluid by the paddling legs, as shown in Figure 11. The minus sign is added here since $\boldsymbol{F}(t)$ represents the hydrodynamic force vector exerted on the paddling legs by the surrounding fluid, while $\boldsymbol{u}(t)$ is its moving velocity vector. Upon a given paddling gait, swing amplitude β_0 and power phase ratio DR, the transient power $P(t)$ is independent of the initial swing angle α_0, as shown in Figure 11a,b. Furthermore, the diagonal and lateral sequence gaits exhibit the same transient power profile. This indicates that, under the same swing amplitude β_0 and power phase ratio DR, the gait sequence, be it diagonal or lateral, has no impact on the power consumption (Figure 11c). Additional power is required when the power phase is shortened in the "Hindering Mode" (DR = 33%) and the "Separate Mode" (DR = 25%) for enhanced propulsion, as shown in Figure 11d. As DR decreases from 50% to 33% and 25% but with $\alpha_0 = 80°$ and $\beta_0 = 40°$ fixed, we integrate the transient power $P(t)$ over a complete paddling cycle, and the resultant requested work are 1.37 J, 2.40 J, and 4.37 J, respectively. As a result, increased energy consumption becomes essential to achieve more effective propulsion. We also present the dimensionless power coefficient $\bar{P}_{avg}$, which is normalized by $\rho_f \bar{U}^3 L_2 W/2$, in Table 4. Again, for a given DR and β_0, $\bar{P}_{avg}$ remains unaffected by α_0. Under fixed β_0,

$\bar{P}_{avg}$ drops discernibly with increasing DR. However, $\bar{P}_{avg}$ with $\beta_0 = 30°$ and $40°$ under the same DR are very close to each other. The excessive power expenditure at smaller DR may impact the long-term performance of quadruped robots, influencing endurance, efficiency, and operational reliability. An appropriate gait strategy is therefore critical to help optimize energy use and extend mission duration. In addition, a lightweight yet durable structural design and a streamlined body shape also help minimize drag, enhance locomotion efficiency and reduce energy loss.

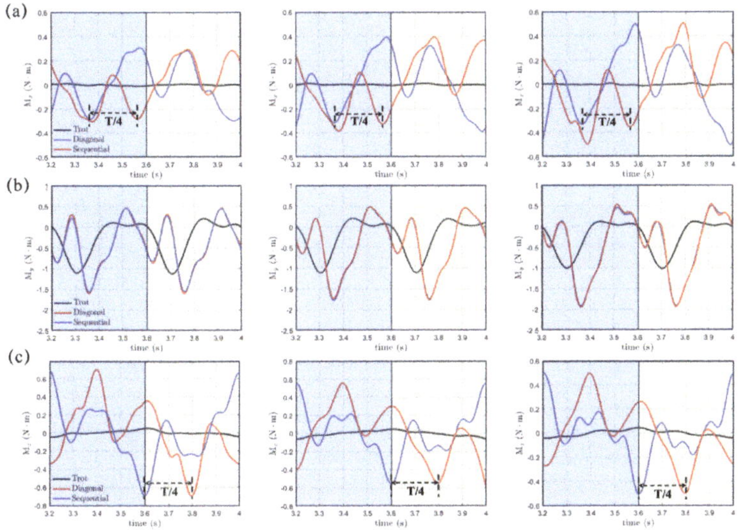

Figure 9. The transient profiles of hydrodynamic moments of (**a**) $M_x(t)$, (**b**) $M_y(t)$ and (**c**) $M_z(t)$ generated in both diagonal and lateral sequence gaits when DR = 33% and comparison with that from the trotting gait.

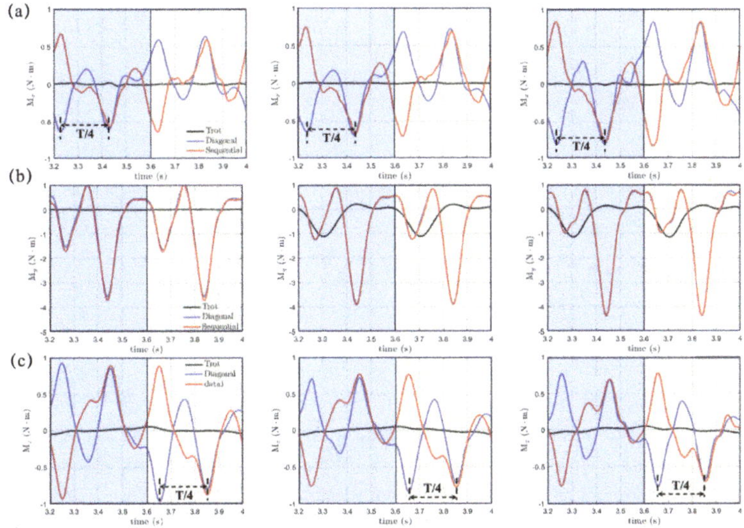

Figure 10. The transient profiles of moments of (**a**) $M_x(t)$, (**b**) $M_y(t)$ and (**c**) $M_z(t)$ generated in both diagonal and lateral sequence gaits when DR = 25% and comparison with that from the trotting gait.

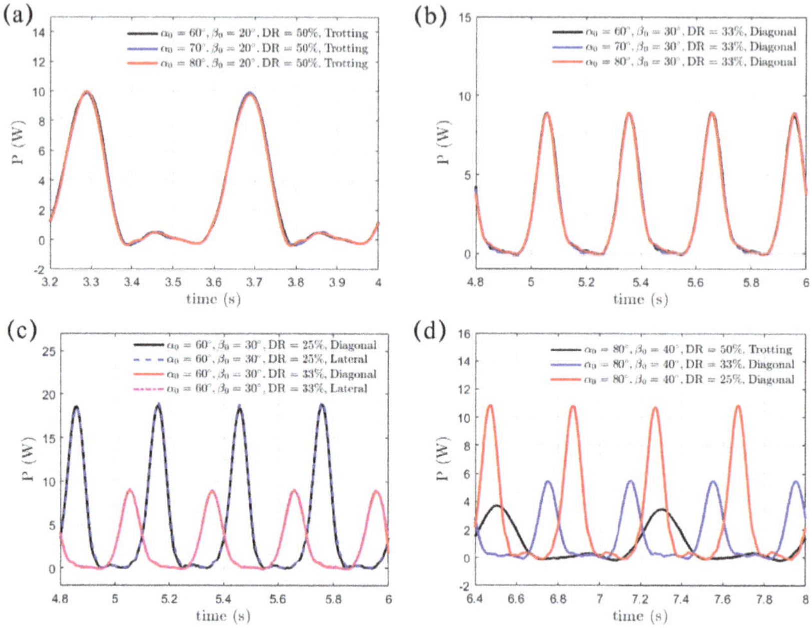

Figure 11. Identical power consumption in the (**a**) trotting gait and (**b**) diagonal sequence gait with varying initial swing angle $\alpha_0 \in [60°, 70°, 80°]$. (**c**) The diagonal and lateral sequence gaits share the same power consumption under fixed swing amplitude β_0 and power phase ratio DR. (**d**) The power expenditure increases with the reducing DR from 50% to 33% and 25%.

Table 4. Values of power coefficient $\bar{P}_{avg}$ under different combinations of initial swing angle and swing amplitude (α_0, β_0).

$\bar{P}_{avg}$	Trotting	Diagonal, DR = 33%	Diagonal, DR = 25%
(60, 20)	3.50	1.72	1.25
(70, 20)	3.50	1.72	1.25
(80, 20)	3.50	1.72	1.25
(60, 30)	4.33	2.41	1.60
(70, 30)	4.33	2.41	1.60
(80, 30)	4.33	2.41	1.60
(60, 40)	4.28	2.38	1.62
(70, 40)	4.28	2.38	1.62
(80, 40)	4.28	2.38	1.62

4. Discussion

In this work, we numerically investigate the effect of gait sequence, initial swing angle, swing amplitude, and power phase ratio on thrust generation of a bio-inspired robotic dog paddling model. For the trotting gait, the thrust engendered by one set of diagonal legs in the power phase is compromised by the other set of diagonal legs. Enhanced thrust is achieved through the reduced power phase in diagonal and lateral sequence gaits. The "Hindering Mode" is observed when the thrust from one paddling leg in the power phase is undermined by the preceding leg leaving its current power phase and entering the following recovery phase; while in the "Separate Mode", the thrust contribution from each of the four individual legs is uniformly distributed within one paddling cycle and is almost independent to each other. In almost all cases, the interaction between limbs positively impacts thrust generation and generates a total thrust that is more than quadruple the thrust of a single leg. When implementing diagonal or lateral sequence gaits for propulsion in real-world robotic systems, particularly for lowerDR, it is essential to develop adequate feedback control strategies based on real-time vibration signals from sensors. These strategies are capable of attenuating the larger oscillatory pitching, rolling, and yawing moments to guarantee stability. Meanwhile, additional energy expenditure, which is independent of both initial swing angle and paddling sequence (whether diagonal or lateral), becomes inevitable.

Supplementary Materials: The following supporting information can be downloaded at: https://www.mdpi.com/article/10.3390/biomimetics10030148/s1.

Author Contributions: Conceptualization, Y.L. and Y.C. (Ye Chen); methodology, C.Z., Y.C. (Ye Chen); software, Y.C. (Ye Chen); validation, Y.W., Y.C. (Yumeng Cai) and B.X.; formal analysis, Y.W. and Y.C. (Yumeng Cai); investigation, Y.W. and Y.C. (Yumeng Cai); resources, Y.C. (Ye Chen); data curation, Y.C. (Ye Chen); writing—original draft preparation, Y.W. and Y.C. (Yumeng Cai); writing—review and editing, C.Z. and Y.C. (Ye Chen); visualization, Y.W. and Y.C. (Yumeng Cai); supervision, Y.C. (Ye Chen); project administration, Y.C. (Ye Chen); funding acquisition, C.Z. and Y.C. (Ye Chen). Supplementary Materials: Y.W. and Y.C. (Yumeng Cai). All authors have read and agreed to the published version of the manuscript.

Funding: The authors acknowledge the support provided by the National Natural Science Foundation of China (Grant No. 12293000 and 12293001).

Institutional Review Board Statement: Not applicable.

Data Availability Statement: The original contributions presented in this study are included in the article/Supplementary Material. Further inquiries can be directed to the corresponding authors.

Conflicts of Interest: The authors declare no conflicts of interest.

Appendix A

The numerical approach and verification details of the sharp-interface immersed-boundary method based on the Cartesian grid for the flow utilized in this work have been expounded previously [39,40]. Successful applications to biological systems involve the fluid-structure interaction of heart valves [41,42] and vocal fold vibrations [43]. Furthermore, a parallel algorithm based on domain decomposition has been implemented to accelerate the costly flow simulation. Some of the specific issues related to the immersed-boundary treatment and its parallelization are briefly discussed here. As illustrated in Figure A1b, all nodes on the Cartesian grid are classified into four groups: fluid nodes, solid nodes, ghost nodes, and hybrid nodes. The ghost nodes are located within the solid domain and are immediately adjacent to the fluid-solid interface, and the hybrid nodes are located within the fluid domain and are immediately adjacent to the fluid-solid interface. The solid nodes are simply dummy nodes, while the fluid nodes anchor the standard 2nd-order finite-difference stencil used in the equation discretization. The identification of node types, the set-up of interpolation and extrapolation stencils for the ghost and hybrid nodes, as well as the computations related to updating these nodes during the solution process, are restricted to the subdomain itself. Therefore, the computational overhead associated with the immersed boundary scales very well with the number of subdomains or with the number of processors. Each subdomain is supplemented with two buffer slices on each side to support the interpolation or extrapolation zone for the ghost and hybrid nodes located near the boundary of the subdomain. These slices of data are conveyed through the Message Passing Interface (MPI) in the parallel implementation. Each subdomain also has a full copy of the unstructured surface mesh of the paddling legs. In this work, the bounding box of the whole flow domain is uniformly partitioned into 256 subdomains by employing the two-dimensional domain-decomposition strategy (16 subdomains along both y and z directions). Thus, a total number of 256 CPU cores is employed.

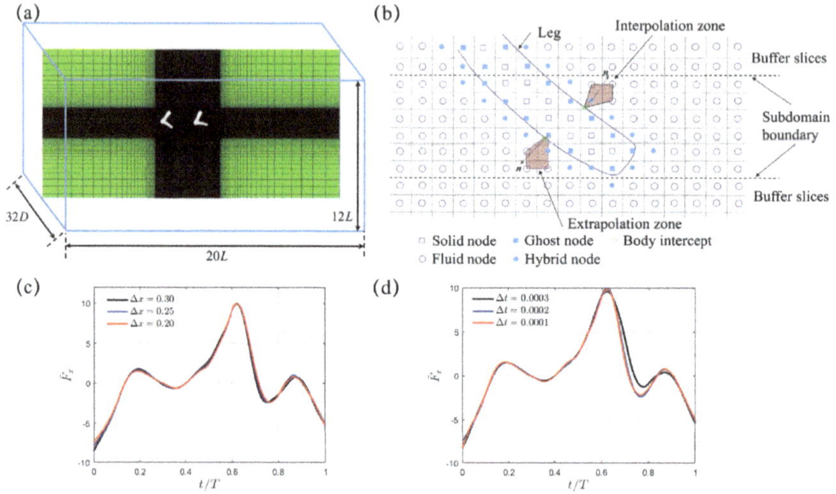

Figure A1. (**a**) Computational domain size and the stretched Cartesian mesh around the paddling legs. (**b**) Schematic of a subdomain with two buffer slices on each side in our parallel implementation. The shaded regions represent the stencil for local interpolation and extrapolation. (**c**) Mesh independency study with $\Delta x = 0.30$ cm, 0.25 cm, and 0.20 cm. (**d**) Time independency study with $\Delta t = 0.0003$ s, 0.0002 s, and 0.0001 s.

Independency studies are carried out for mesh and time-step in the case of a single paddling leg with initial swing angle $\alpha_0 = 70°$ and swing amplitude $\beta_0 = 20°$. For the mesh independency study, three different mesh sizes with $\Delta x = 0.30$ cm ($200 \times 90 \times 160$ with about 288 million mesh points), 0.25 cm ($228 \times 96 \times 180$ with about 394 million points), and 0.20 cm ($270 \times 105 \times 210$ with about 595 million points) are used in the refined region around the paddling leg. The results of the transient propulsion force $\bar{F}_x(t)$ during the first cycle, where $t \in [0, 0.8]$ s, are presented in Figure A1c. We can see that the three mesh sizes can capture the major variation characteristics of $\bar{F}_x(t)$. Figure A1d presents the results of the time-step independence study. Evidently, $\bar{F}_x(t)$ from simulations with time steps $\Delta t = 0.0001$ s and 0.0002 s agree well with each other, while that from a larger time step $\Delta t = 0.0003$ s becomes less satisfactory. As a compromise between computation cost and simulation accuracy, $\Delta x = 0.25$ cm and $\Delta t = 0.0002$ s are adopted in this work.

References

1. Bai, X.J.; Shang, J.Z.; Luo, Z.R.; Jiang, T.; Yin, Q. Development of amphibious biomimetic robots. *J. Zhejiang Univ. Sci. A* **2022**, *23*, 157–187. [CrossRef]
2. Yu, J.; Ding, R.; Yang, Q.; Tan, M.; Wang, W.; Zhang, J. On a bio-inspired amphibious robot capable of multimodal motion. *IEEE/ASME Trans. Mechatron.* **2011**, *17*, 847–856. [CrossRef]
3. Zhang, S.; Zhou, Y.; Xu, M.; Liang, X.; Liu, J.; Yang, J. AmphiHex-I: Locomotory performance in amphibious environments with specially designed transformable flipper legs. *IEEE/ASME Trans. Mechatron.* **2015**, *21*, 1720–1731. [CrossRef]
4. Greiner, H.; Shectman, A.; Won, C.; Elsley, R.; Beith, P. Autonomous legged underwater vehicles for near land warfare. In Proceedings of the Symposium on Autonomous Underwater Vehicle Technology, Monterey, CA, USA, 2–6 June 1996; IEEE: Piscataway, NJ, USA, 1996; pp. 41–48.

5. Saranli, U.; Buehler, M.; Koditschek, D.E. Design, modeling and preliminary control of a compliant hexapod robot. In Proceedings of the 2000 ICRA IEEE International Conference on Robotics and Automation, San Francisco, CA, USA, 24–28 April 2000; IEEE: Piscataway, NJ, USA, 2000; Volume 3, pp. 2589–2596.
6. Prahacs, C.; Saudners, A.; Smith, M.K.; McMordie, D.; Buehler, M. Towards legged amphibious mobile robotics. In Proceedings of the Canadian Engineering Education Association (CEEA), Montreal, QC, Canada, 29–30 July 2004.
7. Zhong, B.; Zhou, Y.; Li, X.; Xu, M.; Zhang, S. Locomotion performance of the amphibious robot on various terrains and underwater with flexible flipper legs. *J. Bionic Eng.* **2016**, *13*, 525–536. [CrossRef]
8. Hirose, S.; Mori, M. Biologically inspired snake-like robots. In Proceedings of the 2004 IEEE International Conference on Robotics and Biomimetics, Shenyang, China, 22–26 August 2004; IEEE: Piscataway, NJ, USA, 2004; pp. 1–7.
9. Hirose, S.; Yamada, H. snake-like robots machine design of biologically inspired robots. *IEEE Robot. Autom. Mag.* **2009**, *16*, 88–98. [CrossRef]
10. Matsuo, T.; Yokoyama, T.; Ueno, D.; Ishii, K. Biomimetic motion control system based on a CPG for an amphibious multi-link mobile robot. *J. Bionic Eng.* **2008**, *5*, 91–97. [CrossRef]
11. Ijspeert, A.J.; Crespi, A.; Ryczko, D.; Cabelguen, J.M. From swimming to walking with a salamander robot driven by a spinal cord model. *Science* **2007**, *315*, 1416–1420. [CrossRef] [PubMed]
12. Xia, M.; Wang, H.; Yin, Q.; Shang, J.; Luo, Z.; Zhu, Q. Design and mechanics of a composite wave-driven soft robotic fin for biomimetic amphibious robot. *J. Bionic Eng.* **2023**, *20*, 934–952. [CrossRef]
13. Fish, F.E. Transitions from drag-based to lift-based propulsion in mammalian swimming. *Am. Zool.* **1996**, *36*, 628–641. [CrossRef]
14. Li, Y.; Fish, F.; Chen, Y.; Ren, T.; Zhou, J. Bio-inspired robotic dog paddling: Kinematic and hydro-dynamic analysis. *Bioinspir. Biomim.* **2019**, *14*, 066008. [CrossRef] [PubMed]
15. Lee, J.; Hwangbo, J.; Wellhausen, L.; Koltun, V.; Hutter, M. Learning quadrupedal locomotion over challenging terrain. *Sci. Robot.* **2020**, *5*, eabc5986. [CrossRef]
16. Tang, Y.; Tan, J.; Harada, T. Learning agile locomotion via adversarial training. In Proceedings of the 2020 IEEE/RSJ International Conference On Intelligent Robots And Systems (IROS), Las Vegas, NV, USA, 24 October 2020–24 January 2021; IEEE: Piscataway, NJ, USA, 2020; pp. 6098–6105.
17. Fu, Z.; Kumar, A.; Malik, J.; Pathak, D. Minimizing energy consumption leads to the emergence of gaits in legged robots. *arXiv* **2021**, arXiv:2111.01674.
18. Choi, S.; Ji, G.; Park, J.; Kim, H.; Mun, J.; Lee, J.H.; Hwangbo, J. Learning quadrupedal locomotion on deformable terrain. *Sci. Robot.* **2023**, *8*, eade2256. [CrossRef]
19. Hao, T.; Xu, D.; Yan, S. Quadrupedal Locomotion in an Energy-efficient Way Based on Reinforcement Learning. *Int. J. Control Autom. Syst.* **2024**, *22*, 1613–1623. [CrossRef]
20. Lee, S.; Nahrendra, I.; Lee, D.; Yu, B.; Oh, M.; Myung, H. DreamFLEX: Learning Fault-Aware Quadrupedal Locomotion Controller for Anomaly Situation in Rough Terrains. *arXiv* **2025**, arXiv:2502.05817.
21. Dudek, G.; Giguere, P.; Prahacs, C.; Saunderson, S.; Sattar, J.; Torres-Mendez, L.A.; Jenkin, M.; German, A.; Hogue, A.; Ripsman, A.; et al. Aqua: An amphibious autonomous robot. *Computer* **2007**, *40*, 46–53. [CrossRef]
22. Zhang, S.; Liang, X.; Xu, L.; Xu, M. Initial development of a novel amphibious robot with transformable fin-leg composite propulsion mechanisms. *J. Bionic Eng.* **2013**, *10*, 434–445. [CrossRef]
23. Wang, C.; Xie, G.; Yin, X.; Li, L.; Wang, L. CPG-based locomotion control of a quadruped amphibious robot. In Proceedings of the 2012 IEEE/ASME International Conference on Advanced Intelligent Mechatronics (AIM), Kaohsiung, Taiwan, 11–14 July 2012; IEEE: Piscataway, NJ, USA, 2012; pp. 1–6.
24. Alben, S. Flapping propulsion using a fin ray. *J. Fluid Mech.* **2012**, *705*, 149–164. [CrossRef]
25. Maertens, A.P.; Gao, A.; Triantafyllou, M.S. Optimal undulatory swimming for a single fish-like body and for a pair of interacting swimmers. *J. Fluid Mech.* **2017**, *813*, 301–345. [CrossRef]
26. Dagenais, P.; Aegerter, C.M. How shape and flapping rate affect the distribution of fluid forces on flexible hydrofoils. *J. Fluid Mech.* **2020**, *901*, A1. [CrossRef]
27. Fernández-Gutiérrez, D.; Van Rees, W.M. Effect of leading-edge curvature actuation on flapping fin performance. *J. Fluid Mech.* **2021**, *921*, A22. [CrossRef]
28. Chao, L.M.; Alam, M.M.; Cheng, L. Hydrodynamic performance of slender swimmer: Effect of travelling wavelength. *J. Fluid Mech.* **2022**, *947*, A8. [CrossRef]
29. Zhang, D.; Huang, Q.G.; Pan, G.; Yang, L.M.; Huang, W.X. Vortex dynamics and hydrodynamic performance enhancement mechanism in batoid fish oscillatory swimming. *J. Fluid Mech.* **2022**, *930*, A28. [CrossRef]
30. Georgiades, C.; Nahon, M.; Buehler, M. Simulation of an underwater hexapod robot. *Ocean Eng.* **2009**, *36*, 39–47. [CrossRef]
31. Chen, G.; Tu, J.; Ti, X.; Wang, Z.; Hu, H. Hydrodynamic model of the beaver-like bendable webbed foot and paddling characteristics under different flow velocities. *Ocean Eng.* **2021**, *234*, 109179. [CrossRef]

32. Zhang, C.; Guy, R.D.; Mulloney, B.; Zhang, Q.; Lewis, T.J. Neural mechanism of optimal limb coordination in crustacean swimming. *Proc. Natl. Acad. Sci. USA* **2014**, *111*, 13840–13845. [CrossRef]
33. Baines, R.; Patiballa, S.K.; Booth, J.; Ramirez, L.; Sipple, T.; Garcia, A.; Fish, F.; Kramer-Bottiglio, R. Multi-environment robotic transitions through adaptive morphogenesis. *Nature* **2022**, *610*, 283–289. [CrossRef] [PubMed]
34. Fish, F.E.; DiNenno, N.K.; Trail, J. The "dog paddle": Stereotypic swimming gait pattern in different dog breeds. *Anat. Rec.* **2021**, *304*, 90–100. [CrossRef] [PubMed]
35. Khalid, M.S.U.; Akhtar, I.; Imtiaz, H.; Dong, H.; Wu, B. On the hydrodynamics and nonlinear interaction between fish in tandem configuration. *Ocean Eng.* **2018**, *157*, 108–120. [CrossRef]
36. Sun, X.; Ji, F.; Zhong, S.; Huang, D. Numerical study of an undulatory airfoil with different leading edge shape in power-extraction regime and propulsive regime. *Renew. Energy* **2020**, *146*, 986–996. [CrossRef]
37. Tian, F.B. Hydrodynamic effects of mucus on swimming performance of an undulatory foil by using the DSD/SST method. *Comput. Mech.* **2020**, *65*, 751–761. [CrossRef]
38. Khalid, M.S.U.; Wang, J.; Dong, H.; Liu, M. Flow transitions and mapping for undulating swimmers. *Phys. Rev. Fluids* **2020**, *5*, 063104. [CrossRef]
39. Mittal, R.; Dong, H.; Bozkurttas, M.; Najjar, F.; Vargas, A.; Von Loebbecke, A. A versatile sharp interface immersed boundary method for incompressible flows with complex boundaries. *J. Comput. Phys.* **2008**, *227*, 4825–4852. [CrossRef] [PubMed]
40. Luo, H.; Dai, H.; de Sousa, P.J.F.; Yin, B. On the numerical oscillation of the direct-forcing immersed-boundary method for moving boundaries. *Comput. Fluids* **2012**, *56*, 61–76. [CrossRef]
41. Chen, Y.; Luo, H. A computational study of the three-dimensional fluid–structure interaction of aortic valve. *J. Fluids Struct.* **2018**, *80*, 332–349. [CrossRef]
42. Chen, Y.; Luo, H. Pressure distribution over the leaflets and effect of bending stiffness on fluid–structure interaction of the aortic valve. *J. Fluid Mech.* **2020**, *883*, A52. [CrossRef]
43. Tian, F.B.; Dai, H.; Luo, H.; Doyle, J.F.; Rousseau, B. Fluid–structure interaction involving large deformations: 3D simulations and applications to biological systems. *J. Comput. Phys.* **2014**, *258*, 451–469. [CrossRef] [PubMed]

biomimetics

MDPI

Article

Development of a Wire-Driven Robotic Fish Based on Double Sine Mechanism

Qian Yang, Qixin Wang, Zihao Cao, Zeyue Zhao, Ye Chen * and Yong Zhong *

Shien-Ming Wu School of Intelligent Engineering, Guangzhou International Campus, South China University of Technology, Guangzhou 510640, China; 202320160054@mail.scut.edu.cn (Q.Y.)
* Correspondence: yechen@scut.edu.cn (Y.C.); zhongyong@scut.edu.cn (Y.Z.)

Abstract: Wire-driven robotic fish can effectively simulate the movement of real fish, but research on high-frequency wire-driven robotic fish is limited. This paper introduces the development of wire-driven robotic fish based on a double-sine mechanism. The appearance of the fish body is designed based on the morphology of tuna, and a mechanism that can support the high-frequency movement of the wire-driven mechanism is designed. The swimming speed and turning performance of the robotic fish are experimentally tested at various swing frequencies. The experimental results show that within the range of 1 to 4 Hz, the swimming speed of the robotic fish with different tail stiffness increases as the frequency increases. However, when the frequency exceeds 4 Hz, the swimming speed decreases. The tail joint with lower stiffness performs better at low frequencies, but as frequency increases, higher stiffness results in better swimming performance. Experimental tests show that the turning radius increases with higher swing frequencies and lower stiffness, resulting in a larger turning radius. This experiment will help to improve the application of high-frequency wire-driven mechanisms in the study of robot fish movement and carry out more in-depth bionic research in the future.

Keywords: robotic fish; wire-driven; double sine mechanism; different tail stiffness; various swimming frequencies

1. Introduction

After millions of years of evolution, fish have acquired excellent swimming abilities in water, which has aroused great interest among scholars of bionics and oceanography. They have tried to incorporate the fish's superb swimming skills into robotic fish. With continuous advancements in science and technology, research on robotic fish has matured. Researchers at the Massachusetts Institute of Technology in the United States successfully developed the world's first bionic robotic tuna (Robo Tuna) in 1994 after observing the swimming of tuna for a long time [1]. As motors are widely used in robotic fish, Liang et al. [2] developed a tail fin propeller driven by a servo motor to verify the high maneuverability and reliability of fishtail fin propulsion. With the maturity of bionic technology, more and more robotic fish have been designed and developed to pursue a swimming posture closer to that of real fish. Research on robotic fish is no longer limited to the multi-link design driven by motors and propellers [3,4]. Although this design is effective, its mechanism is complex and inefficient, and the rigid connection between the joints leads to a low fit between the fish body swing and the swimming of real fish. In order to be more similar to real fish, Zhong proposed a robotic fish based on a wire drive mechanism [5,6]. The wire-driven mechanism can not only improve the swimming of the robotic fish and the swimming of real fish but also reduce the difficulty and complexity of control. However, due to its structural reasons, this traditional wire-driven robotic fish makes it difficult to achieve high-frequency swimming.

Although the movement of robotic fish based on wire-driven mechanisms is very similar to that of real fish, most studies have only explored its movement at low frequencies, which usually results in insufficient propulsion in the water. High-frequency movement can better simulate the rapid swimming and complex movements of fish and improve the swimming efficiency of robotic fish. Zhang et al. [7] designed a robotic fish with high-frequency vibration characteristics based on the characteristics of electromagnetically driven machinery and tuna imitation. The results showed that increasing the tail-flicking frequency can improve the swimming speed of the robotic fish to a certain extent. However, electromagnetic drive is not only difficult to control but also consumes a significant amount of energy. Yu et al. [8] developed a two-joint bionic robotic fish driven by a single motor. It is driven at high frequency by an eccentric wheel structure, with a swing frequency of more than 9 Hz and a maximum swimming speed of 3.07 BL/s. However, its flexibility is poorer than that of multi-joint robotic fish. Lauder's team [9] continued to optimize the platform to achieve three different numbers of joint configurations. The experiment compared robotic fish with two, three, and four fish joints and analyzed their swimming performance. The results showed that increasing the number of joints in the fish's body can improve swimming performance, but this will also increase the difficulty of control and reduce its flexibility. Although there have been many studies on the high-frequency swimming of robotic fish, most of them are multi-link or single-link mechanisms, and there are few studies on high-frequency wire-driven robotic fish, so we explored this field.

Liao [10] proposed a robotic fish equipped with a wire-driven dual-elastic tail with energy storage and passive flexibility, capable of achieving high-frequency swings of 12 Hz, thereby improving swimming performance. To further advance the research on high-frequency wire-driven robotic fish, Liao et al. [11] also designed an elastic robotic fish based on wire-driven technology, where the tail is actuated by a wire system controlled by dual servomotors to simulate natural fish movements. Studies have shown that at high frequencies, the amplitude has a more significant effect on improving swimming speed. However, the design of the elastic fishtail makes the drive structure more complicated, and frequent high-frequency swings easily lead to structural deformation. In addition, in 2022, S.C. van den Berg et al. [12] proposed a high-speed swimming bionic robotic fish OpenFish based on a cable mechanism. The system drives the cable through the continuous rotation of the DC motor, driving the coordinated movement of the flexible tail on both sides and improving propulsion efficiency during high-speed swimming. However, due to the limitations of flexible tail structure and cable material, OpenFish still faces challenges in achieving high frequency and high precision control, especially in complex water flow environments, and its control response is relatively limited.

The experimental results in this paper show that increasing the frequency can increase the swimming speed of the robotic fish to a certain extent, but too high a frequency will reduce the speed of the robotic fish [13]. The robotic fish with the lowest tail stiffness at 1 Hz has the best swimming performance. As the frequency increases to 4 Hz, the swimming speed of the fully rigid fishtail increases to the fastest. The turning performance of the robotic fish is better at low frequencies, and the stronger the stiffness of the robotic fishtail, the smaller the turning radius.

2. The Design and Modeling of the Robotic Fish

Inspired by biology, most bionic robotic fish imitate the design and swimming style of real fish to improve propulsion efficiency and flexibility. However, most current research on bionic robotic fish cannot take into account both bionic effects and high-frequency oscillations. Few people study the high-frequency motion of robotic fish while ensuring the same flexibility as real fish. Although there is still a long way to go to achieve the high-frequency oscillation movement posture of real fish, this paper has achieved the combination of the two to a certain extent by studying the high-frequency movement of tuna and proposing novel contents.

2.1. Mechanical Design of the Robotic Fish

The robotic fish designed in this paper adopts the (Body and/or Caudal Fin propulsion) BCF model of the tuna family. Figure 1a shows a robotic fish designed in imitation of tuna [14]. It consists of three parts: head, body, and tail. The head length is 201 mm, the body length is 129 mm, the tail length is 100 mm, and the total weight is 1610 g, as shown in Table 1. The head was manufactured using 3D printing technology and houses the control panel, servo, motor, battery, encoder, synchronous pulley, and transmission mechanism. Counterweights were added inside to balance gravity and buoyancy. The brushless motor was used as the power source, and the rotation of the motor was output to the synchronous pulley through a double sine mechanism. The synchronous pulley drives the synchronous belt to move, thereby realizing the swing of the fishtail. The body contains an active wire-driven mechanism with four joints and synchronous belts. One end of the synchronous belt is connected to the pulley, while the other end is fixed to the last joint of the body. When the synchronous pulley rotates, the synchronous belt is rotated one circle, the other side is extended, and the tail fin on the seabed is applied to collect. The fishtail is connected to the body through the tail joint. The tail connection joint can be equipped with torsion springs of different wire diameters to change the tail stiffness. Figure 1b is the visual model.

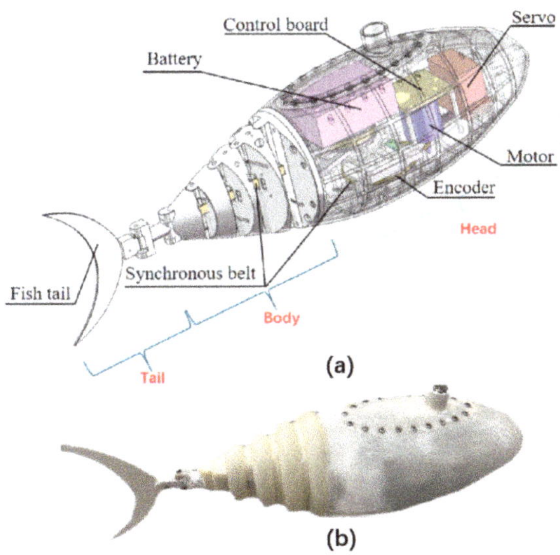

Figure 1. Overview and actual image of the robotic fish: (**a**) the overview design of the robotic fish; (**b**) robotic fish physical model.

Table 1. Specifications of the robotic fish prototype.

	Size (mm)
Overall length of the robotic fish	430 × 102 × 140 (Length × Width × Height)
Head length	201
Body length	129
Tail length	100

In order to be closer to the swimming posture of real fish, we estimate the oscillating swimming of fish as a circular arc segment, as shown in Figure 2b. From the driving perspective, the curvature of the body arc segment was obtained by limiting the stolen. Figure 2a shows the fracture and resection arrangement of bony fish. The joints are used to simulate the vertebrae, and the axis is rotated at the connecting joints to constrain the joint movement [15]. Since the resection can only bear tension, the complex bending of the fish body on both sides requires the joint action of the sculpture: lateral contraction, as shown in the left side of the figure, the fish body bends, and vice versa. The structure of the fish and its body drive mechanism are very similar to the bionic wire drive mechanism, which can easily construct the body curve. As shown in Figure 2c, a synchronous belt is used to replace the resection of the fish. The two synchronous belts are stuck by the bearing and the toothed support wheel when passing through the vertebrae, ensuring that the joints can rotate synchronously without skipping teeth. At the same time, it also ensures that the synchronous belt on the steering wheel does not slip when the fish body swings at high speed, effectively increasing the stability of the robotic fish body when moving. In addition, the fish body is waterproofed by wrapping the soft rubber material fish skin prepared by 3D printing technology on the outside of the fish body [16]. The fish skin is fixed to the base of the robotic fish head and the last joint of the fish body with silicone rubber.

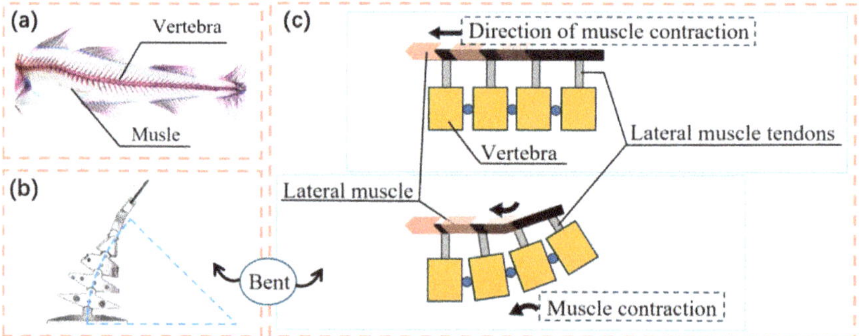

Figure 2. Overview of the skeleton structure and muscle arrangement of fish: (a) skeleton and muscle arrangement; (b) schematic diagram of the bending of the robotic fish's body swing; (c) the spine bends due to muscle contraction.

Compared to traditional multi-link robotic fish driven by motors, wire-driven robotic fish better mimics the swimming posture of real fish. However, due to mechanical strength limitations and swing stability concerns, most research has focused on low-frequency motion. To enable high-frequency swinging, this paper designs a transmission mechanism based on a double sine mechanism. As shown in Figure 3a, the transmission mechanism consists of a fixed base, a linear guide rail, a dual-sine mechanism, and a synchronous wheel. Two linear slide rails are installed under the head of the robotic fish. The fixed base is connected to the slide rails, and the fixed base can be moved forward and backward by rotating the servo. Figure 3b is a structural diagram of the dual-sine mechanism. In principle, the dual-sine mechanism is composed of two oppositely placed sine mechanisms.

By swapping the input and output of the two and then combining them, it can be ensured that the input of the motor can be output to the synchronous pulley according to the transmission ratio. Figure 3c shows a sine mechanism, which consists of a crank AB, a slider B, and vertically intersecting slides BCD. When crank AB rotates, slider B makes a circular motion. Since the vertically intersecting slides offset the horizontal motion, the motion trajectory output from the C end changes with time as a sine change, that is, a sine motion in the vertical direction [17]. By using the output of a sine mechanism as the input of another sine mechanism, a double sine mechanism can be formed. The schematic diagram of the double sine mechanism is shown in Figure 3d. The motor signal is input from the A end, driving the sine wheel to rotate. The slide CD perpendicular to the slider B can only reciprocate in one direction under the restriction of the limit device. At the same time, the slide rail where slider B′ is located is also perpendicular to CD and parallel to the slide rail where slider B is located. Due to the different transmission ratios, slider B′ is driven by CD to reciprocate in an incomplete circle at an initially determined middle position, thereby outputting a rotation signal to the A′ end. The synchronous pulley is fixedly connected to A′, so when the motor signal is input, the synchronous pulley also reciprocates.

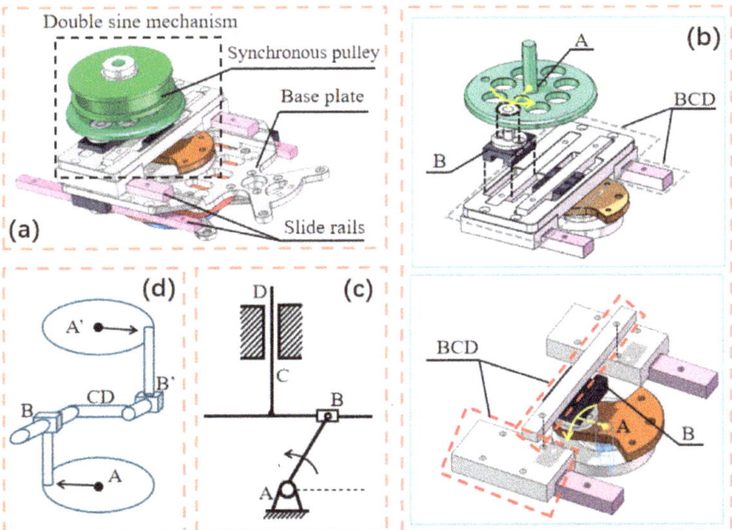

Figure 3. Double sine mechanism structure and schematic diagram: (**a**) overall schematic diagram of transmission mechanism (**b**) physical picture of double sine mechanism; (**c**) sine mechanism schematic diagram; (**d**) double sine mechanism schematic diagram.

Although the output of a single sine structure can ensure that the speed is symmetrical compared to the center point, the lateral deflection force received by the synchronous pulley is always applied to a single side of the limit device. When the movement is long-term or high-frequency, the output end of the sine mechanism will be deformed due to excessive lateral force during the movement and additional vibration caused by the body, resulting in asymmetric output of the sine mechanism, thereby affecting the fishtail swing. The dual sine mechanism is designed to solve this problem. Taking Figure 3d as an example, when slider B moves along the slide rail, the output end is connected to the other slide rail. At this time, the force applied to the limit device is dispersed by the movement of the slider B′, reducing the local force and making the output movement more stable.

As can be seen from the body part of Figure 1, the synchronous belt replaces the traditional steel rope as the "muscle" of the robotic fish, which is used to drive the body to swing. The output end of the double sine mechanism is connected to the synchronous pulley in Figure 3a, and the synchronous belt is fixed on the synchronous pulley. The body part of the robotic fish contains four joints. There are holes on both sides of each joint to pass the synchronous belt. Synchronous gears are installed inside the joint holes to engage the synchronous belt. This design ensures that the robotic fish can swing at high frequencies without breaking the line or slipping the teeth. The experimental test results show that the structure can withstand a maximum of 24 Hz swing while the transmission mechanism is still stable, and the synchronization does not loosen or break. In contrast, the robotic fish driven by steel rope will break when the swing frequency exceeds 12 Hz. When the sine mechanism is used as the transmission mechanism, the synchronous wheel will loosen when the swing frequency exceeds 15 Hz, resulting in a decrease in the swing angle of the robotic fish. The experimental results show that the double sine mechanism can increase the stability of the robotic fish during high-frequency movement while reducing the wear of the internal transmission mechanism under high-frequency movement [18].

2.2. Electronics and Control System

All electronic devices are installed in the head of the robotic fish. The power source is a 22.2 V 6S lithium battery (850 mAh), which powers the motor control module, microcontroller, and wireless module via a step-down converter. The system uses a 5 V Savox SW-1210SG brushless motor, while the auxiliary steering system is powered by a KST X20-7512 servo. The sensor system is composed of an AS5047P encoder, which can control the motor speed according to the feedback signal to achieve speed control or turning movement of the robotic fish. The Arduino Seeed Studio XIAO SAMD21 serves as the central computer, receiving signals from the wireless module, processing sensor data, and sending control commands to the motor and servo. When the robotic fish is turned on, the servo drives the fixed bottom plate back to the calibration start position, and the remote control handle can send a frequency signal. When the wireless module receives the corresponding remote control signal, the motor will rotate according to the frequency, and the robotic fishtail swings its body at this frequency to swim straight. The servo signal and the motor signal do not interfere with each other. This is because the turning signal is only sent to the steering auxiliary servo, which is used to adjust the swimming direction. When the external device sends the servo signal, the steering auxiliary servo drives the bottom plate to move a certain distance, causing the middle position of the robotic fish to deflect. Figure 4 is the control system framework.

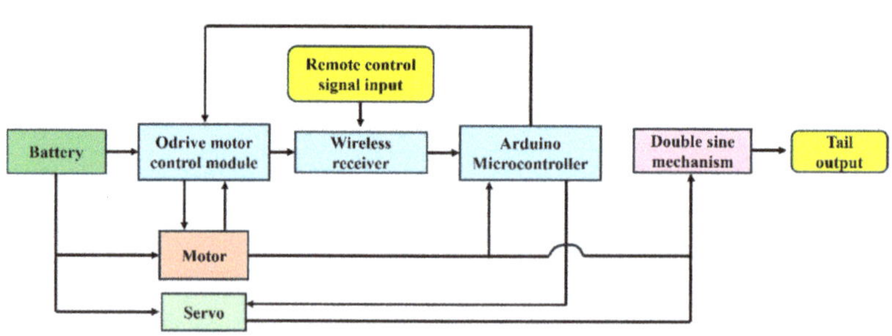

Figure 4. Block diagram of electromechanical control system.

3. Modeling and Design of Robotic Fish

3.1. Kinematic Model of Robotic Fish Swimming

Figure 5 is a cross-sectional view of adjacent joints when the robotic fish is swimming, and its body is bent. In this figure, d is the vertical distance between the synchronous belts, H is the thickness of the joints, h_0 is the distance between the joints in the initial state, φ is the angle of the fish body relative to the central axis, N is the number of joints of the fish body, and w_l and w_r are the left and right distances between adjacent joint holes after the fish body swings [19,20]. Because of the existence of the joint torsion spring, each joint can rotate evenly to the same side under the drive of the synchronous belt, so it can be assumed that the rotation angle of each joint is the same. φ/N is the rotation angle of each joint. From this, the length change in the two synchronous belts can be calculated as follows:

$$\Delta w_l = d\sin\frac{\varphi}{2N} - 2h_0\sin^2\frac{\varphi}{4N} \tag{1}$$

$$\Delta w_r = -d\sin\frac{\varphi}{2N} - 2h_0\sin^2\frac{\varphi}{4N} \tag{2}$$

where Δw_l and Δw_r are the variable lengths of the left and right synchronous belts, respectively. Since the rotation angle φ/N between each joint is very small and can be almost ignored, the binomial in the above formula can be omitted, and the entire expression can be simplified to the following:

$$\Delta w_l = d\sin\frac{\varphi}{2N} \tag{3}$$

$$\Delta w_r = -d\sin\frac{\varphi}{2N} \tag{4}$$

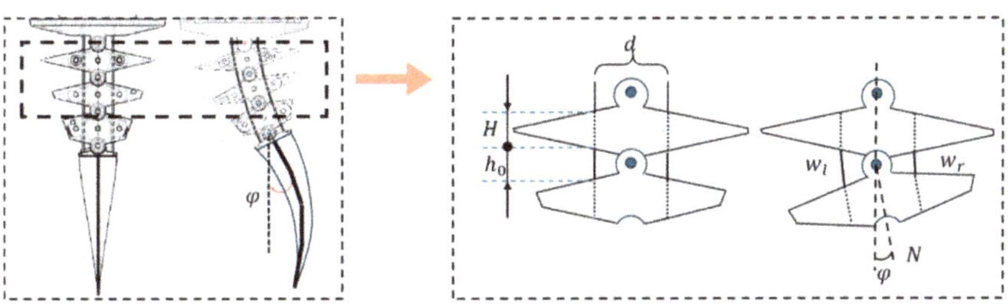

Figure 5. Schematic diagram of the robotic fish's body swing.

When the upper synchronous pulley rotates by an angle of α driven by the double sine mechanism, the length Δl of the synchronous belt rotating around the rotating axis of the upper synchronous pulley is as follows:

$$\Delta l = 2\pi l\frac{\alpha}{360^\circ} \tag{5}$$

In Formula (5), l is the radius of the upper synchronous belt pulley. Driven by the rotation of the upper synchronous belt pulley, the length of the synchronous belt on the right side of the fish body becomes shorter, while the length of the synchronous belt on the left side becomes longer. From (3) and (4), it can be seen that the shortened length of the synchronous belt on the right side is equal to the increased length of the synchronous belt on the left side. Because the synchronous belt passes through the idler wheel and meshes, it can be determined that the synchronous belt does not deform under the action of tension; that is, the total length does not change. Therefore, the following formula is obtained:

$$\Delta l = N|\Delta w_l| = N|\Delta w_r| \tag{6}$$

Substituting Formulas (3) and (5) into Formula (6), we can obtain the following:

$$\varphi = 2N\sin^{-1}\frac{\pi l\alpha}{180^{\circ}Nd} \tag{7}$$

where φ is the swing angle of the fish body, α is the rotation angle of the synchronous pulley. Since the inverse sine function in (7) is small, it can be simplified to the following:

$$\varphi = \frac{2l\alpha}{d} \tag{8}$$

Since the motor output angle is β, r: R is the ratio of the input axis offset to the output axis offset of the double sine mechanism, and the input gear ratio of the motor and the double sine mechanism is $m : M$, the synchronous belt rotation length ΔL of the sine mechanism input can be obtained:

$$\Delta L = \sin\frac{m\beta}{M} \tag{9}$$

According to the transmission characteristics of the double sine mechanism, the rotation angle of the upper synchronous pulley can be obtained as follows:

$$\alpha = \sin^{-1}\left(\frac{r}{R}\sin\frac{m\beta}{M}\right) \tag{10}$$

Substituting Equation (10) into Equation (8), the relationship between the motor rotation angle and the machine tail swing angle can be obtained after simplification:

$$\varphi = \frac{2l}{d}\sin^{-1}\left(\frac{r}{R}\sin\frac{m\beta}{M}\right) \tag{11}$$

Table 2 provides the transmission ratio between the synchronous pulley and the double sine mechanism. Substituting these data into Equation (11), when the β angle reaches its maximum, we can calculate that the maximum swing angle φ of the robotic fish is 45°.

Table 2. Parameter ratio of high-speed transmission mechanism.

Parameter	Actual Value	Parameter Ratio
$r : R$	1 cm:2 cm	1:2
$m : M$		1:6
$l : d$	1.32 cm:1.76 cm	3:4

When the robotic fish turns, the steering gear provides a transfer deflection angle to drive the base plate of the double sine mechanism to move, and the moving distance is x (x is the absolute value of the distance from the calibration starting position of the base plate. When the base plate moves forward relative to the starting position, the robotic fish turns left; when the base plate moves backward relative to the starting position, the robotic fish turns right.) At this time, the tail swing angle of the robotic fish can be rewritten as follows:

$$\varphi = \frac{2l}{d}\sin^{-1}\left(\frac{r}{R}\sin\frac{m\beta}{M} + x\right) \tag{12}$$

When the robotic fish receives a turning signal, the servo drives the bottom plate of the double sine mechanism to move, thereby providing a deflection angle. The new center position has an additional angle compared to the original fish body. The sensor feedback system obtains the current swing direction and position of the fishtail and switches the speed value at the maximum left and right swing position of the fishtail. The speeds generated by the left and right motors when working alternately are different, so a speed difference is generated, thereby providing the deflection force required for steering. The robotic fish in this article achieves differential turning based on this.

3.2. Hydrodynamics Modeling of Robotic Fish

To study the motion performance of the robotic fish, we modeled its movement in two modes: straight swimming and turning. The modeling during the swimming process includes two parts: the head and the tail. The fish's head is modeled as a rigid body, with hydrodynamic forces and thrust from the tail treated as external forces. The dynamic model of the horizontal plane is established according to the Newton–Euler equation. We model the tail swing using a pseudo-rigid body approach to simulate the deformation during movement. The Morison [21] equation is then applied to calculate the hydrodynamic forces generated by the tail's motion.

In order to facilitate the analysis of the movement of the robotic fish underwater, a spatial motion coordinate system is established, as shown in Figure 6. The coordinate system includes the inertial coordinate system OXYZ and the body coordinate system oxyz. Assuming that the center of gravity of the head coincides with point c,$v = (v_1, v_2, v_3)^T$ and $\Omega = (\Omega_1, \Omega_2, \Omega_3)^T$ represent the speed and angular velocity of the robotic fish in the body coordinate system, respectively. v_1, v_2, v_3 are the velocity components along the x, y, and z axes respectively; $\Omega_1, \Omega_2, \Omega_3$ are the angular velocity components of the x, y, and z axes, respectively. The transformation relationship between the body coordinate system and the inertial coordinate system is expressed by the Euler angle, which includes the pitch angle θ, the yaw angle ψ, and the roll angle ϕ. The pitch angle θ is defined as the angle between the ox axis and the OXY plane, which is positive when the ox axis is facing upward. The yaw angle ψ is defined as the angle between the projection of the ox axis on the OXY plane and the OX axis, which is positive when the projection line is between the OX axis and the OY axis. The roll angle ϕ is defined as the angle between the oxz plane and the longitudinal plane where the ox axis is located. It rotates around the ox axis according to the right-hand rule. When the finger passes through the longitudinal plane first, ϕ is positive. α is the angle of attack of the v velocity vector relative to the x-axis.

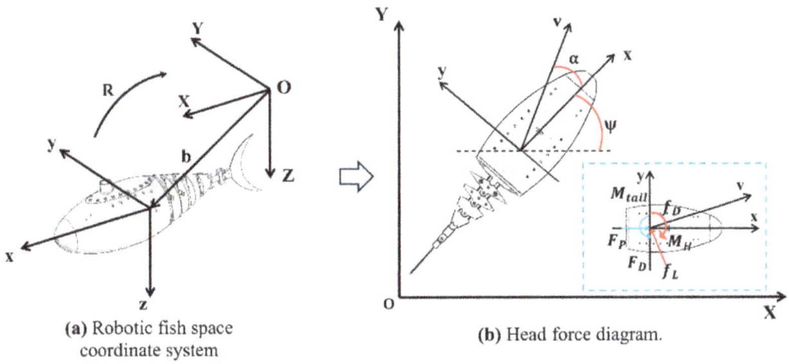

Figure 6. Robotic fish space motion coordinates.

The positional relationship between the body coordinate system and the inertial coordinate system is represented by the vector b, and R is the rotation matrix from the body coordinate system to the inertial coordinate system. The modeling of the gliding robot fish in this paper mainly focuses on its movement on the horizontal plane OXY without considering the floating and diving movements. The coordinate relationship of the OXY plane is shown in Figure 6b.

When the robotic fish is in water, its head can be regarded as a rigid body. The motion of a rigid body in water can be expressed by the Kirchoff equation, so there is the following formula:

$$\dot{P} = P \times \Omega + F_e \tag{13}$$

$$\dot{H} = H \times \Omega + P \times v + M_e \tag{14}$$

where P is the total momentum of the robotic fish head, H is the total angular momentum of the head, and the external force and torque applied to the center of gravity of the head are $F_e = [F_x, F_y, F_z]^T$ and $M_e = [M_x, M_y, M_z]^T$, respectively.

Since we are focusing on the motion of the plane and not on the vertical motion of the robotic fish, we do not consider the effect of the change in the center of gravity of the tail on the center of gravity of the head during the swinging process. The robotic fish is affected by the combined forces from fluid interaction and tail fin motion. Therefore, this paper also needs to consider the effects of the drag f_D, lift f_L, hydrodynamic torque M_H, water flow impact force F_D, thrust F_p, and thrust torque M_{tail}, generated by the tail fin swing. In summary, the swimming dynamics equation of the robotic fish can be simplified as follows:

$$m_h \dot{v_1} = m_h v_2 \Omega_3 + F_x \tag{15}$$

$$m_h \dot{v_2} = -m_h v_1 \Omega_3 + F_y \tag{16}$$

$$J_3 \dot{\Omega}_3 = M_{tail} + M_H \tag{17}$$

where m_h is the weight of the robotic fish head, and J_3 is the tensile inertia of the z-axis at the center of gravity of the robotic fish. At the same time, the external force and torque can be expressed by the content given in Figure 6.

$$F_x = F_P + f_L \sin\alpha - f_D \cos\alpha \tag{18}$$

$$F_y = F_D - f_L \cos\alpha - f_D \sin\alpha \tag{19}$$

The calculation formula of the hydrodynamic force on the head of the robotic fish during movement is as follows:

$$f_L = \frac{1}{2}\rho_w C_{HL} A_H v^2 \alpha_L \tag{20}$$

$$f_D = \frac{1}{2}\rho_w C_{HD} A_H v^2 \tag{21}$$

$$M_H = -C_{HM} {\Omega_3}^2 \mathrm{sgn}(\Omega_3) \tag{22}$$

Among them, C_{HL}, C_{HD} and C_{HM} are the lift coefficient, drag coefficient, and torque coefficient of the robotic fish head, respectively; α_L represents the average angle of attack of the fish's body during swimming and swinging; A_H is the surface area of the robotic fish head; and ρ_w is the water density.

Figure 7 is the tail kinematic model of the robot fish.The thrust of the robotic fish's forward movement and the torque when turning are both generated by the tail swing. The force analysis of the tail swing is crucial to the kinematic mechanics modeling of the robotic fish. The model is shown in Figure 6. This paper adopts the 1R tail rigid body model, which consists of two rigid rods connected by a hinge and a torsion spring. To simplify the calculation, L_c is the length of the active tail section, L_f is the length of the passive tail section, K_1 is the torsion spring constant, T_1 is the torque generated by the torsion spring,

M_1 is the hydrodynamic torque that causes the tail to bend, and φ_2 is the deflection angle of the second tail section.

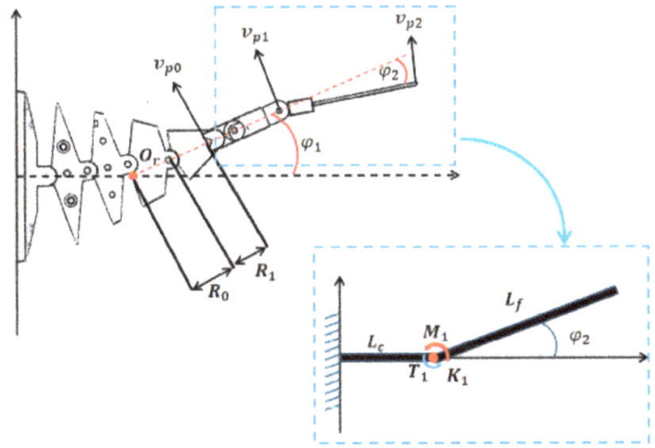

Figure 7. Tail kinematics model.

The torque T_1 generated when the torsion spring deflects φ_2 can be approximately calculated as follows:

$$T_1 = K_1\varphi_2 \tag{23}$$

Under the combined effect of the hydrodynamic torque and the torsion spring, the deflection acceleration of the second tail section is as follows:

$$\ddot{\varphi_2} = \frac{(T_1 + M_1)}{J_c} \tag{24}$$

where J_c is the moment of inertia of the second tail model, and the torsion spring constant can be obtained by calculating the following equation:

$$K_1 = \frac{Ed^4}{10.8DN} \tag{25}$$

where E is the elastic modulus (200 GPa for steel), d is the wire diameter, D is the average diameter, and N is the number of turns.

In the process of driving the wire to drive the joint to swing, in order to obtain the swing speed of the end joint, the center line of the end joint can be extended to intersect with the x-axis of the fuselage. The intersection O_r is used as the rotation center of the end joint; φ_1 is the deflection angle of the end join; v_{p0}, v_{p1}, v_{p2} correspond to the speed of P_0, P_1, P_2 points, respectively, and R_0, R_1 are parameters for calculating the rotation radius, where R_1 is a constant, and R_0 needs to be calculated based on the current deflection angle φ_1. According to the geometric relationship, the following can be calculated:

$$R_0 = \begin{cases} \frac{H_0\left(\sin\frac{\varphi_1}{4} + \sin\frac{2\varphi_1}{4} + \sin\frac{3\varphi_1}{4}\right)}{\sin\varphi_1} & \varphi_1 \neq 0 \\ 3H_0 & \varphi_1 = 0 \end{cases} \tag{26}$$

The speed of any point between the angular velocities $\omega_t = \frac{d\varphi_1}{dt} = \dot{\varphi_1}$, P_0, and P_1 of the tail swing can be expressed as follows:

$$V_{P1} = \omega_t \times r_{pm}. \tag{27}$$

where r_{pl} is the position vector between O_r and point P_1, the velocity magnitude at P_1 is $V_{P1} = \dot{\varphi_1}(R_0 + R_1 + l)$, and l is the distance between P_0 and P_1. According to the velocity relationship between the rods, the velocity of any point P_m between P_1 and the end of the tail fin consists of two parts: the velocity at point P_1 and the linear velocity around point P_1, which can be expressed as follows:

$$V_{pm} = V_{P1} + \omega_c \times r_{pm} \tag{28}$$

where ω_c is the bending angular velocity of the hinge of the pseudo-rigid body model, its magnitude is $\omega_c = \dot{\varphi}_2$, and r_{pm} is the position vector from point P_1 to P_m. According to the Morison equation, the hydrodynamic forces on the flexible middle section and the tail fin during swinging mainly include drag force and inertia force, where the drag force is the force on the tail caused by the velocity of the water flow particle and the inertia force is the force on the tail caused by the acceleration of the water flow particle. The hydrodynamic force on any point n of the flexible middle section and the tail fin can be expressed as follows:

$$f(n) = -\frac{1}{2}\rho_w C_d V_n |V_n| d(n) - \frac{1}{4}\pi d^2(n)\rho_w C_m \dot{V}_n \tag{29}$$

where C_d and C_m are the drag coefficient and inertia coefficient, respectively, V_n is the velocity at point n, and $d(n)$ is the chord length of the section at point n of the tail. The hydrodynamic force on the entire tail can be obtained by calculating the sum of the hydrodynamic forces from P_0 to P_1 and from P_1 to P_2.

$$F_{tail} = \begin{pmatrix} F_P \\ F_D \end{pmatrix} = \int_0^{L_c+L_f} f(x)dx \tag{30}$$

The torque M_{tail} generated by the tail swing is as follows:

$$M_{tail} = \int_0^{L_c+L_f} r_x \times f(x)dx \tag{31}$$

where r_x is the position vector from the coordinate origin o of the robotic fish body to the tail action point.

4. Simulation and Experimental Research

In order to verify the swimming performance of the transmission structure designed in this paper on the robotic fish, a series of underwater swimming experiments at different frequencies were conducted on robotic fish with different tail stiffness. The tail connection joint can replace the torsion spring to change the tail stiffness of the robotic fish. This experiment was equipped with 0.8 mm torsion spring wire diameter, 1.0 mm torsion spring wire diameter, and full rigid fishtail (hereinafter collectively referred to as 0.8 mm stiffness fishtail, 1.0 mm stiffness fishtail, and full rigidity fishtail). The whole experiment was mainly divided into two parts: a straight swimming experiment and a turning swimming experiment. In addition, the swimming experiment was also to further verify the accuracy of the established mathematical model. The size of the pool for the experiment was 2 m × 3 m, and the depth was 40 cm. The experiment was carried out at room temperature of 25 °C. A camera was installed above the pool, as shown in Figure 8, to upload the picture of the robotic fish swimming to the computer monitoring interface. This paper uses MATLAB as the simulation platform to implement the dynamic model simulation, with the environment defined as a flow environment. In conjunction with the dynamic modeling in Chapter 3.2, the simulation fully accounts for the dynamic effects of water flow on the robot fish's head and tail fins.

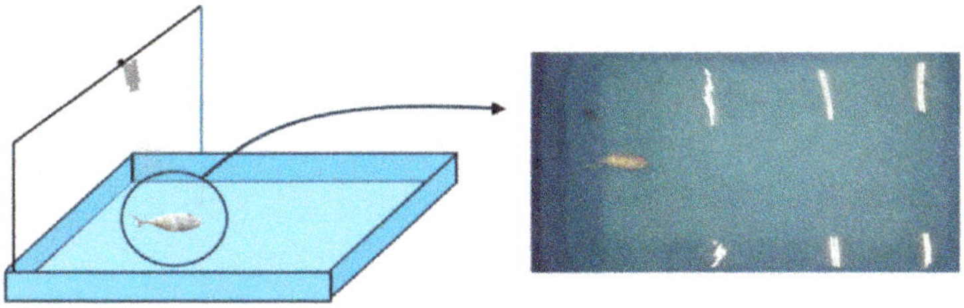

Figure 8. Robotic fish swimming experimental platform.

Both the forward swimming and turning experiments assessed the swimming performance of three groups of robotic fish with different tail stiffness at various frequencies [22]. The forward swimming experiment measured the swimming speed of the robotic fish with different tail stiffnesses at frequencies ranging from 1 to 7 Hz, while the turning experiment measured the turning radius of the robotic fish at frequencies ranging from 1 to 3 Hz.

4.1. Forward Swimming

The linear swimming part mainly explores the effect of different tail stiffness and frequency on the swimming speed of the robotic fish. In order to refine the experimental process, the experiment tested three groups of robotic fish with different tail stiffness in the range of 1–7 Hz. Three experiments were conducted at each frequency [20], and the average of the three experimental results was taken as the experimental result, with a total of 21 groups of experimental results. Figure 9 is a process diagram of the robotic fish swimming in a straight line. When the different parameters were tested experimentally, the 21 different parameter groups were simulated at the same time. The simulation result and actual trajectory of one linear swimming are shown in Figure 10.

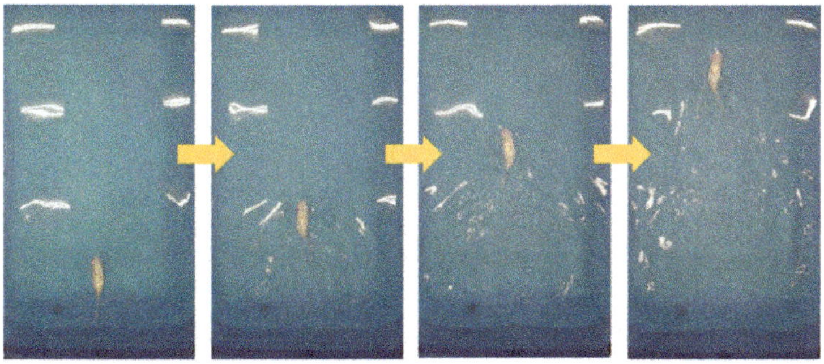

Figure 9. Forward swimming process.

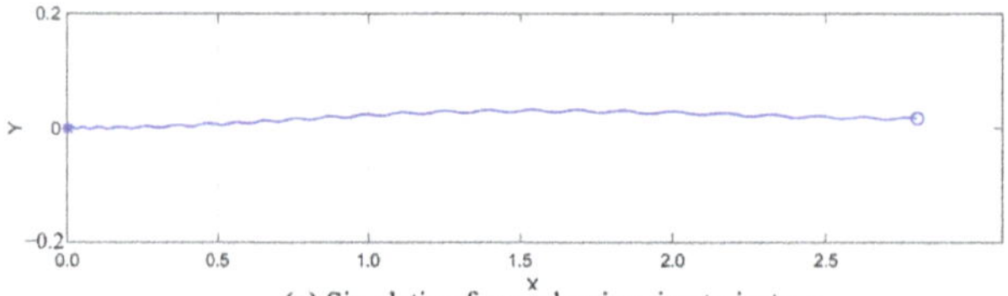

(a) Simulation forward swimming trajectory

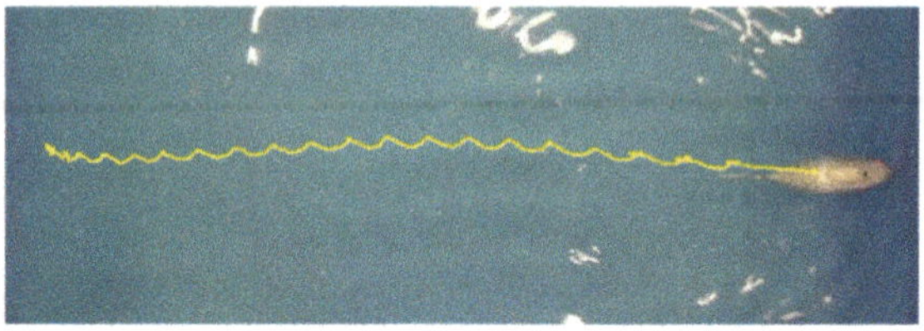

(b) Experimental forward swimming trajectory

Figure 10. Simulation and experimental motion trajectory.

Figure 11a,b show the yaw angle and forward swimming speed obtained from the simulation, respectively. While the simulation of the yaw angle deviates from the actual yaw angle observed in the swimming experiment, this discrepancy arises because the robot fish's tail does not remain perfectly centered when it starts swinging, leading to an asymmetric swing waveform. On the other hand, the simulation of swimming speed closely matches the experimental swimming speed for this group.

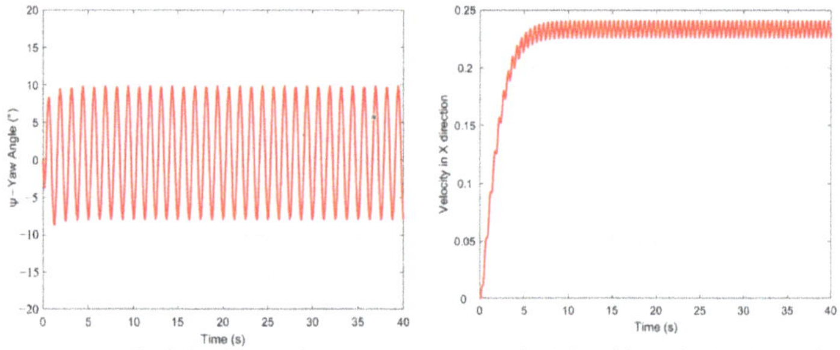

(a) Simulation of yaw angle (b) Simulation of forward swimming speed

Figure 11. Simulation of yaw angle and forward swimming speed.

From the comparison chart of simulation results and experiments, it can be found that the straight swimming trajectory of the robotic fish is consistent with the simulation results; the overall deviation of the yaw angle is not large, and it corresponds to the swimming trajectory. In addition, this experiment intercepted three groups of video frames of robotic fish with different tail stiffness flapping in the water to verify whether the maximum swing angle of the robotic fish swimming in a straight line is consistent with the kinematic model. In order to reduce the interference of water resistance during high-frequency swinging, taking the swing frequency of 2 Hz as an example, Figure 12 is an image of a full rigid fishtail in a complete swing cycle. Six frames of images were intercepted within T, namely 0, T/6, 2T/6, 3T/6, 4T/6, and 5T/6. In each frame of the image, the red solid line is the fish body swing curve, and the yellow dotted line is the median direction. The experimental results show that the maximum swing angle of straight swimming is about 45°, which is consistent with the result calculated by Formula (11). For the 0.8 mm stiffness fishtail and 1.0 mm stiffness fishtail, due to the influence of water resistance on both sides, when the active section body bends, the torsion spring cannot generate sufficient restoring force in time, resulting in the fishtail swing angle being less than 45°, but the bending angle of the joint at the end of the active section body is still 45°.

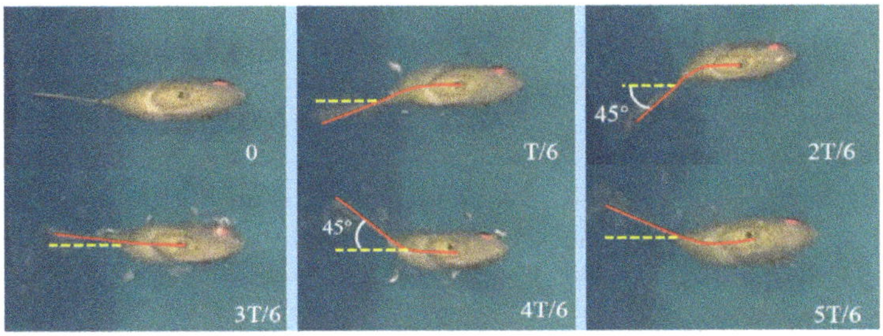

Figure 12. Image of a full rigid fishtail during a complete swing cycle.

In addition, the experimental measurement results of the swimming speed of three groups of fishtails with different stiffness at different frequencies are shown in Figure 13. The experimental results show that with the increase in frequency, the swimming speed of the three groups of robotic fish all tend to increase first and then increase. When the frequency reaches 4 Hz, the swimming speed begins to decrease. The reason is that the stiffness of the passive joint of the fishtail is relatively low. Before the swing frequency reaches 4 Hz, the swing amplitude of the tail fin is large, the propulsion efficiency is high, and the swimming speed is positively correlated with the frequency. When the swing frequency of the fishtail exceeds 4 Hz, the passive joint of the fishtail cannot respond in time, and a waveform will appear, resulting in a significant decrease in the swing amplitude of the tail fin. When the frequency continues to increase, the swing amplitude of the end of the fishtail continues to decrease, resulting in a negative correlation between the swimming speed and the frequency at this time. At the same time, when it reaches 7 Hz, the active joint of the robotic fish does not respond in time, and the swimming speed of the robotic fish drops rapidly at this time, so it is not within the scope of the study. Due to the existence of the active joint torsion spring, a higher frequency response can be achieved by replacing the joint torsion spring with a larger stiffness, but this is not discussed in this experiment.

Figure 13d is a comparison of the average values of the experimental results. When the swing frequency is 1 Hz, the swimming performance of the robotic fish using a 0.8 mm fishtail joint torsion spring is the best, with a speed of 0.34 times the body length, while the swimming performance of the robotic fish using a fully rigid fishtail joint is the worst, with a speed of only 0.285 times the body length. This is due to the existence of inertia and water resistance; the 0.8 mm torsion spring can undergo a large deformation, so the fishtail swing angle of the robotic fish is larger when it swings. Similarly, the swimming speed of the robotic fish using a 1.0 mm fishtail joint torsion spring is also slightly higher than that of the robotic fish using a fully rigid fishtail joint. However, as the swing frequency gradually increases to 3 Hz, the swimming performance of the robotic fish using a fully rigid fishtail joint gradually becomes better than the other two groups of rigid fishtails. This is because the passive section of the fishtail with too low stiffness will cause a slower response as the frequency increases, thereby affecting the swimming speed. When the frequency reaches 4 Hz, the swimming speed of the robotic fish reaches its maximum, which is 0.71 times the body length. When the frequency exceeds 4 Hz, the speed begins to decrease, and the swimming speed is negatively correlated with the increase in frequency. Since the 0.8 mm fishtail joint torsion spring is more easily deformed, the passive section of the fishtail cannot respond in time at high frequencies, and the swimming speed slows down rapidly. This is also the reason why the swimming speed of the robotic fish using this set of rigid fishtails in high-frequency mode lags far behind the other two sets of rigid fishtails. Although the overall swimming speed of the robotic fish slows down with the increase in frequency, when the frequency exceeds 5 Hz, the swimming speed of the robotic fish using a 1.0 mm fishtail joint torsion spring is better than that of the robotic fish using a fully rigid fishtail joint. Analysis shows that when the frequency is too high, the swing of the active section of the fishtail begins to fail to respond quickly, resulting in a large water resistance of the fully rigid fishtail, which leads to a large decrease in swimming speed. However, due to the characteristics of the torsion spring, the passive section of the fishtail of the robotic fish using the 1.0 mm fishtail joint torsion spring will still swing outward for a distance along the original swing direction under the action of inertia, making its swing angle larger, so the water resistance is relatively small, and the swimming speed decreases less. When the swing frequency is higher than 7 Hz, the swimming speed of the robotic fish continues to decrease and is unstable, so no more test records are made.

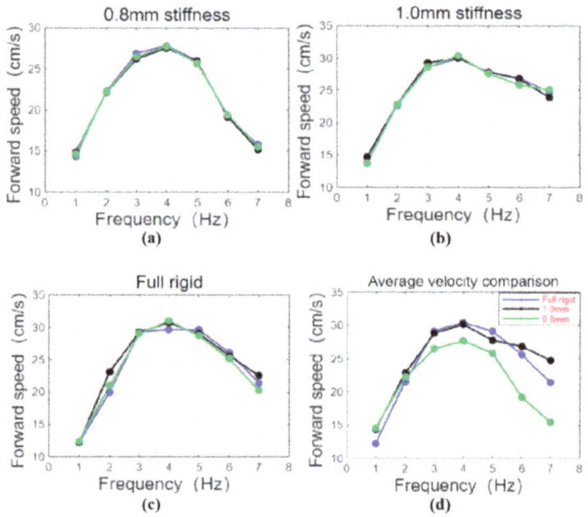

Figure 13. Results of the swimming speed of fishtails with different stiffness at different frequencies: (**a**–**c**) the experimental results of the forward swimming speed of the fishtail with 0.8 mm stiffness, 1.0 mm stiffness, and full stiffness at a frequency of 1–7 Hz; (**d**) comparison of average speeds of robotic fish with different tail stiffness.

4.2. Turning Swimming

Turning is another important part of the robotic fish swimming mode. The robotic fish designed in this paper relies on the turning auxiliary servo to control the movement of the fixed bottom plate to change the middle position of the body swing. Specifically, the output of the double sine mechanism is offset by the movement of the bottom plate, resulting in a change in the standard sine motion, thereby changing the symmetry center of the sine motion. At the same time, the program controls the two sides of the robotic fish's motion in the middle position to produce differential motion, and the two work together to complete the robotic fish's turning. The accuracy of the swimming motion model is verified by comparing the experimental results with the swimming simulation results.

The experiment tested the turning of three groups of robotic fish with different tail stiffness in the range of 1–3 Hz. Three experiments were conducted at each frequency, and the average of the three experimental results was taken as the experimental result. The experimental results were taken at intervals of 0.5 Hz, and a total of 15 groups of experimental results were obtained. Figure 14 is a process diagram of the robotic fish turning and swimming. When the different parameters were tested experimentally, the 15 different parameter groups were simulated at the same time. The simulation results and actual trajectories of one-turning swimming are shown in Figure 15.

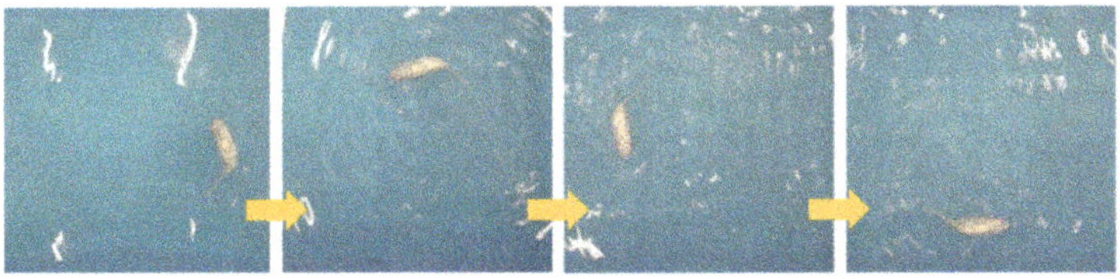

Figure 14. Process diagram of the robotic fish turning and swimming.

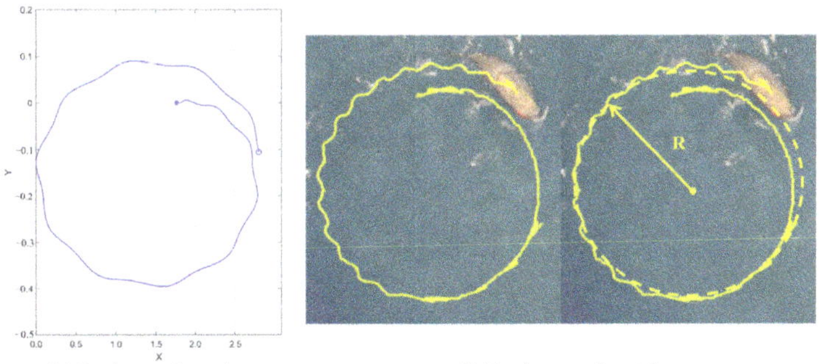

(a) Turning motion trajectory simulation

(b) Turning experimental motion trajectory

Figure 15. Turning experiment and simulation trajectory.

The simulation results are consistent with the actual experimental test results. Overall, the turning performance of the robotic fish designed in this paper is highly correlated with the results calculated by the dynamic model, which also proves the accuracy of the swimming model to a certain extent. Figure 16 is a schematic diagram of the robotic fish turning comparison under the line drive frequency of 1~3 Hz. It can be seen that the high-frequency line drive robotic fish designed in this paper can achieve a good turning effect. The three sets of experimental data all show that the turning radius increases with the increase in the robotic fish swing frequency. Among them, the robotic fish using a fully rigid fishtail joint has the smallest turning radius at a frequency of 1 Hz, which is 21.75 cm, about 0.51 body lengths. The robotic fish using a 0.8 mm fishtail joint torsion spring has the largest turning radius. This is because during differential steering, the lower the stiffness of the tail joint, the easier it is to sweep out a larger steering angle at a certain steering speed [23,24], thereby increasing the forward speed of the robotic fish and resulting in a larger swimming radius of the robotic fish [25–27]. High-frequency swinging increases the forward speed of the robotic fish, resulting in a larger turning radius. Therefore, the higher the swing frequency, the larger the turning radius; the lower the swing frequency, the smaller the turning radius. However, when the frequency exceeds 2.5 Hz, the turning radius of the robotic fish using a 0.8 mm fishtail joint torsion spring decreases. This phenomenon is caused by the low stiffness of the tail. When the frequency is too high, the passive section of the fishtail cannot respond in time, so the steering angle swept by the fishtail decreases, resulting in a decrease in the turning radius. At the same time, due to the presence of the variable central axis mechanism, the steering effect of the robotic fish is significantly improved compared to the traditional motor-driven differential steering robotic fish, but there is still a gap compared to the traditional servo-driven robotic fish. Figure 17 shows the turning results of fishtails with different stiffness at different frequencies. From the turning comparison experiment in the figure, we can more intuitively obtain the same conclusion as the experimental data. In addition, the experimental results of turning radius are more directly expressed in Table 3.

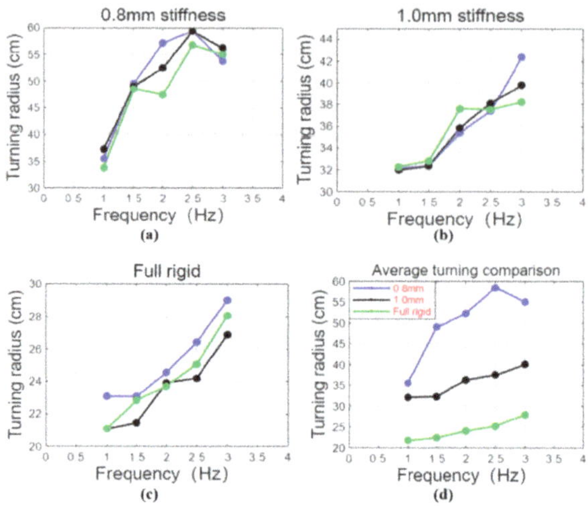

Figure 16. Results of turning radius of fishtail with different stiffness at different frequencies: (**a–c**) the experimental results of the turning radius of the fishtail with 0.8 mm stiffness, 1.0 mm stiffness, and full rigidity at 1–3 Hz frequency; (**d**) comparison of turning radius of robotic fish with different tail stiffness.

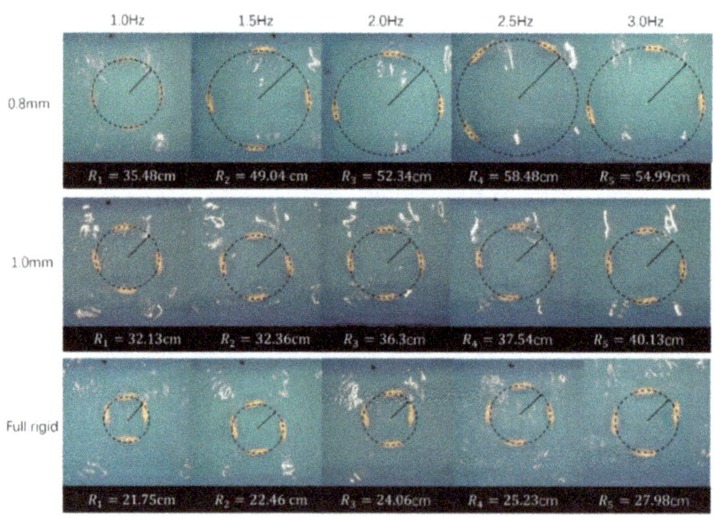

Figure 17. Turning results of fishtails with different stiffness at different frequency.

Table 3. Radius of robotic fish with different tail stiffness at different frequencies.

Tail Stiffness \ Frequency	1.0 Hz	1.5 Hz	2.0 Hz	2.5 Hz	3.0 Hz
0.8 mm	35.48 cm	49.04 cm	52.34 cm	8.48 cm	54.99 cm
1.0 mm	32.13 cm	32.36 cm	36.30 cm	37.54 cm	40.13 cm
Full rigid	21.75 cm	22.46 cm	24.06 cm	25.23 cm	27.98 cm

5. Conclusions and Future Works

Inspired by the crank rocker, this paper designs a transmission mechanism called a double sine mechanism. Compared with the traditional wire-driven robotic fish, the mechanism designed in this paper can withstand greater lateral forces and ensure the stability of the robotic fish when swimming at high frequencies, with a maximum frequency of more than 7 Hz. The tail connector of the robotic fish uses torsion springs with three different diameters to simulate experiments under different tail stiffness. The experimental results show the following:

- As the tail swing frequency changes, the swimming speed change trends of the three groups of robotic fish with different tail stiffness are consistent. When the tail swing frequency is lower than 4 Hz, the linear swimming speed of the robotic fish increases with the increase in frequency, and when the tail swing frequency exceeds 4 Hz, its linear swimming speed decreases with the increase in frequency. When the frequency exceeds 7 Hz, the tail material reaches its limit and cannot swim normally.
- When the tail swing frequency is lower than 4 Hz, the swimming speed of the robotic fish with tail stiffness of 0.8 and 1.0 (the diameters of the torsion spring of the tail connector are 0.8 mm and 1.0 mm) gradually becomes weaker than that of the robotic fish with full rigid tail. This is because as the swing frequency increases, the response of the passive section of the tail with too low stiffness also slows down, thereby affecting the swimming speed.
- When the tail swing frequency exceeds 5 Hz, although the overall swimming speed tends to decrease, the swimming speed of the robotic fish with a tail stiffness of 1.0 is better than that of the robotic fish with a fully rigid tail. This is because when the frequency is too high, the active section of the tail cannot respond quickly, causing the fully rigid tail to be subject to greater water resistance. Due to the characteristics of the torsion spring, the passive section of the tail of the robotic fish with a tail stiffness of 1.0 will continue to swing outward for a distance in the original swing direction under the action of inertia, increasing its swing angle and reducing the influence of water resistance. Therefore, the swimming speed decreases relatively less.
- The turning radius of the robotic fish increases with the increase in frequency, and the weaker the tail stiffness, the larger the turning radius. This is because the lower the tail stiffness during turning, the easier it is to sweep a larger turning angle, resulting in a larger swimming radius. High-frequency swinging increases the swimming speed of the robotic fish, which also leads to an increase in the turning radius.

Although this study reveals the effect of different tail stiffness on swimming speed and agility under high-frequency motion, there is still a long way to go to catch up with real fish, including the turning mechanism, and there is still a gap compared with the turning of traditional servo-driven robotic fish. Future research will focus on the turning of the robotic fish in high-frequency mode.

Author Contributions: All authors contributed to the conceptualization and discussion of the paper. Y.Z. led the development of the paper's overarching framework and supervised the project. Y.C. performed a thorough review and revision of the manuscript. Z.Z. designed and built the prototype. Q.Y., Q.W. and Z.C. were responsible for the theoretical derivation, experimental design, and manuscript writing. All authors have read and agreed to the published version of the manuscript.

Funding: This work was supported by GDNRC [2023]33, the National Natural Science Foundation of China under Grant 62473161, the Natural Science Foundation of Guangdong Province (2024A1515012582), Guangzhou applied Basic Research project 2024A04J9942, and the GJYC program of Guangzhou (2024D03J0005).

Institutional Review Board Statement: Not applicable.

Data Availability Statement: No new data were created or analyzed in this study. Data sharing is not applicable to this article.

Conflicts of Interest: The authors declare that they have no known competing financial interests or personal relationships that could have appeared to influence the work reported in this paper.

References

1. Triantafyllou, M.S.; Triantafyllou, G.S. An Efficient Swimming Machine. *Sci. Am.* **1995**, *272*, 64–70. [CrossRef]
2. Liang, J.; Wang, T.; Wen, L. Development of a two-joint robotic fish for real-world exploration. *J. Field Robot.* **2011**, *28*, 70–79. [CrossRef]
3. Wen, L.; Wang, T.; Wu, G.; Liang, J.; Wang, C. Novel Method for the Modeling and Control Investigation of Efficient Swimming for Robotic Fish. *IEEE Trans. Ind. Electron.* **2012**, *59*, 3176–3188. [CrossRef]
4. Li, L.; Wang, C.; Xie, G. Modeling of a carangiform-like robotic fish for both forward and backward swimming: Based on the fixed point. *2014 IEEE Int. Conf. Robot. Autom. (ICRA)* **2014**, 800–805.
5. Zhong, Y.; Li, Z.; Du, R. A Novel Robot Fish with Wire-Driven Active Body and Compliant Tail. *IEEE/ASME Trans. Mechatron.* **2017**, *22*, 1633–1643. [CrossRef]
6. Zhong, Y.; Song, J.; Yu, H.; Du, R. Toward a Transform Method from Lighthill Fish Swimming Model to Biomimetic Robot Fish. *IEEE Robot. Autom. Lett.* **2018**, *3*, 2632–2639. [CrossRef]
7. Zhang, B.; Chen, Y.; Wang, Z.; Ma, H. Research and Experiment on a Bionic Fish Based on High-Frequency Vibration Characteristics. *Biomimetics.* **2023**, *8*, 253. [CrossRef]
8. Yu, J.; Zhang, C.; Liu, L. Design and Control of a Single-Motor-Actuated Robotic Fish Capable of Fast Swimming and Maneuverability. *IEEE/ASME Trans. Mechatron.* **2016**, *21*, 1711–1719. [CrossRef]
9. White, C.H.; Lauder, G.V.; Bart-Smith, H. Tunabot Flex: A tuna-inspired robot with body flexibility improves high-performance swimming. *Bioinspiration Biomim.* **2021**, *16*, 026019. [CrossRef]
10. Liao, X.; Zhou, C.; Wang, J.; Tan, M. A Wire-Driven Dual Elastic Fishtail with Energy Storing and Passive Flexibility. *IEEE/ASME Trans. Mechatron.* **2024**, *29*, 1914–1925. [CrossRef]
11. Liao, X.; Zhou, C.; Wang, J.; Fan, J.; Zhang, Z. A Wire-driven Elastic Robotic Fish and its Design and CPG-Based Control. *J. Intell. Robot. Syst.* **2022**, *107*, 4. [CrossRef]
12. van den Berg, S.C.; Scharff, R.B.; Rusák, Z.; Wu, J. OpenFish: Biomimetic design of a soft robotic fish for high speed locomotion. *HardwareX* **2022**, *12*, e00320. [CrossRef] [PubMed]
13. Chen, D.; Wu, Z.; Meng, Y.; Tan, M.; Yu, J. Development of a High-Speed Swimming Robot with the Capability of Fish-Like Leaping. *IEEE/ASME Trans. Mechatron.* **2022**, *27*, 3579–3589. [CrossRef]
14. Chen, B.; Jiang, H. Swimming Performance of a Tensegrity Robotic Fish. *Soft Robot.* **2019**, *6*, 520–531. [CrossRef] [PubMed]
15. Patterson, C. Bony Fishes. *Short Courses Paleontol.* **1994**, *7*, 57–84. [CrossRef]
16. Fiazza, C.; Salumäe, T.; Listak, M.; Kulikovskis, G.; Templeton, R.; Akanyeti, O.; Megill, W.; Fiorini, P.; Kruusmaa, M. Biomimetic mechanical design for soft-bodied underwater vehicles. In Proceedings of the OCEANS'10 IEEE SYDNEY, Sydney, NSW, Australia, 24–27 May 2010; pp. 7–10.
17. Liu, S.; Liu, C.; Liang, Y.; Ren, L.; Ren, L. Tunable Stiffness Caudal Peduncle Leads to Higher Swimming Speed Without Extra Energy. *IEEE Robot. Autom. Lett.* **2023**, *8*, 5886–5893. [CrossRef]
18. Wen, L.; Liang, J.; Wu, G.; Li, J. Hydrodynamic Experimental Investigation on Efficient Swimming of Robotic Fish Using Self-propelled Method. In Proceedings of the Twentieth International Offshore and Polar Engineering Conference, Beijing, China, 20–25 June 2010. ISOPE-I-10-166.

19. Qiu, C.; Wu, Z.; Wang, J.; Tan, M.; Yu, J. Locomotion Optimization of a Tendon-Driven Robotic Fish with Variable Passive Tail Fin. *IEEE Trans. Ind. Electron.* **2023**, *70*, 4983–4992. [CrossRef]
20. Wang, Q.; Hong, Z.; Zhong, Y. Learn to swim: Online motion control of an underactuated robotic eel based on deep reinforcement learning. *Biomim. Intell. Robot.* **2022**, *2*, 100066. [CrossRef]
21. Morison, J.R.; O'Brien, M.P.; Johnson, J.W.; Schaaf, S.A. The Force Exerted by Surface Waves on Piles. *J. Pet. Technol.* **1950**, *2*, 149–154. [CrossRef]
22. Clapham, R.J.; Hu, H. iSplash: Realizing Fast Carangiform Swimming to Outperform a Real Fish. In *Robot Fish: Bio-inspired Fishlike Underwater Robots*; Du, R., Li, Z., Youcef-Toumi, Valdivia, P., Alvarado, Y., Eds.; Springer: Berlin, Heidelberg, 2015; pp. 193–218.
23. Xie, F.; Li, Z.; Ding, Y.; Zhong, Y.; Du, R. An Experimental Study on the Fish Body Flapping Patterns by Using a Biomimetic Robot Fish. *IEEE Robot. Autom. Lett.* **2020**, *5*, 64–71. [CrossRef]
24. Liao, B.; Li, Z.; Du, R. Robot fish with a novel biomimetic wire-driven flapping propulsor. *Adv. Robot.* **2014**, *28*, 339–349. [CrossRef]
25. Feilich, K.L.; Lauder, G.V. Passive mechanical models of fish caudal fins: Effects of shape and stiffness on self-propulsion. *Bioinspiration Biomim.* **2015**, *10*, 036002. [CrossRef] [PubMed]
26. Yu, J.; Su, Z.; Wu, Z.; Tan, M. Development of a Fast-Swimming Dolphin Robot Capable of Leaping. *IEEE/ASME Trans. Mechatron.* **2016**, *21*, 2307–2316. [CrossRef]
27. Yu, J.; Wu, Z.; Wang, M.; Tan, M. CPG Network Optimization for a Biomimetic Robotic Fish via PSO. *IEEE Trans. Neural Netw. Learn. Syst.* **2016**, *27*, 1962–1968. [CrossRef]

biomimetics

MDPI

Article

A Pneumatic Soft Glove System Based on Bidirectional Bending Functionality for Rehabilitation

Xiaohui Wang [1,2], Qinkun Cheng [1,2], Zhifeng Wang [1,2], Yongxu Lu [1,2], Zhaowei Zhang [1,*] and Xingang Zhao [1,*]

[1] State Key Laboratory of Robotics, Shenyang Institute of Automation, Chinese Academy of Sciences, Shenyang 110016, China; xiaohuiw@mail.ustc.edu.cn (X.W.); chengqinkun@mail.ustc.edu.cn (Q.C.); wangzhifeng@sia.cn (Z.W.); luyongxu@sia.cn (Y.L.)

[2] University of Chinese Academy of Sciences, Beijing 100049, China

* Correspondence: zhangzhaowei@sia.cn (Z.Z.); zhaoxingang@sia.cn (X.Z.)

Abstract: Stroke-related hand dysfunction significantly limits the ability to perform daily activities. Pneumatic soft gloves can provide rehabilitation training and support for individuals with impaired hand function, enhancing their independence. This paper presents a novel pneumatic soft robotic system for hand rehabilitation featuring bidirectional bending actuators. The system comprises a pneumatic soft glove and a pneumatic control platform, enabling various rehabilitation gestures and assisting with finger grasping. The main bending module of the pneumatic soft actuator features a three-stage cavity structure, allowing for a wider range of finger rehabilitation training gestures and greater bending angles. The reverse-bending module uses a trapezoidal cavity design to enhance the reverse-bending capability, effectively facilitating finger extension motion. The pneumatic control platform is simple to set up, but effectively controls the actuators of the soft glove, which enables both main and reverse bending. This allows individuals with hand impairments to perform various gestures and grasp different objects. Experiments demonstrate that the pneumatic soft glove has a measurable load capacity. Additionally, the pneumatic soft glove system is capable of executing single-finger movements, a variety of rehabilitation gestures, and the ability to grasp different objects. This functionality is highly beneficial for the rehabilitation of individuals with hand impairments.

Keywords: pneumatic soft glove; bidirectional bending actuator; rehabilitation gestures; assisted grip

1. Introduction

With the increasing global aging population, stroke has become the third-leading cause of death and disability worldwide [1]. Approximately 55% to 75% of stroke survivors experience persistent motor impairments in the upper limb and hand [2,3]. Stroke patients often struggle to control the affected areas of their body, significantly impacting their quality of life, ability to perform activities of daily living (ADLs), and even their psychological well-being, ultimately reducing their independence [4,5]. Among the various body parts involved, the hand plays a crucial role in physical interactions between humans and their environment [6,7], which has led to an increasing demand for hand rehabilitation among patients. However, traditional manual rehabilitation methods are costly, inefficient, and vary widely in quality, with a severe shortage of therapists [8,9]. As a result, robot-assisted rehabilitation has garnered increasing attention as a promising approach for supporting hand function recovery in stroke patients [10]. Studies have shown that robot-assisted repetitive task training can effectively aid in the recovery of hand function [11–13].

 https://doi.org/10.3390/biomimetics10030129

Traditional hand rehabilitation robots typically use rigid structures. These rigid robots were among the first to be developed for rehabilitation and assistive applications, as their rigid mechanical drive systems can transmit high forces with high precision [14]. Rigid robots for hand rehabilitation primarily use motors to drive finger joints. The average weight of these devices is approximately 500 g. They are relatively large compared with the size of the human hand, making them uncomfortable to wear and difficult to carry. Additionally, they pose safety risks, potentially causing further injury to patients [15,16]. Rigid structures reduce the biomimetic quality of robots, limiting their therapeutic potential. To overcome these limitations, soft actuators made from hyperelastic materials have emerged as novel types of continuous biomimetic robots applied in finger rehabilitation and assistive movements [17]. Among these, pneumatic soft hand rehabilitation robots have become a key focus of research because of their light weight, safety, flexibility, portability, and excellent biocompatibility [18,19], making them highly suitable for hand rehabilitation and assistance [20].

Polygerinos et al. introduced a pneumatic network actuator made from hyperelastic materials featuring an integrated channel design that incorporates a soft actuator into an open glove made of flexible material. Their study demonstrated the actuator's ease of use and feasibility of fist formation [21]. However, this type of soft actuator can only produce curved contours resembling arcs, limiting its ability to replicate the complex curvature of finger flexion. Connolly et al. designed a fiber-reinforced soft actuator that mimics biological joints by extending, expanding, twisting, and pressurizing sections made of different cylindrical elastic tubes [22]. Although the actuator's bending appears to simulate the flexion of three joints, it is actually controlled by a single actuation, meaning all joints must bend simultaneously. Feng et al. introduced a pneumatic soft actuator with a three-stage cavity structure that can independently drive the three joints of a finger [23]. However, this actuator is limited to flexion movement and cannot facilitate the extension movement of the finger. Yap et al. introduced a stiffness-adjustable pneumatic soft actuator designed to mimic the natural flexion of the human hand under pneumatic pressure [24]. On the basis of the anatomical structure of the human finger, J. Wang et al. developed a segmented pneumatic soft actuator that conforms to the finger's contours, and embedded an internal bladder into the cavity to improve airtightness [25]. However, both of these soft actuators are unable to support finger extension movement. F. Wang et al. introduced a bidirectional pneumatic soft rehabilitation glove that combines both rehabilitation and assistive functions. The main body of the soft actuator, responsible for both main and reverse bending, uses rectangular solids. The joints of the main bending module feature trapezoidal air cavities, while the reverse-bending module uses triangular air cavities at its joints. This glove mimics human finger joints and offers four distinct rehabilitation postures [26]. However, the design of the main bending module limits the range of both finger flexion and extension angles. Simultaneously, the design of the reverse-bending module affects the reverse-bending angle of the finger and results in a smaller fingertip force during reverse bending.

Soft actuators have attracted substantial interest; however, their fluid-driven systems have been relatively overlooked [27,28]. Pneumatic actuation in soft hand robots is typically powered by compressed air, with air pumps serving as the system's air source, and various valves control the direction of airflow [29–31]. As soft hand rehabilitation robot systems become more complex, there is an increasing demand for precise control of actuator movements. In this context, relays and solenoid valves play crucial roles [32,33]. Several studies on soft hand rehabilitation robots have discussed the use of solenoid valve-based systems [34–36]. Skorina et al. designed a valve system where the inlet is directly connected to a compressed air pump, the outlet is linked to the soft actuator, and the exhaust port is open to the atmosphere [37]. T. Wang et al. employed a dual two-position, two-way solenoid valve (2/2-way solenoid valve) system to reduce energy consumption and increase the valve lifespan [38]. Young et al. used 26 switch valves and 5 proportional valves to control a pneumatic soft hand robot, enabling simultaneous control of multiple actuators [31]. Overall, there are few studies on control platforms specifically designed for managing the bidirectional bending of soft actuators.

This paper presents a bidirectional pneumatic soft glove system, comprising both a pneumatic soft glove and a pneumatic control platform. The main bending module of the bidirectional pneumatic soft glove features a three-stage cavity structure that mimics the finger joints. To allow reverse bending without interfering with the main bending movement, the reverse-bending module uses a trapezoidal cavity design. Additionally, a simple and easily replicable programmable pneumatic control platform was designed to manage the complex movements of the soft glove. First, an analysis of the soft actuator revealed that the pneumatic soft glove has a measurable load capacity. Then, through rehabilitation experiments in which the glove was worn on both a prosthetic hand model and a human hand, its ability to perform various rehabilitation gestures and assist patients in grasping different objects was demonstrated. This pneumatic soft glove system enables both finger flexion and extension, supporting individuals with hand impairments in achieving greater independence in ADLs.

2. Design and Fabrication of the Soft Finger Actuator

2.1. The Structure Design of the Actuator

Owing to the complex interactions between tendons, ligaments, muscles, and bones, the human hand has at least 20 degrees of freedom [39], enabling it to perform intricate movements. The design of our pneumatic soft actuator is inspired by the soft actuator developed by Polygerinos et al. and J. Wang et al. [21,25]. To enable finger extension, a reverse-bending module was added. The design is inspired by J. Wang et al., drawing on their key features and principles. This adaptation makes our reverse-bending module better aligned with the human hand, improving the fit of the soft glove and the hand's natural movement. Building on the design by Polygerinos et al., we have improved the structure and internal cross-sectional shape of the air cavity in the main bending module, as well as the spacing between the cavities, etc. These modifications enhance the bending angle of the main bending module and provide sufficient space for the reverse bending. Following the anatomical structure of the human hand, a pneumatic network with three joints was designed, making the actuator more closely resemble the human hand (excluding the thumb), as shown in Figure 1a. On the basis of the structure and length of the thumb, a pneumatic network with two joints was designed, as shown in Figure 1b.

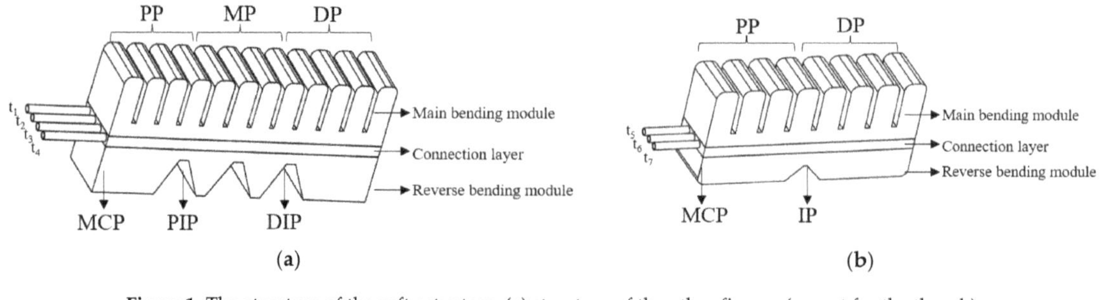

Figure 1. The structure of the soft actuators. (**a**) structure of the other fingers (except for the thumb); (**b**) structure of the thumb.

Except for the thumb, the soft actuator for each finger consists of three phalanges—the distal phalanx (DP), middle phalanx (MP), and proximal phalanx (PP)—corresponding to the distal interphalangeal (DIP), proximal interphalangeal (PIP), and metacarpophalangeal (MCP) joints, respectively. The main bending module features a three-stage cavity structure, with each phalanx consisting of four air cavities, for a total of twelve cavities. Each air cavity features a chamfered cross-sectional design [23], with an axial wall thickness of 2 mm and a lateral wall thickness of 3 mm. This design ensures a maximum bending angle during axial flexion, meeting the requirements for the angle of finger flexion while also providing sufficient fingertip force. The soft actuator can perform single-joint movements or multi-joint coordinated motions. By adjusting the airflow through different combinations of the t_1, t_2, and t_3 tubes, various rehabilitation gestures can be achieved. Without affecting the main bending module, the reverse-bending module, which is designed with a trapezoidal cross section, provides greater space for reverse bending, compared to the rectangular cross-sectional design by F. Wang et al. [26]. By inflating the t_4 tube, the reverse-bending module generates adequate force for finger extension while preventing any secondary injury to the finger.

The thumb consists of the distal phalanx (DP) and proximal phalanx (PP), corresponding to the interphalangeal joint (IP) and metacarpophalangeal joint (MCP) of the thumb. The main bending module of the thumb contains eight air cavities distributed across two phalanges. The reverse-bending module is also designed with trapezoidal cross-sectional air cavities. By adjusting the airflow through different combinations of the t_5 and t_6 tubes, different joint flexions of the thumb can be achieved. By inflating the t_7 tube, extension movement of the thumb can be achieved. Without compromising the degrees of freedom of the finger, the thumb plays a crucial role in grasping objects of various shapes.

Therefore, the designed pneumatic soft glove can perform a wider range of rehabilitation gestures and assist patients with hand impairments in grasping tasks during daily activities.

2.2. Structural Theoretical Analysis

Pneumatic soft actuators are made of silicone rubber and exhibit nonlinear characteristics. When appropriate pressure is applied, the actuator can extend to several times its original size. Once the pressure value drops to zero, it returns to its initial state. Abaqus simulations of the soft actuator were conducted via the Yeoh model for third-order hyperelastic materials to analyze its mathematical behavior and complete the fabrication of the soft actuator. The geometric parameters and values of the soft actuator are provided in Table A1.

2.2.1. Bending Angle of the Main Bending Module

The deformation of a single air cavity in the main bending module, along with its re-lated parameters, is shown in Figure 2. The initial length of a single air cavity is a, whereas its extended length after deformation is b. The height and width of the cavity are denoted h and c, respectively. The total bending angle after deformation is θ. There-fore, the deformation can be expressed in terms of known parameters such as a, c and h.

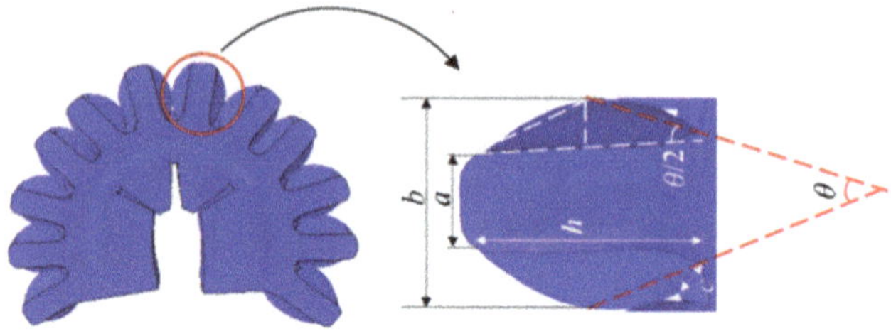

Figure 2. Parameters of a single air cavity of the main bending module after expansion.

After pressurization, the deformed section of the air cavity can be approximated as an isosceles triangle. By drawing a perpendicular line within the triangle, a smaller triangle containing angle $\theta/2$ is formed. For the smaller triangle, the lengths of the perpendicular sides are denoted $h/2$, $(a-b)/2$. According to the principle of equal areas in triangles, the following relationship is derived:

$$\sin\frac{\theta}{2}=\frac{b-a}{\sqrt{(b-a)^2+h^2}} \tag{1}$$

From this, the following equation for θ can be derived:

$$\theta=\arcsin\left(2\frac{(b-a)h}{(b-a)^2+h^2}\right)+\varepsilon, \tag{2}$$

The error due to factors such as the environment, material properties, and fabrication precision is represented by ε.

Since the final posture of the soft actuator results from the coupling between individual air cavities [23], the geometric relationship between the internal air pressure and the bending angle of each cavity can be inferred. It is assumed that the gas behaves as an ideal gas and that the ambient temperature remains constant throughout the experiment. The soft actuator deforms and bends upon inflation. According to Avogadro's hypothesis, at constant temperature (T) and pressure, the volume of gas in a single cavity is directly proportional to its mass. Within the elastic deformation stage, the stress–strain relationship of Ecoflex 00-50 follows Hooke's law, where Young's modulus (E), shear modulus (G), and Poisson's ratio (μ) remain constant [40,41]. Theoretically, the elastic constants E, G, and μ of an isotropic material have the following relationship:

$$E=2G(1+\mu) \tag{3}$$

The change in volume during the deformation process is referred to as the bulk modulus (K):

$$K=\frac{\Delta P}{\Delta V/V_0}=\frac{Pahc}{\Delta V} \tag{4}$$

The relationship between the bulk modulus, tensile modulus, and Poisson's ratio is ex- pressed as follows:

$$E=3K(1-2\mu) \tag{5}$$

where ΔP represents the variation in pressure, P is the pressure, ΔV represents the variation in gas volume, V represents the initial value of the gas volume, and the values of E and G are shown in Table A1.

When the cavity inflates, it can be approximated as a semi-ellipsoid, as shown in Figure 2. Therefore, the incremental volume of a single cavity can be expressed as:

$$V = \frac{1}{3}\pi hc(b-a) \tag{6}$$

From Equations (3)–(6), we obtain:

$$\frac{\pi hc(b-a)}{3} = Pahc\frac{9G-3E}{EG}, \tag{7}$$

thus:

$$b-a = \frac{3Pa(9G-3E)}{\pi EG} \tag{8}$$

Substituting Equation (8) into Equation (2) yields the relationship between the bending angle θ and the input air pressure P:

$$\theta = \arcsin\left(\frac{2Pkh}{P^2k^2+h^2}\right) + \varepsilon, \tag{9}$$

where:

$$k = \frac{3a(9G-3E)}{\pi EG} \tag{10}$$

2.2.2. Bending Angle of the Reverse-Bending Module

The central section of the reverse-bending module is designed with an air cavity structure in the shape of an isosceles trapezoid, with identical base angles, as shown in Figure 1a. The deformed state of a single trapezoidal air cavity after inflation and its related parameters are shown in Figure 3. The trapezoidal module has a short side of a_1, base length of b_1, height of h_1, and width of c. The angle of the base of the trapezoid is α, and half of the bending angle after deformation is φ_1, where φ_1 is the difference between the angle γ between the original trapezoidal hypotenuse and the vertical direction, and the angle δ between the vertex of the expanded triangle on the arc and the vertical direction, i.e., $\varphi_1 = \gamma - \delta$.

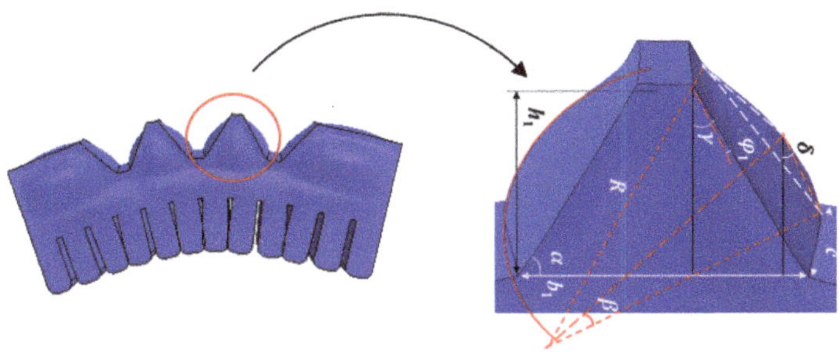

Figure 3. Parameters of a single air cavity of the reverse-bending module after expansion.

After pressurization, part of the air cavity approximates a circular arc. Let the radius of the resulting sector be r, the corresponding central angle be 2β, and the length of the trapezoid's slanted side be t. The following relationship can be derived:

$$t = \frac{h_1}{\sin\alpha} \tag{11}$$

$$\beta = 2\varphi_1 \tag{12}$$

$$r = \frac{t}{2\sin\beta} \tag{13}$$

After inflation, the chord volume on one side, denoted V_s, is given by:

$$V_s = V_S - V_T = \beta r^2 c - \frac{t}{2} cr\cos\beta \tag{14}$$

where V_S represents the volume of the sector formed by the circular arc and V_T denotes the volume of the triangular prism formed by the intersection of the sector and the trapezoid.

The volume of the original trapezoid is V_t:

$$V_t = \frac{1}{2}(a_1 + b_1)h_1 c \tag{15}$$

When the cavity inflates, the incremental volume of a single cavity, denoted V', can be expressed as:

$$V' = 2V_s = 2\beta r^2 c - ctr\cos\beta \tag{16}$$

Similarly, on the basis of the relationship between the elastic parameters E, G, and the bulk modulus K during the material's elastic deformation stage, the following expression is obtained:

$$2(V_S - V_T) = P' V_t \frac{9G - 3E}{EG} \tag{17}$$

Substituting Equations (15) and (16) into (17), we obtain:

$$2\beta r^2 - tr\cos\beta = \frac{P' h_1 (a_1 + b_1)(9G - 3E)}{2EG} \tag{18}$$

Substituting Equations (11) and (13) into (18), we obtain:

$$\frac{\beta - \cos\beta\sin\beta}{\sin^2\beta} = k', \tag{19}$$

where:

$$k' = \frac{P'(a_1 + b_1)(9G - 3E)}{EGh_1}\sin^2\alpha \tag{20}$$

where P' is the pressure and the values of E, G and the other parameters are shown in Table A1.

When the inflation pressure is constant, k' can be treated as a constant, yielding the following expression:

$$\frac{\beta - \cos\beta\sin\beta}{\sin^2\beta} = k' \tag{21}$$

$$\beta - \cos\beta\sin\beta = k'\sin^2\beta \tag{22}$$

Taking the derivative of both sides of Equation (22) gives:

$$1 - (\cos^2\beta - \sin^2\beta) = 2k'\sin\beta\cos\beta \tag{23}$$

$$\beta = \arctan k' \tag{24}$$

Substituting Equation (12) into Equation (24), we obtain half of the bending angle after deformation, denoted φ_1:

$$\varphi_1 = \frac{\arctan k'}{2} \tag{25}$$

Finally, we obtain the relationship between the bending angle φ of a single deformed trapezoidal gas chamber and the input air pressure P':

$$\varphi = \arctan(\frac{P'(a_1 + b_1)(9G - 3E)}{EGh_1} \sin^2 \alpha + \varepsilon \quad (26)$$

For the irregular trapezoidal sides of the reverse-bending module, the bending angle of a single irregular trapezoid after inflation is φ_1.

When the input air pressure is within the safe operating range of the soft actuator, this theoretical model can be used to calculate the initial input pressure. By adjusting the initial pressure, the desired finger bending angle can be achieved.

2.3. Fabrication of the Single-Finger Actuator

The soft actuator is made primarily of Ecoflex 00-50 silicone (Smooth-On Inc., Macungie, PA, USA), as shown in Figure 4a. The silicone was mixed at a 1A:1B weight ratio, thoroughly stirred, and then placed in a vacuum pump (Danhao oil-free vacuum pump) to eliminate air bubbles (Figure 4b) [42]. Afterward, the mixture was poured into a mold and left to cure for 3 h before demolding. After demolding, the layers were bonded together via a silicone adhesive (Smooth-On Silpoxy) and left to cure for one hour to ensure thorough bonding. To facilitate demolding, a release agent (Smooth-On Release 200) was evenly sprayed onto the inner surface of the mold 5 min before pouring. The material properties of Ecoflex 00-50 silicone are summarized in Table 1.

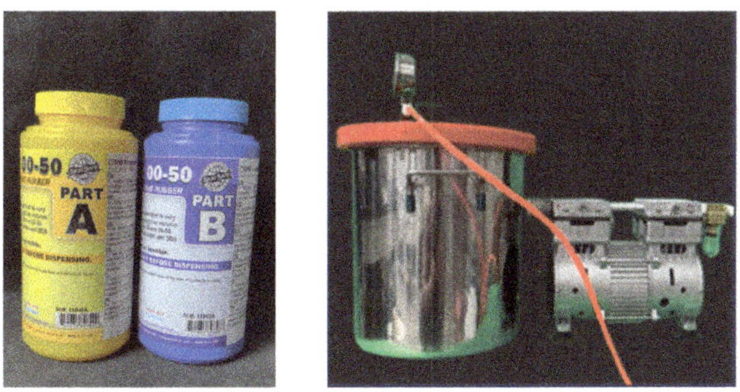

(a) (b)

Figure 4. (**a**) Material of soft actuator-Ecoflex 00-50; (**b**) vacuum pump.

Table 1. The material parameters of Ecoflex 00-50.

Material Characteristics	Ecoflex 00-50
Specific Gravity	1.07
Cure Time	3 h
Shore Hardness	00-50
Tensile Strength	2.172 MPa
100% Modulus	0.083 MPa
Elongation at Break %	980%

The molds for the soft actuators were made from polylactic acid (PLA) and were printed with a Bambu Lab 3D printer, as shown in Figure 5. In this study, the middle finger is used as a representative example of the other fingers. Figure 5a,b display the molds for the middle finger and the thumb, respectively. In Figure 5a, the molds for the main bending module are labeled M-Mm1 and M-Mm2, the intermediate connecting layer module molds are labeled M-Cm1 and M-Cm2, and the reverse-bending module molds are labeled M-Rm1 and M-Rm2. In Figure 5b, the molds for the main bending module are labeled T-Mm1 and T-Mm2, the intermediate connecting layer module molds are labeled T-Cm1 and T-Cm2, and the reverse-bending module molds are labeled T-Rm1 and T-Rm2.

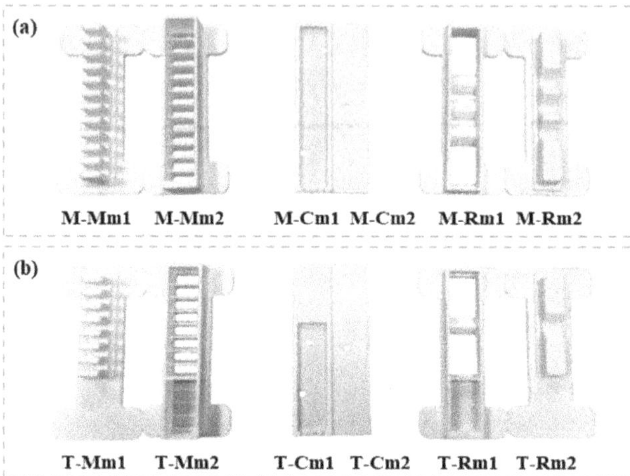

Figure 5. Finger molds. (**a**) Mold for index finger, middle finger, ring finger, and little finger; (**b**) mold for thumb.

The fabrication process for the soft actuator of the middle finger is shown in Figure 6, which can be explained in the following three steps.

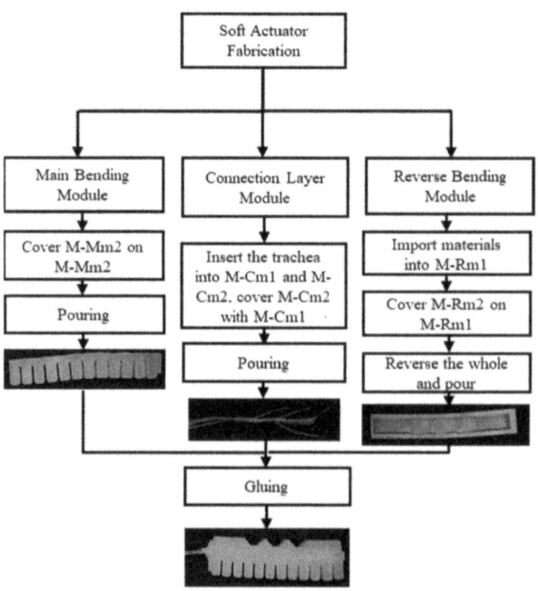

Figure 6. Fabrication process diagram of the middle finger of the soft actuator.

Step I: Make the main bending and reverse-bending modules. The main bending module is formed by combining molds M-Mm1 and M-Mm2, followed by casting. To prevent air bubbles or incomplete filling in the trapezoidal cavities during the casting of the reverse-bending module, we employed a two-step casting process. First, mold M-Rm1 is placed on a thin slice, and silicone material is poured into it. Then, mold M-Rm2 is securely positioned inside mold M-Rm1. Finally, the entire mold is inverted, and additional material is added to ensure full encapsulation.

Step II: Fabricating the intermediate layer and securing the soft tubing. The soft tube is inserted into the holes of molds M-Cm1 and M-Cm2, with one end extending approximately 15 cm beyond the lower edge of mold M-Cm1. Silicone material is then used to fill mold M-Cm1, after which mold M-Cm2 is placed over M-Cm1 and securely fixed.

Step III: After demolding, the main bending and reverse-bending sections are bonded together through the intermediate layer, completing the fabrication of the soft actuator.

The fabrication process for the thumb soft actuator follows the same steps as those described above.

3. Design of the Pneumatic Control Platform for Soft Glove

Many soft robots have been developed to emulate the movement of human fingers [43], including pneumatic soft gloves. As a key component of soft robotics, the pneumatic system plays a critical role. In particular, pneumatic systems based on pumps and electromagnetic valves demonstrate excellent performance [44]. To achieve our design requirements with a simple and controllable pneumatic circuit, we developed a programmable pneumatic control platform, as illustrated in Figure 7b. This platform, which is based on a combination of pumps and electromagnetic valves, enables a soft robotic glove with bidirectional bending capability to assist patients with hand impairments in rehabilitation tasks. The model and origin of the components used in the pneumatic control platform are detailed in Table A2.

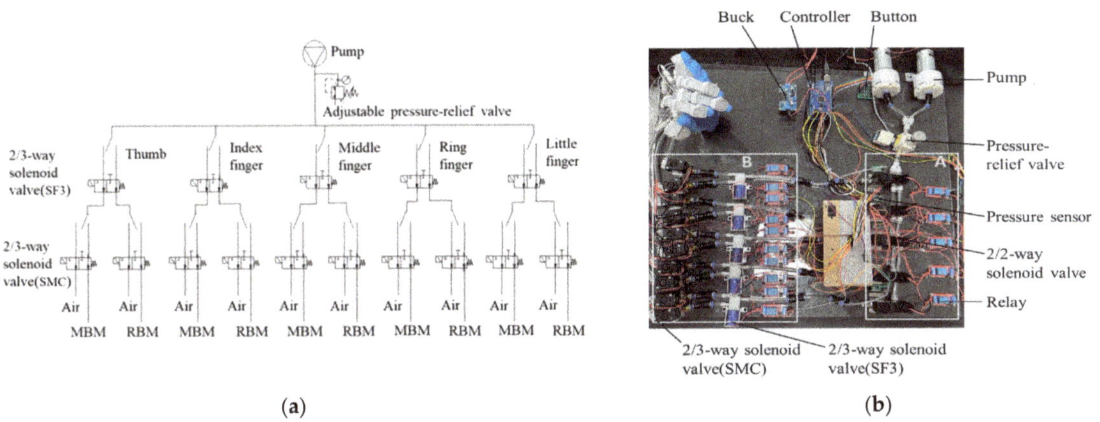

Figure 7. Design of the pneumatic system. (**a**) pneumatic circuit diagram; (**b**) pneumatic control platform.

The working principle of the pneumatic system is illustrated in Figure 7a. The system consists of a main airway and a branch airway, as shown in Figure 7b(A,B).

The pneumatic control platform uses a micro air pump as the air source, with a maximum flow rate of 15 L/min and an output pressure range of −70 to 90 kPa. A pressure-relief valve is used as a mechanical safeguard to prevent excessive pressure and allows for adjustment of the input pressure. A DC-DC buck converter (input: 6.5–40 V, output: 5 V/3 A) provides +5 V to power the Arduino controller. The desired rehabilitation gesture or assistive function is selected via a 4 × 4 matrix keypad. The Arduino controller then sends signals to control the relays, which regulate the two-position, two-way solenoid valves (2/2-way solenoid valves), enabling the selection of the main airway for the corresponding finger. During this process, an air pressure sensor (0–100 kPa) measures the pressure in the airway, and the feedback data are used to adjust the inflation pressure through the pressure regulation module. Finally, the relays in the branch circuit control the two-position, three-way solenoid valves (SF3 and SMC) (2/3-way solenoid valves) to select either the main bending module (MBM) or reverse-bending module (RBM) of soft actuators, enabling human finger flexion or extension, thereby completing the rehabilitation exercises of the pneumatic soft glove. The system also realizes the maintaining of the rehabilitation gesture of the soft glove. During the hold state, the air pump and relays in the main airway are powered off, whereas the two-position, three-way solenoid valve (SMC) in the branch airway operates to maintain the gesture. After the hold state, the components in the branch circuit are de-energized. The 2/3 solenoid valves (SMC) are switched to vent to the atmosphere (Air), releasing the gas from the soft actuator.

Experiments demonstrated the practicality of the pneumatic control platform. Additionally, its versatility and scalability offer valuable insights for the drive and control of soft robots.

4. Experimental Verification

To validate the ability of the pneumatic soft glove in performing various rehabilitation gestures and assisting with grasping tasks, we conducted experiments with both a single soft actuator and a soft glove.

4.1. *Characteristic Testing of the Single-Finger Actuator*

4.1.1. Bending Angle Measurement

To assess the impact of gravity on the bending angle and input pressure, tests were conducted in three orientations: horizontal, against the vertical direction of gravity (against gravity), and the direction of gravity (with gravity). As the soft actuator reaches its maximum bending angle without distortion, the air pressure value is continuously monitored and recorded in real-time via an air pressure sensor. In each direction, multiple experiments were conducted under each entity model, and the average values were taken. Kalman filtering was then applied for optimization to obtain the pressure values, which is the optimal air pressure value.

Figure 8 shows the entity models and air pressure values of both the main bending of the soft actuator's finger joints and the overall reverse bending of the soft actuator in the horizontal direction. Figure 8a shows the entity models of the soft actuator in different bending states. M1–M5 represent the middle finger's DP, MP, PP, and full joint in both main bending (MB-FJ) and reverse bending (RB-FJ), respectively. Similarly, T1–T4 represent the thumb's DP, PP, and full joint in both main bending and reverse bending, respectively. The air pressure values corresponding to M1–M5 for the middle finger and T1–T4 for the thumb are shown in Figure 8b.

Figures 9 and 10 illustrate the entity models and air pressure values of both the main bending of the soft actuator's finger joints and the overall reverse bending of the soft actuator in both the direction of against gravity and the direction of gravity, respectively. The arrangement and correspondence of these images follow the same pattern as those in Figure 8.

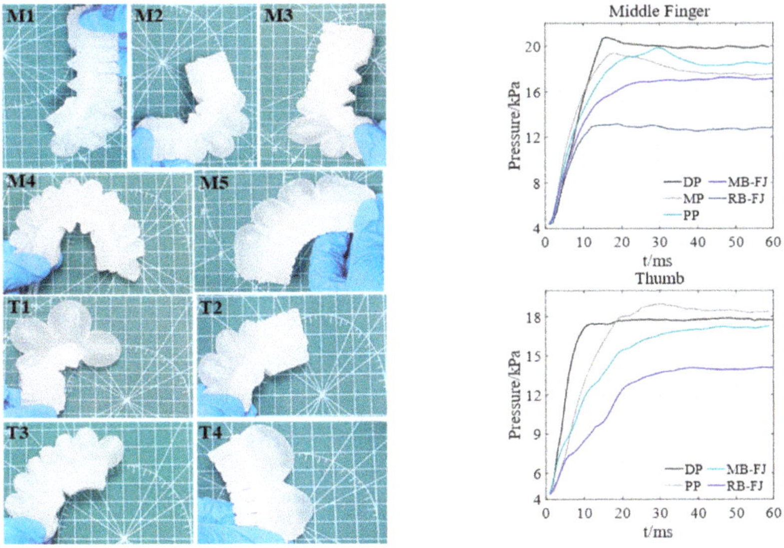

(**a**) (**b**)

Figure 8. The main bending of the soft actuator's finger joints and the overall reverse bending of the soft actuator in the horizontal direction. (**a**) entity models of the middle finger and thumb during bending; (**b**) air pressure values of the bending of the middle finger and thumb.

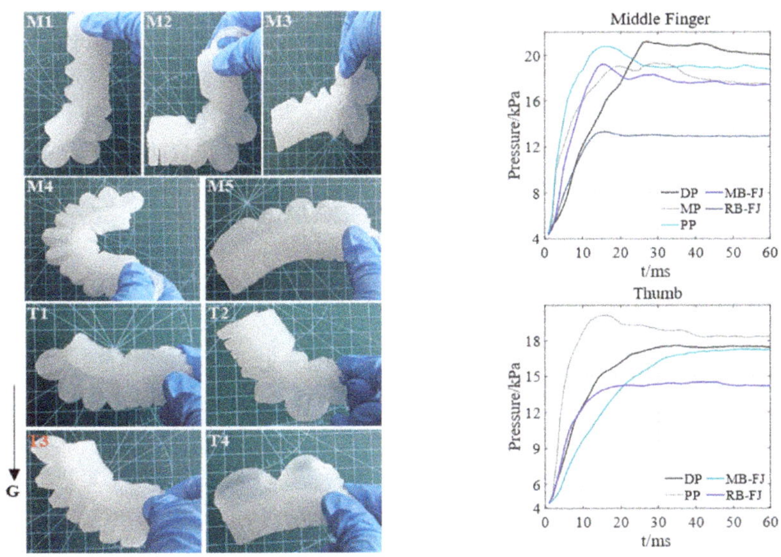

(**a**) (**b**)

Figure 9. The main bending of the soft actuator's finger joints and the overall reverse bending of the soft actuator in the direction of against gravity. (**a**) entity models of the middle finger and thumb during bending; (**b**) air pressure values of the bending of the middle finger and thumb.

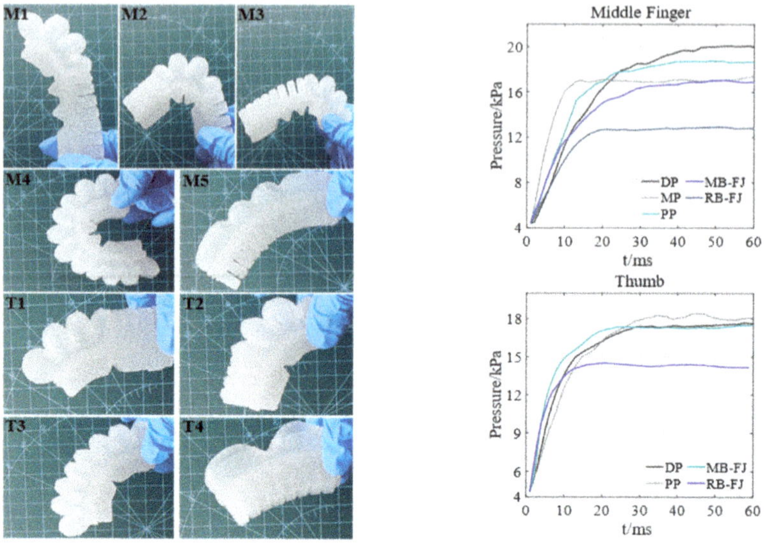

(a) (b)

Figure 10. The main bending of the soft actuator's finger joints and the overall reverse bending of the soft actuator in the direction of gravity: (**a**) entity models of the middle finger and thumb during bending; (**b**) air pressure values of the bending of the middle finger and thumb.

By comparing Figures 8b, 9b and 10b, the optimal air pressure values for both the main bending of the soft actuator's finger joints of the middle finger and thumb and the overall reverse bending of the soft actuator's middle finger and thumb are obtained. The air pressure values at the maximum main and reverse-bending angles of the soft actuator are independent of its orientation. Therefore, the effect of gravity on the soft actuator can be considered negligible.

At the optimal air pressure, we select the Yeoh model of a third-order hyperelastic material to describe the mathematical model of the soft actuator, as illustrated in Figure 11. The parameters of the Yeoh hyperelastic strain energy potential are as follows: C10 = 0.11, C20 = 0.02, and C30 = 6.07×10^{-5} [45]. The density of Ecoflex 00-50 is 1070 kg/m^3.

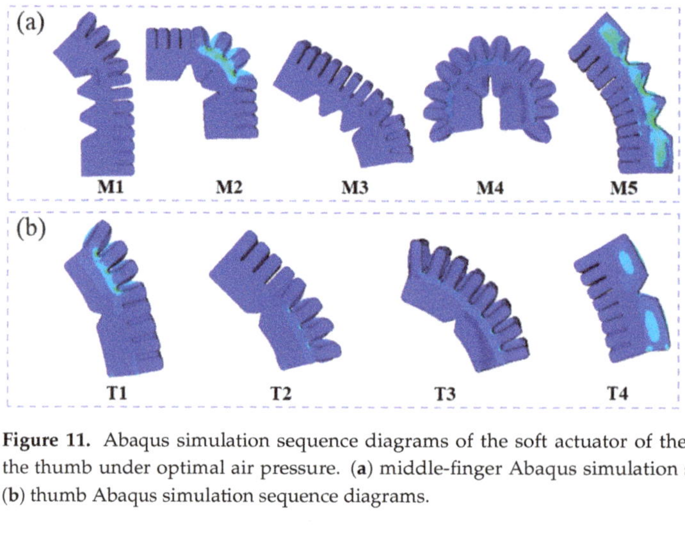

Figure 11. Abaqus simulation sequence diagrams of the soft actuator of the middle finger and the thumb under optimal air pressure. (**a**) middle-finger Abaqus simulation sequence diagrams; (**b**) thumb Abaqus simulation sequence diagrams.

The bending angles of both the entity model (EM) shown in Figure 10a and the Abaqus simulation model (ASM) of the soft actuators are shown in Figure 12. The simulation models of the soft actuators, obtained through finite element analysis in Abaqus, closely match the entity models. For the thumb, the soft actuator can reach a maximum main bending angle of 68° and a reverse-bending angle of 23°. The middle finger can achieve a maximum main bending angle of 180° and a maximum reverse-bending angle of 50°. The main and reverse-bending angles of each joint in soft actuators meet the movement requirements for hand function in daily human activities [46]. Subsequently, all experiments were conducted in the direction aligned with gravity.

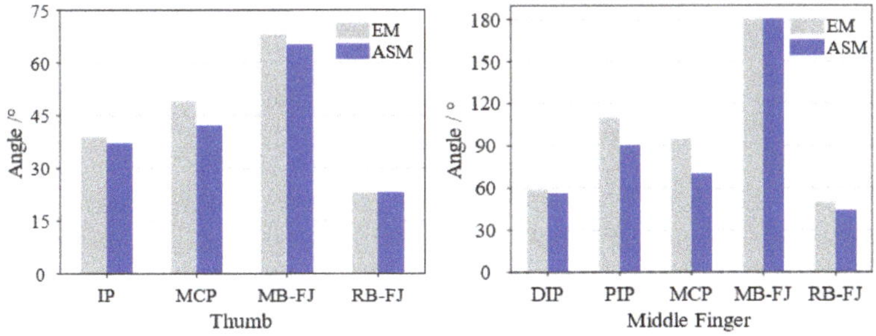

Figure 12. Bending angle values of the entity model and Abaqus simulation model of the soft actuators.

Using the middle finger as an example, the bending angles of both the EM and theoretical model (TM) of the soft actuator are shown in Figure 13, considering both the main and reverse bending under different air pressures. In the main bending condition, the deviation between the theoretical and entity models for the bending angle is less than 2.25%, based on the bending angle of the entity model. For the reverse bending, the value of the entity model is lower than the value of the theoretical model, with an error of less than 5%.

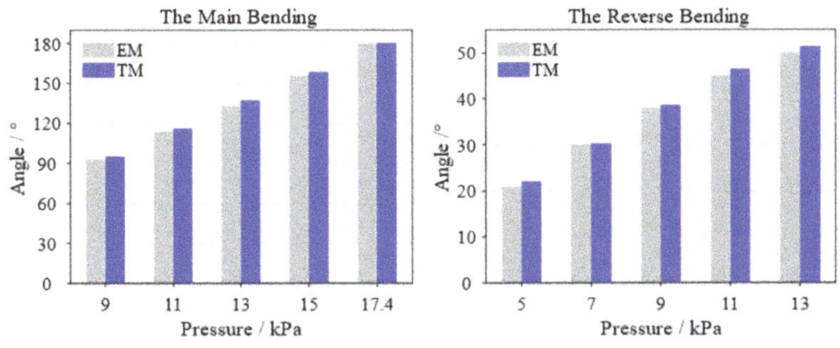

Figure 13. Bending angle values of the entity model and theoretical model of the soft actuator of the middle finger.

4.1.2. Tip Force Measurement

Functional grasping requires sufficient fingertip force and proper directional control, which are crucial for activating the relevant hand muscles and are essential for rehabilitation [47]. Therefore, the fingertip force was measured for different soft actuators via a force gauge (M5-100, MARK-10 Series 5, Copiague, NY, USA), as shown in Figure 14.

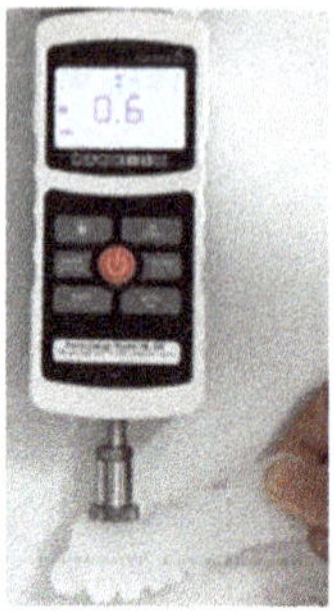

Figure 14. Fingertip force measuring device.

The air pressure within the soft actuator was varied from 20 kPa to 70 kPa in 10 kPa increments. The fingertip force for both main bending (MBFT) and reverse bending (RBFT) of the thumb and middle finger was measured at each air pressure level. The results are shown in Figure 15. Compared with the initial air pressure of 20 kPa, when the starting air pressure was increased to 70 kPa, the fingertip force for the main bending of the thumb increased to 2.35 N, a 273% increase. The fingertip force for reverse bending of the thumb reached 1.15 N, representing an 85.5% increase. The fingertip force for the main bending of the middle finger reached 1.62 N, a 211.5% increase. The fingertip force for reverse bending of the middle finger was 1.05 N, a 94% increase. The experimental results indicate that as air pressure is applied, the fingertip force increases, demonstrating the load-bearing capacity of the soft actuator.

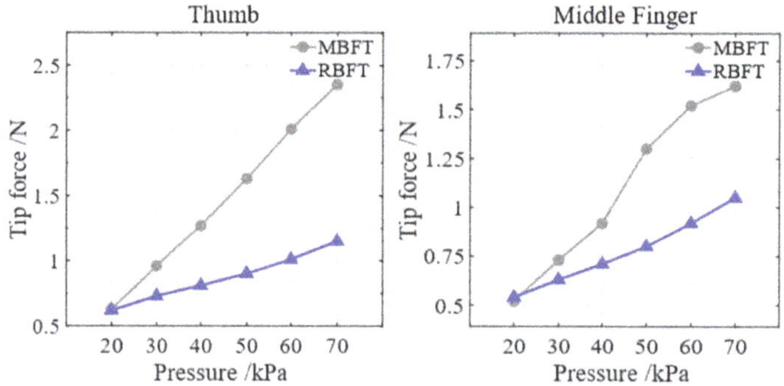

Figure 15. Fingertip forces for main and reverse bending of the soft actuator at different air pressures.

Additionally, at an air pressure of 70 kPa, the maximum force of the joints (DP, PP) of the thumb and the maximum fingertip force of the overall reverse bending of the thumb in the soft actuator were measured. The maximum force of the joints (DP, MP and PP) of the middle finger and the maximum fingertip force of the overall reverse bending of the middle finger in the soft actuator were also measured. As shown in Figure 16, six experiments were conducted. The maximum force at each joint of the soft actuator was consistently greater than the maximum fingertip force generated during reverse bending. The average forces were as follows: for the thumb, 2.37 N at the DP, 1.95 N at the PP, and 1 N for the reverse-bending fingertip force; for the middle finger, 1.4 N at the DP, 2.25 N at the MP, 2.55 N at the PP, and 0.95 N for the reverse-bending fingertip force. The reverse fingertip force of the soft actuator designed by F. Wang et al. is 0.6 N under an air pressure of 70 kPa, and the main fingertip force of the soft actuator is 1.59 N under an air pressure of 130 kPa [26].

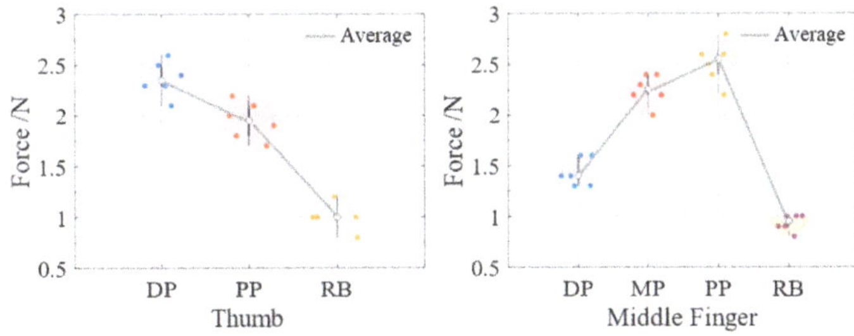

Figure 16. Forces at each joint and reverse-bending fingertip forces of the soft actuators under air pressure of 70 kPa.

4.2. Soft Glove Testing

The soft glove was constructed by attaching the soft actuators to a nitrile glove via nylon cords. To evaluate the performance of the pneumatic soft glove, we created a prosthetic hand model, as shown in Figure 17a. On the basis of the proportions and joint range of motion (ROM) of a human hand, a joint opening of 60° and a connection thickness of 1 mm were designed. The model was then 3D-printed using thermoplastic urethane (TPU) material. The soft robotic glove was worn on the prosthetic hand, with a thin-film pressure sensor (RX-G0505M, Roxifsr, Changzhou, China) placed inside the glove, as shown in Figure 17b. The thin-film pressure sensor is illustrated in Figure 17c.

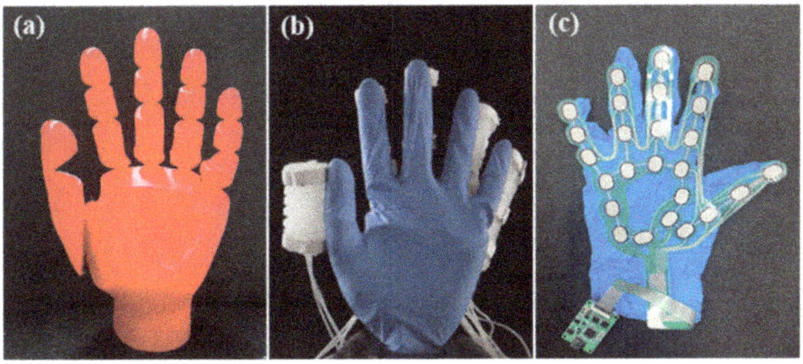

Figure 17. The prosthetic model and initial state of wearing the soft glove. (**a**) the prosthetic model; (**b**) the initial state of the prosthetic model wearing the soft glove; (**c**) the thin-film pressure sensor.

4.2.1. Range-of-Motion Test for the Fingers in the Soft Glove

To evaluate the impact of the soft glove on the finger range of motion, the glove was worn on both the prosthetic hand model and a human hand for comparison. The flexion angles of the index, middle, and ring fingers on both the prosthetic hand and the human hand were measured at different air pressures.

The flexion states of the index finger on the prosthetic hand model (Figure 18a) and the human hand (Figure 18b) at air pressures of 9 kPa, 11.5 kPa, 13 kPa, 15 kPa, and 18 kPa are shown in Figure 18. Similarly, the flexion states of the middle finger on the prosthetic hand model (Figure 19a) and the human hand (Figure 19b) at air pressures of 9 kPa, 11.5 kPa, 13 kPa, 15 kPa, and 18 kPa are shown in Figure 19. Finally, the flexion states of the ring finger on the prosthetic hand model (Figure 20a) and the human hand (Figure 20b) at air pressures of 9 kPa, 11.5 kPa, 13 kPa, 15 kPa, and 18 kPa are shown in Figure 20.

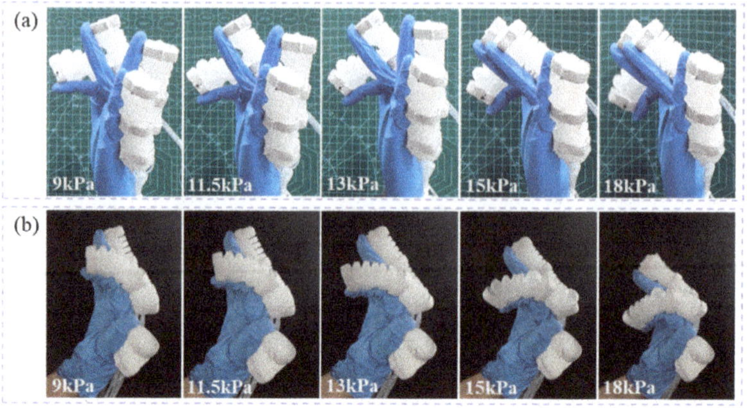

Figure 18. The flexion states of the index finger under different air pressures. (**a**) the prosthetic hand model; (**b**) the human hand.

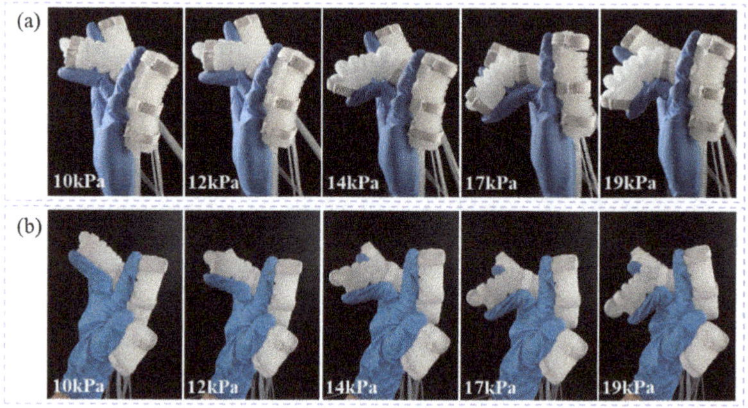

Figure 19. The flexion states of the middle finger under different air pressures. (**a**) the prosthetic hand model; (**b**) the human hand.

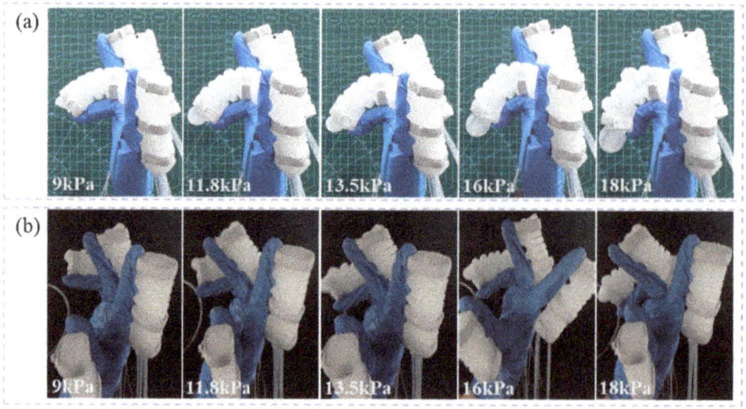

Figure 20. The flexion states of the ring finger under different air pressures. (**a**) the prosthetic hand model; (**b**) the human hand.

The finger flexion angles (the angle between the base and tip of the finger) for both the prosthetic hand model and the human hand are plotted in Figure 21. The experimental results demonstrate that the soft glove allows the index, middle, and ring fingers to achieve flexion angles of 101°, 112°, and 112°, respectively, indicating its strong practicality and biomimetic performance.

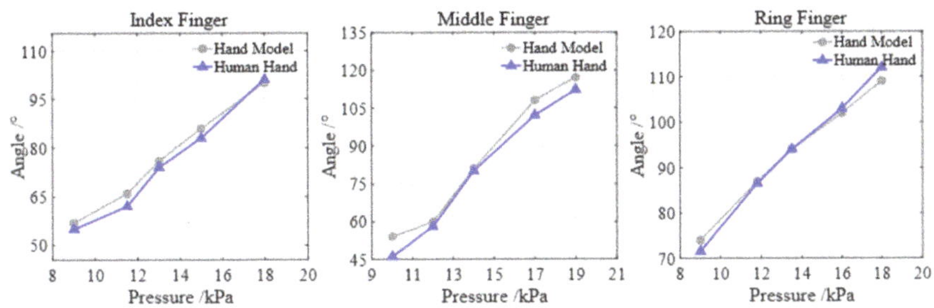

Figure 21. Finger flexion angles of both the prosthetic hand model and the human hand.

4.2.2. Rehabilitation Gesture Testing of the Soft Glove

Under the control of the pneumatic platform, both the prosthetic hand model and the human hand, worn with the soft glove, are capable of performing various rehabilitation gestures, as shown in Figure 22. The initial state of the prosthetic hand model wearing the soft glove is shown in Figure 17b. With the assistance of the soft glove, both the prosthetic hand model (Figure 22a) and the human hand (Figure 22b) successfully performed various gestures, including "1", "2", "3", "4", "5", "love", and "fist." The experimental results demonstrate that the pneumatic soft glove has the potential to support patients with hand impairments in training for a variety of rehabilitation gestures.

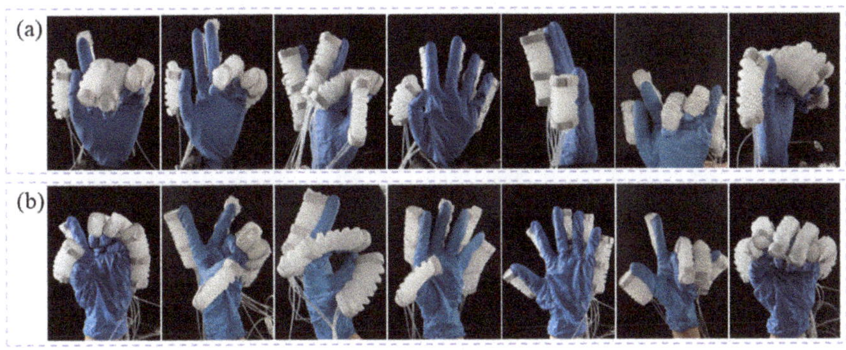

Figure 22. Soft glove-assisted rehabilitation gestures. (**a**) the prosthetic hand model; (**b**) human hand.

4.2.3. Test of the Auxiliary Grip

The soft glove assists the prosthetic hand in grasping objects of varying weights and shapes, as shown in Figure 23. The objects grasped included a rectangular box (15.4 g), foam (3.8 g), a blackboard eraser (18.5 g), and a PLA mold (19.2 g). The experimental results demonstrate that the pneumatic soft glove can assist the hand to grasp objects.

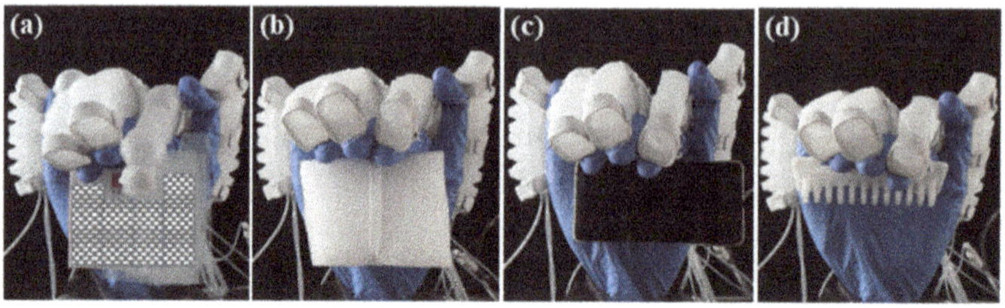

Figure 23. Gripping objects by prosthetic hand with soft glove. (**a**) a rectangular box; (**b**) a foam; (**c**) a blackboard eraser; (**d**) a PLA mold.

By using a pneumatic soft glove to assist with daily tasks, meaningful grasping exercises can be designed to enhance hand function in stroke patients [48]. Two different grasping techniques, full grip (FG) and flat pinch (FP) were employed to handle objects of varying shapes and weights. The pneumatic soft glove assisted the human hand in grasping tasks involving a blackboard eraser (18.5 g), U-disk box (139.4 g), glasses case (38.4 g), and coil (22.6 g), as shown in Figure 24. Figure 24a shows the FG scenes, whereas Figure 24b shows the FP scenes. The experimental results indicate that the soft robotic glove can grasp objects weighing at least 139.4 g.

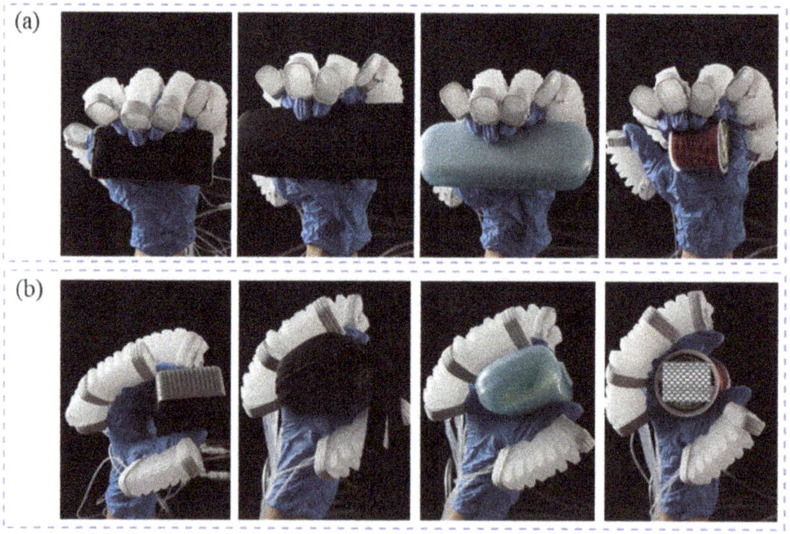

Figure 24. Grasping objects with two grasping postures by human hand wearing soft glove. (**a**) full grip; (**b**) flat pinch.

The fingertip forces of all five fingers were measured by a thin-film pressure sensor during the grasping of different objects with the two grip techniques, as shown in Figure 25. The experimental results indicate that the pneumatic soft glove can assist patients with hand impairments in performing various daily tasks.

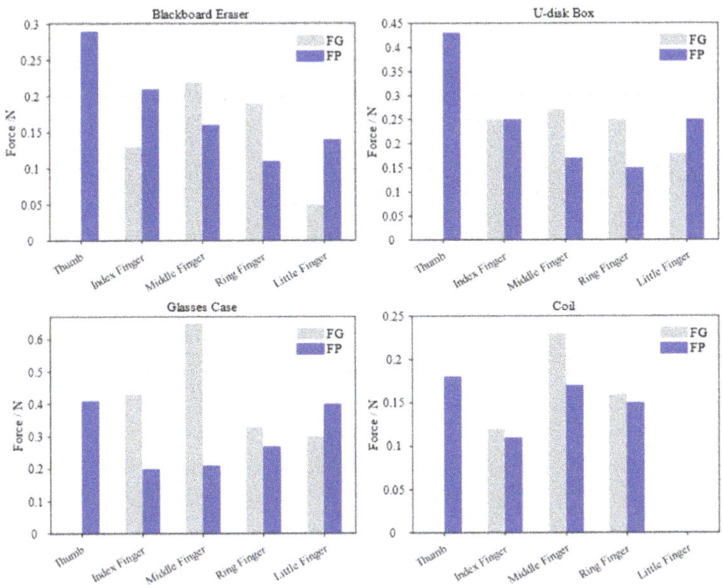

Figure 25. Five-finger tip forces through two grasp types.

5. Conclusions

This work presents a novel pneumatic soft glove with bidirectional bending functionality. It can perform various rehabilitation gestures and grasping tasks. An expandable pneumatic control platform is developed, enabling the soft glove to assist patients with hand impairments in rehabilitation exercises and daily activities. Experiments show that the soft actuator possesses a measurable load capacity and that the effect of gravity on the bending angle and air pressure of the soft actuator is negligible, enabling not only finger flexion and extension but also grasping functionality. The maximum main and reverse-bending angles of the soft actuator are sufficient to meet the requirements of daily human activities. In addition, a prosthetic hand model was designed and a soft glove was tested on both the prosthetic hand and the human hand, verifying that the soft glove system is capable of enabling individual finger movement, performing various rehabilitation gestures, and assisting with grasping tasks. The experimental results indicate that the pneumatic soft glove has the potential to support patients with hand impairments in various rehabilitation exercises and enable them to grasp different objects to complete daily tasks.

In future work, we will focus on designing a versatile pneumatic soft glove with bidirectional bending capabilities and exploring new materials or structures to increase the grasping force, thereby enabling the completion of a broader range of daily tasks. Furthermore, computer vision will be incorporated into the motion data collection of the soft glove to obtain more precise and comprehensive experimental results. Finally, electromyographic (EMG) signals will be utilized for detecting user intent and controlling the movements of the soft glove, improving the accuracy of the movements of glove and providing better support for stroke patients with hand impairments.

Author Contributions: Conceptualization, X.W. and Q.C.; methodology, X.W., Q.C., Z.W. and Z.Z.; software, X.W. and Y.L.; validation, X.W., Q.C., Z.W. and Y.L.; formal analysis, X.W., Q.C. and X.Z.; investigation, X.W.; resources, X.W., Q.C. and Z.Z.; data curation, X.W.; writing—original draft preparation, X.W.; writing—review and editing, X.W., Q.C., Z.Z. and X.Z.; visualization, X.W. and

Q.C.; supervision, project administration, funding acquisition, Z.Z. All authors have read and agreed to the published version of the manuscript.

Funding: This research was funded by the Strategic Priority Research Program on Space Science of the Chinese Academy of Sciences (grant XDA1502030505), the Youth Innovation Promotion Association of the Chinese Academy of Sciences (202202), and the State Key Laboratory of Robotics (2024-Z20).

Institutional Review Board Statement: Not applicable.

Data Availability Statement: The data that support the findings of this study are available from the authors upon reasonable request.

Conflicts of Interest: The authors declare no conflicts of interest.

Appendix A

Table A1. Parameters of the soft actuator.

Values	Parameters
$a = 3.6$ mm	Length of a single air chamber of the main bending module
$h = 9.75$ mm	Height of the air chamber of the main bending module
$c = 10$ mm	Effective width of the soft actuator
$a_1 = 1.88$ mm	Length of the short side of the isosceles trapezoid of the reverse-bending module
$b_1 = 10.8$ mm	Length of the long side of the isosceles trapezoid of the reverse-bending module
$h_1 = 5.24$ mm	Height of the air cavity in the reverse-bending module
$\alpha = 49.5°$	Bottom corner of a trapezoid
$T = 23$ °C	Normal temperature
$E = 0.1$ MPa	Young's modulus of Ecoflex 00-50
$G = 0.044$ MPa	Shear modulus of Ecoflex 00-50

Table A2. Model and origin of components for pneumatic control platform.

Components and Parts	Model	Brand	Place of Origin
Pressure-relief valve	IR2020	SMC	Tokyo, Japan
Buck	LM2596S	-	China
Controller	A Tmega2560	KEYES	Shenzhen, China
2/2-way solenoid valve	2V025	AirTac	Taiwan, China
Air pressure sensor	MSP20	-	China
2/3-way solenoid valve	SF3	PIN YA	Shenzhen, China
2/3-way solenoid valve	VT307	SMC	Tokyo, Japan

References

1. Feigin, V.L.; Brainin, M.; Norrving, B.; Martins, S.; Sacco, R.L.; Hacke, W.; Fisher, M.; Pandian, J.; Lindsay, P. World Stroke Organization (WSO): Global Stroke Fact Sheet 2022. *Int. J. Stroke* **2022**, *17*, 18–29. [CrossRef] [PubMed]
2. Cassidy, J.M.; Mark, J.I.; Cramer, S.C. Functional connectivity drives stroke recovery: Shifting the paradigm from correlation to causation. *Brain* **2022**, *145*, 1211–1228. [CrossRef] [PubMed]
3. Feigin, V.L.; Stark, B.A.; Johnson, C.O.; Roth, G.A.; Bisignano, C.; Abady, G.G. Global, regional, and national burden of stroke and its risk factors, 1990–2019: A systematic analysis for the Global Burden of Disease Study 2019. *Lancet Neurol.* **2021**, *20*, 795–820. [CrossRef]
4. Kamper, D.G. Restoration of hand function in stroke and spinal cord injury. In *Neurorehabilitation Technology*; Springer: Berlin/Heidelberg, Germany, 2016; pp. 311–331.
5. Bützer, T.; Lambercy, O.; Arata, J.; Gassert, R. Fully Wearable Actuated Soft Exoskeleton for Grasping Assistance in Everyday Activities. *Soft Robot.* **2021**, *8*, 128–143. [CrossRef]
6. Bae, J.H.; Kang, S.H.; Seo, K.M.; Kim, D.-K.; Shin, H.I.; Shin, H.E. Relationship between grip and pinch strength and activities of daily living in stroke patients. *Ann. Rehabil. Med.* **2015**, *39*, 752–762. [CrossRef]

7. Lee, J.J.; Shin, J.-H. Predicting clinically significant improvement after robot-assisted upper limb rehabilitation in subacute and chronic stroke. *Front. Neurol.* **2021**, *12*, 668923. [CrossRef]
8. Maciejasz, P.; Eschweiler, J.; Gerlach-Hahn, K.; Jansen-Troy, A.; Leonhardt, S. A survey on robotic devices for upper limb rehabilitation. *J. Neuroeng. Rehabil.* **2014**, *11*, 3. [CrossRef]
9. Wang, X.J.; Cheng, Y.; Zheng, H.D.; Li, Y.H.; Wang, C.D. Design and optimization of actuator for multi-joint soft rehabilitation glove. *Ind. Robot* **2021**, *48*, 877–890. [CrossRef]
10. Lambercy, O.; Ranzani, R.; Gassert, R. Robot-assisted rehabilitation of hand function. In *Rehabilitation Robotics*; Academic Press: Cambridge, MA, USA, 2018; pp. 205–225.
11. Wolf, S.L.; Winstein, C.J.; Miller, J.P.; Taub, E.; Uswatte, G.; Morris, D.; Giuliani, C.; Light, K.E.; Nichols-Larsen, D. Effect of constraint-induced movement therapy on upper extremity function 3 to 9 months after stroke: The EXCITE randomized clinical trial. *JAMA* **2006**, *296*, 2095–2104. [CrossRef]
12. Liu, C.; Guo, K.; Lu, J.; Yang, H. A review on the application of intelligent control strategies for post-stroke hand rehabilitation machines. *Adv. Mech. Eng.* **2023**, *15*, 16878132221148018. [CrossRef]
13. Kutner, N.G.; Zhang, R.; Butler, A.J.; Wolf, S.L.; Alberts, J.L. Quality-of-life change associated with robotic-assisted therapy to improve hand motor function in patients with subacute stroke: A randomized clinical trial. *Phys. Ther.* **2010**, *90*, 493–504. [CrossRef] [PubMed]
14. Noronha, B.; Accoto, D. Exoskeletal Devices for Hand Assistance and Rehabilitation: A Comprehensive Analysis of State-of-the-Art Technologies. *IEEE Trans. Med. Robot. Bionics* **2021**, *3*, 525–538. [CrossRef]
15. Tran, P.; Jeong, S.; Herrin, K.; Desai, J. A review: Hand exoskeleton systems, clinical rehabilitation practices, and future prospects. *IEEE Trans. Med. Robot. Bionics* **2021**, *3*, 606–622. [CrossRef]
16. Almenara, M.; Cempini, M.; Gómez, C.; Cortese, M.; Martín, C.; Medina, J.; Vitiello, N.; Opisso, E. Usability test of a hand exoskeleton for activities of daily living: An example of user-centered design. *Disabil. Rehabil. Assist. Technol.* **2017**, *12*, 84–96. [CrossRef]
17. Zhang, Z.; Calderon, A.D.; Huang, X.; Wu, G.; Liang, C. Design and Driving Performance Study of Soft Actuators for Hand Rehabilitation Training. *Med. Devices* **2024**, *17*, 237–260. [CrossRef]
18. Pan, M.; Yuan, C.; Liang, X.; Dong, T.; Liu, T.; Zhang, J.; Zou, J.; Yang, H.; Bowen, C. Soft Actuators and Robotic Devices for Rehabilitation and Assistance. *Adv. Intell. Syst.* **2021**, *4*, 2100140. [CrossRef]
19. Yap, H.K.; Lim, J.H.; Nasrallah, F.; Cho Hong Goh, J.; Yeow, C.H. Characterisation and evaluation of soft elastomeric actuators for hand assistive and rehabilitation applications. *J. Med. Eng. Technol.* **2016**, *40*, 199–209. [CrossRef]
20. Tang, D.; Lv, X.; Zhang, Y.; Qi, L.; Shen, C.; Shen, W. A Review on Soft Exoskeletons for Hand Rehabilitation. *Recent Pat. Eng.* **2024**, *18*, 52–73. [CrossRef]
21. Polygerinos, P.; Lyne, S.; Wang, Z.; Nicolini, L.F.; Mosadegh, B.; Whitesides, G.M.; Walsh, C.J. Towards a soft pneumatic glove for hand rehabilitation. In Proceedings of the 2013 IEEE/RSJ International Conference on Intelligent Robots and Systems, Tokyo, Japan, 3–7 November 2013; pp. 1512–1517.
22. Connolly, F.; Walsh, C.J.; Bertoldi, K. Automatic design of fiber-reinforced soft actuators for trajectory matching. *Proc. Natl. Acad. Sci. USA* **2017**, *114*, 51–56. [CrossRef]
23. Feng, N.; Shi, Q.; Wang, H.; Gong, J.; Liu, C.; Lu, Z. A soft robotic hand: Design, analysis, sEMG control, and experiment. *Int. J. Adv. Manuf. Technol.* **2018**, *97*, 319–333. [CrossRef]
24. Yap, H.K.; Lim, J.H.; Nasrallah, F.; Goh, J.C.; Yeow, R.C. A soft exoskeleton for hand assistive and rehabilitation application using pneumatic actuators with variable stiffness. In Proceedings of the 2015 IEEE International Conference on Robotics and Automation (ICRA), Seattle, WA, USA, 26–30 May 2015; pp. 4967–4972.
25. Wang, J.B.; Fei, Y.Q.; Pang, W. Design, Modeling, and Testing of a Soft Pneumatic Glove with Segmented PneuNets Bending Actuators. *IEEE/ASME Trans. Mechatron.* **2019**, *24*, 990–1001. [CrossRef]
26. Wang, F.; Chen, Y.; Wang, Y.; Liu, Z.; Tian, Y.; Zhang, D. A soft pneumatic glove with multiple rehabilitation postures and assisted grasping modes. *Sens. Actuators A Phys.* **2022**, *347*, 113978. [CrossRef]
27. Xavier, M.S.; Fleming, A.J.; Yong, Y.K. Modelling and Simulation of Pneumatic Sources for Soft Robotic Applications. In Proceedings of the 2020 IEEE/ASME International Conference on Advanced Intelligent Mechatronics (AIM), Boston, MA, USA, 6–9 July 2020; pp. 916–921.
28. Joshi, S.; Paik, J. Pneumatic Supply System Parameter Optimization for Soft Actuators. *Soft Robot.* **2021**, *8*, 152–163. [CrossRef] [PubMed]
29. Jung, Y.; Kwon, K.; Lee, J.; Ko, S.H. Untethered soft actuators for soft standalone robotics. *Nat. Commun.* **2024**, *15*, 3510. [CrossRef]
30. Gerges, F.; Desai, J.; Watkins, J.; Burugupally, S.P. Master-Slave Control for a Pneumatically Actuated low Pressure Soft Robotic Glove to Facilitate Bilateral Training for Stroke Patients. In Proceedings of the 2020 23rd International Symposium on Measurement and Control in Robotics (ISMCR), Budapest, Hungary, 22–24 October 2020.

31. Young, T.R.; Xavier, M.S.; Yong, Y.K.; Fleming, A.J. A Control and Drive System for Pneumatic Soft Robots: PneuSoRD. In Proceedings of the 2021 IEEE/RSJ International Conference on Intelligent Robots and Systems (IROS), Prague, Czech Republic, 28–30 September 2021; pp. 2822–2829.
32. Abidi, H.; Gerboni, G.; Brancadoro, M.; Fras, J.; Diodato, A.; Cianchetti, M.; Wurdemann, H.A.; Althoefer, K.; Menciassi, A. Highly dexterous 2-module soft robot for intra-organ navigation in minimally invasive surgery. *Int. J. Med. Robot. Comput. Assist. Surg.* **2018**, *14*, e1875. [CrossRef]
33. Khan, A.H.; Li, S. Sliding Mode Control with PID Sliding Surface for Active Vibration Damping of Pneumatically Actuated Soft Robots. *IEEE Access* **2020**, *8*, 88793–88800. [CrossRef]
34. Skorina, E.H.; Luo, M.; Tao, W.; Chen, F.; Fu, J.; Onal, C.D. Adapting to Flexibility: Model Reference Adaptive Control of Soft Bending Actuators. *IEEE Robot. Autom. Lett.* **2017**, *2*, 964–970. [CrossRef]
35. Alves, S.; Babcinschi, M.; Silva, A.; Neto, D.; Fonseca, D.; Neto, P. Integrated Design Fabrication and Control of a Bioinspired Multimaterial Soft Robotic Hand. *Cyborg Bionic Syst.* **2023**, *4*, 0051. [CrossRef]
36. Chen, P.; Ding, Q.; Liu, Y.; Deng, Z.; Huang, J. Programmable Pressure Control in Pneumatic Soft Robots With 2-Way 2-State Solenoid Valves. *IEEE Robot. Autom. Lett.* **2024**, *9*, 6448–6455. [CrossRef]
37. Skorina, E.H.; Luo, M.; Ozel, S.; Chen, F.; Tao, W.; Onal, C.D. Feedforward augmented sliding mode motion control of antagonistic soft pneumatic actuators. In Proceedings of the 2015 IEEE International Conference on Robotics and Automation (ICRA), Seattle, WA, USA, 26–30 May 2015; pp. 2544–2549.
38. Wang, T.; Zhang, Y.; Chen, Z.; Zhu, S. Parameter Identification and Model-Based Nonlinear Robust Control of Fluidic Soft Bending Actuators. *IEEE/ASME Trans. Mechatron.* **2019**, *24*, 1346–1355. [CrossRef]
39. Iqbal, M.; Imtiaz, J.; Mahmood, A. Robust tracking control of a minimal realization model of an impaired human hand for anthropomorphic coordination. *Heliyon* **2024**, *10*, e26941. [CrossRef] [PubMed]
40. Vaicekauskaite, J.; Mazurek, P.; Vudayagiri, S.; Skov, A.L. Mapping the mechanical and electrical properties of commercial silicone elastomer formulations for stretchable transducers. *J. Mater. Chem. C* **2020**, *8*, 1273–1279. [CrossRef]
41. Lavazza, J.; Contino, M.; Marano, C. Strain rate, temperature and deformation state effect on Ecoflex 00-50 silicone mechanical behaviour. *Mech. Mater.* **2023**, *178*, 104560. [CrossRef]
42. Kulkarni, P. Centrifugal Forming and Mechanical Properties of Silicone-Based Elastomers for Soft Robotic Actuators. Master's Thesis, Rutgers The State University of New Jersey, New Brunswick, NJ, USA, 2015.
43. Kokubu, S.; Vinocour, P.E.T.; Yu, W. Development and evaluation of fiber reinforced modular soft actuators and an individualized soft rehabilitation glove. *Robot. Auton. Syst.* **2024**, *171*, 104571. [CrossRef]
44. Zhang, B.; Chen, J.; Ma, X.; Wu, Y.; Zhang, X.; Liao, H. Pneumatic System Capable of Supplying Programmable Pressure States for Soft Robots. *Soft Robot.* **2022**, *9*, 1001–1013. [CrossRef]
45. Marechal, L.; Balland, P.; Lindenroth, L.; Petrou, F.; Kontovounisios, C.; Bello, F. Toward a Common Framework and Database of Materials for Soft Robotics. *Soft Robot.* **2021**, *8*, 284–297. [CrossRef]
46. Bain, G.I.; Polites, N.; Higgs, B.G.; Heptinstall, R.J.; McGrath, A.M. The functional range of motion of the finger joints. *J. Hand Surg. Eur. Vol.* **2014**, *40*, 406–411. [CrossRef]
47. Valero-Cuevas, F.J.; Zajac, F.E.; Burgar, C.G. Large index-fingertip forces are produced by subject-independent patterns of muscle excitation. *J. Biomech.* **1998**, *31*, 693–703. [CrossRef]
48. Tang, Z.Q.; Heung, H.L.; Shi, X.Q.; Tong, R.K.Y.; Li, Z. Probabilistic Model-Based Learning Control of a Soft Pneumatic Glove for Hand Rehabilitation. *IEEE Trans. Biomed. Eng.* **2022**, *69*, 1016–1028. [CrossRef]

biomimetics MDPI

Article

An Octopus-Inspired Soft Pneumatic Robotic Arm

Emmanouil Papadakis [1,*], Dimitris P. Tsakiris [1] and Michael Sfakiotakis [2]

[1] Institute of Computer Science, Foundation for Research and Technology–Hellas, GR-70013 Heraklion, Greece; tsakiris@ics.forth.gr

[2] Department of Electrical and Computer Engineering, Hellenic Mediterranean University, GR-71410 Heraklion, Greece; msfak@hmu.gr

* Correspondence: manospapad@ics.forth.gr

Abstract: This paper addresses the design, development, control, and experimental evaluation of a soft robot arm whose actuation is inspired by the muscular structure of the octopus arm, one of the most agile biological manipulators. The robot arm is made of soft silicone and thus possesses enhanced compliance, which is beneficial in a variety of applications where the arm may come into contact with delicate features of its environment. The arm is composed of three elongated segments arranged in series, each one of which contains several pneumatically actuated chambers embedded in its silicone body, which may induce various types of deformations of the segment. By combining the segment deformations, and by imitating the antagonistic muscle group functionality of the octopus, the robot arm can bend in various directions, increase or decrease its length, as well as twist around its central axis. This is one of the few octopus-inspired soft robotic arms where twisting is replicated in its motion characteristics, thus greatly expanding the arm's potential applications. We present the design process and the development steps of the soft arm, where the molding of two-part silicone of low hardness in 3d-printed molds is employed. In addition, we present the control methodology and the experimental evaluation of both a standalone segment and the entire three-segment arm. This experimental evaluation involves model-free closed-loop control schemes, exploiting visual feedback from a pair of external cameras in order to reconstruct in real time the shape of the soft arm and the pose of its tip.

Keywords: soft robotics; biomimetics; bio-inspired robots; robotic arm; pneumatic actuation; robot control; arm twisting

1. Introduction

The development of robotic systems that offer high flexibility and adaptability, especially in delicate interaction environments, represents a key challenge in robotics research. Conventional robotic systems, characterized by their rigid structures, often encounter difficulties in complex tasks that demand a blend of strength, precision, and adaptability. This challenge is underscored in domains such as underwater exploration, agricultural automation, and medical surgery, where the ability to navigate and manipulate within constrained and/or delicate environments is paramount.

Contemporary endeavors in robotics have progressively embraced bio-inspired designs to overcome the limitations inherent to rigidity, drawing inspiration from the natural world. The muscular hydrostat system of the octopus arm, capable of complex locomotion and manipulation tasks, stands out as a particularly compelling blueprint for soft robotic applications. Early pioneering work in this domain, such as the research by Walker et al. [1], investigated continuum robot arms inspired by cephalopods. This exploration was further expanded upon by studies like [2,3], which demonstrated the potential of leveraging biological principles to enhance robotic flexibility and dexterity by mimicking the cephalopod's morphology. In [4–6], soft robotic grippers and a robotic octopus were developed, exemplifying the feasibility and efficiency of biomimetic approaches in promoting robotic adaptability.

Advancements in material science and novel actuation mechanisms have played a pivotal role in the evolution of soft robotics. The integration of soft sensors within robotic systems [7], alongside foundational work on the design and fabrication of soft robots [8], has paved the way for the development of robots capable of safely interacting with humans and delicate objects. Moreover, the creation of soft robotic fish capable of autonomous underwater navigation [9] underscores the application-specific advantages of soft robotics in challenging environments.

The field has continued to evolve with recent studies pioneering the use of innovative materials and structures to more closely mimic biological functions [10–15]. This has opened new avenues for creating more versatile and safer robotic interactions across a broad spectrum of applications, from healthcare to environmental monitoring. Particularly noteworthy is the series of soft robot arms for minimally invasive surgery developed by Menciassi and colleagues (see [16] and references therein). Their work presents the design, fabrication, characterization, and testing of a three-module soft manipulator inspired by the octopus arms and actuated by embedded fluidic actuators, which allow omnidirectional bending and elongation. Furthermore, the stiffness of the manipulator is controlled via a granular-jamming-based stiffening mechanism integrated along the length of the arm. Other arm designs of the same group exploit tendons for tuning the arm stiffness [17]. The development of soft actuators, as explored in [18–21], further illustrates the potential of soft robotics to transform traditional approaches to robotic design and application.

In addition to material innovations, control strategies for soft robotic systems have seen rapid advancements. Research focusing on the integration of sensory feedback and biomimetic design principles [22–26] has contributed significantly to our understanding of how soft robotic systems can be designed, controlled, and applied in real-world scenarios with a level of sophistication and efficiency akin to their biological counterparts. Together, these studies form a robust foundation for the future of soft robotics, pointing the way toward systems that are not only adaptable and safe but also capable of complex autonomous decision-making and sophisticated environmental interactions. Performance is a critical metric in the field of soft robotics, encompassing attributes such as dexterity, force output, motion range, adaptability, and responsiveness to varying tasks and environments. Metrics such as energy efficiency, actuation speed, load capacity, and control accuracy are commonly used to evaluate the effectiveness of soft robots. By benchmarking these performance indicators [27], soft robotic systems can be optimized for specific use cases, paving the way for advances in functionality and real-world applicability.

Adding twisting ability to a soft robotic arm significantly enhances its versatility and functionality, making it more capable of mimicking the complex motion capabilities of biological systems. While bending and elongation allow for basic directional control and reach, twisting enables the arm to perform additional tasks that require rotational movement. This is particularly important for grasping and manipulating objects with precision, as twisting can adjust the orientation of the end-effector or wrapped arm to better interact with an object. Moreover, twisting greatly expands the arm's range of potential applications. For example, in minimally invasive surgery, twisting motions are essential for navigating confined spaces and reorienting instruments without requiring external rotation mechanisms. Similarly, in underwater robotics or search-and-rescue scenarios, the ability to twist can allow for intricate interactions with irregularly shaped objects or environments. Twisting also facilitates more natural and efficient movement patterns, reducing the need for complex kinematic configurations or additional joints, therefore simplifying the system while enhancing its adaptability.

Despite these potential advantages, only a limited number of existing soft robotic arm prototypes provide for twisting motions. One such prototype, proposed in [28], utilizes a total of 32 pneumatically driven McKibben actuators to emulate the octopus arm musculature. Four of these actuators are helically wound (two in the clockwise and two in the counterclockwise direction) and can provide torsional motions of the arm through appropriate pressurization. However, the helical actuators form the outer layer of the arm, making them susceptible to breaking if sufficient contact forces are exerted, e.g., from an environment containing sharp edges. In [29], the previous arm is used in conjunction with flexible strain sensors providing information on its bending to implement master-slave feedback control of the arm. In [30], a single-segment pneumatic soft robotic wrist is presented, based on four parallel soft helical actuators, able to generate both bending and twisting movements. In [31], an innovative modular magnetically actuated robot arm based on reconfigurable bistable Kresling origami modules is presented, which can stretch, fold, bend, and twist in various directions. This arm has the advantages of scalability and reconfigurability, but its need for precise magnetic actuation can be seen as a limitation.

In this context, our approach in this paper builds upon these foundational studies and extends our prior work on octopus-inspired robotic systems [3–5], aiming to bridge the gap between bio-inspired design principles and the practical requirements of robotic applications. By focusing on the octopus arm's muscular structure, we have developed a modular soft pneumatic robotic arm that not only demonstrates enhanced compliance and maneuverability but also introduces the ability for twisting movements. The robot arm features linear air chambers mimicking the longitudinal muscles of the octopus arm, enabling bending and elongation, while helical air chambers, replicating the oblique muscles of the octopus, enable twisting motions—a novel feature, not frequently occurring in similar soft robot arms. Compressed air is delivered through silicone tubes routed through the center of the arm, leading to a robust design capable of complex, multi-segment motions. Furthermore, rigid 3d-printed rings were employed to constrain the radial expansion of the arm, similarly to the transverse muscles of the octopus, thus ensuring precise, directed movement, as bending and twisting movements are decomposed and can be controlled independently of each other. The arm also features a conical, overlapping design of its three segments in order to avoid bending dead zones, with appropriately shaped chambers for uniform actuation. The arm prototype was fabricated by casting silicone in 3d printed molds, an involved multi-step process, often prone to inconsistencies. The arm is controlled via simple, model-free closed-loop control schemes, whose efficiency was demonstrated by tip positioning experiments. This experimental evaluation demonstrated the functionality and advantages of the developed soft robot arm.

2. Octopus Arm Actuation and Control

The main characteristics of the octopus arm muscular and motion control systems are presented here, to the extent that they are relevant to the design and functionality of our soft robot arm. For a more detailed overview of the octopus arm biomechanics, see, e.g., [32]. The octopus is an ideal animal model for studying elongated flexible appendages, including the generation and control of their movement, as well as the generation of related complex behaviors. The muscular system of the octopus arm consists of a densely packed three-dimensional array of muscle fibers and connective tissue, which extends for the full length of the arm and surrounds a central axial nerve cord. This generates both the forces required for arm movement and deformation, as well as the stiffness variations needed for skeletal-like support. Such a muscular system is called *muscular hydrostat*, and its main characteristic is that it maintains the arm volume constant. Thus, any change in one dimension of the arm will be compensated for by a change in at least one other dimension [33,34].

The muscular tissue of each arm is organized into three major groups (Figure 1): The first consists of the longitudinal muscles (LM), oriented parallel to the long axis of the arm and placed in a cross pattern surrounding the central core of the arm. The second group consists of the transverse muscles (TM), which are oriented perpendicular to the long axis of the arm, directly surround the central axial nerve cord, and are surrounded by the longitudinal muscles. The third group consists of the helical or oblique muscles, namely three pairs of helicoidally arranged muscles, the external (EOM), medial (MOM), and internal (IOM) muscles, arranged as a clockwise and a counterclockwise helix along the arm.

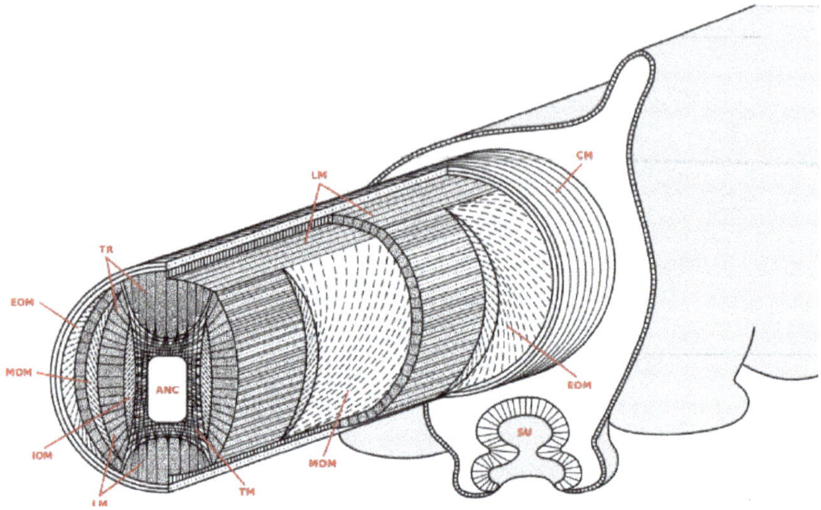

Figure 1. Octopus vulgaris arm muscle system. Image source: [35].

A wide variety of, possibly highly localized, arm movements are feasible. The longitudinal muscles are responsible for shortening the arm, the transverse muscles are responsible for its elongation, and the oblique muscles are responsible for torsion. When the transverse muscles are activated, the diameter of the arm is shortened, and since its volume is constant, the arm is elongated. The longitudinal muscles act antagonistically to the transverse ones by shortening the arm and re-extending the transverse muscle fibers. Simultaneous contraction of the longitudinal and the transverse muscles results in stiffening of the arm. Furthermore, select longitudinal and transverse muscles may be simultaneously contracted to bend the arm in all possible directions: as the longitudinal muscles are selectively contracted, the transverse muscles also need to contract in order to resist the arm shortening that would result from this [33,36]. Finally, the oblique muscles are employed in order to twist the arm along its axis, which is accomplished by the controlled activation of these muscle groups, while the direction of torsion, clockwise or counterclockwise, depends on the handedness of the actuated oblique muscle group [35]. These arm deformations can be combined to create more complex movements.

The virtually infinite degrees-of-freedom (dof), which characterize the flexible octopus arms, make efficient control and arm behavior generation quite challenging [37–39]. From the engineering point of view, the special strategies and shortcuts employed by these robust and adaptable biological mechanisms may provide valuable hints for addressing corresponding problems in robotic systems [1]. Such strategies include the exploitation of basic "hardwired" motion patterns, like those unveiled in detailed studies of the reaching and fetching primitive movements of the octopus arm. Reaching for a target is implemented by forming a bend near the base of the arm and propagating it towards its tip via a stiffening wave of muscle contraction [39–41]. Fetching is implemented by forming a 3-joint-like structure with the arm and rotating the joint-like bends so that the tip reaches the mouth [42]. Control of both of these movements exploits a drastic reduction of the infinite-dof problem via the exploitation of a few basic motion patterns to the control of just three variables [38,39,42,43]. Detailed modeling of the octopus muscular system and computational investigation of its stereotypical behaviors clarified the dynamics and neuronal control mechanisms producing these behaviors [40,41,44–47].

3. Prototype Design and Development

Drawing inspiration from the octopus, we propose a soft biomimetic arm that employs embedded, pneumatically actuated chambers instead of muscles in a configuration analogous to that of the octopus arm muscles to achieve similar motion capabilities.

3.1. Design of the Soft Arm Prototypes

Two soft biomimetic arm prototypes were designed and built within the scope of this work: A single-segment arm and a three-segment arm. The former was developed as a preliminary proof-of-concept prototype to study the design, fabrication, and functional characteristics of the proposed concept. Based on these, the three-segment arm was subsequently developed as a demonstrator prototype of a more integrated and capable system. The arm prototypes are fabricated from a soft, compliant, and elastic material, with a series of linear- and helical-shaped embedded pneumatic chambers. Using solenoid valves to control the flow of compressed air in these chambers allows for bending, extension, and twisting motions of the arms through the elongation of the appropriate, in each case, chambers.

In particular, the single-segment arm (length: 50 mm, weight: 250 g) utilizes four linear chambers, arranged symmetrically at 90° intervals and running along the segment, that recreate the functionality of the longitudinal muscles of the octopus arm. Additionally, a set of helical chambers are placed in the outer surface of the segment and function as the oblique muscles of the octopus arm, providing twisting motions in both the clockwise and counterclockwise directions. For design simplicity, we are not implementing the transverse muscles since the elongation of the arm can be achieved by the simultaneous inflation of all four linear chambers. A series of rigid rings are used to restrict the radial expansion of the chambers when pressurized to implement the desired motions.

The three-segment arm, which is designed along similar principles, consists of three segments, each housing four linear chambers as per above, thus providing the ability to bend in three different areas along the length of the arm simultaneously and independently from each other. It is noted that the placement of the linear chambers is rotated by 45° for consecutive segments. In addition, four helical chambers, two clockwise and two counterclockwise, encapsulate the three segments and can be used to twist the arm as a whole. Pressurized air is supplied to each segment via ⌀4 mm silicone tubes (⌀2 mm inner). In order for the tubes to reach all segments, they must pass through each preceding segment. Hence, eight tubes pass through the first (base) segment and four through the second (mid) one. This requirement, combined with the desire for the arm to have a smooth, continuous outer surface, led to designing the arm to be of a conical shape, with a wide base that gradually tapers at the arm's tip. As a result, the three individual segments forming the arm are also conical. Because of this conical shape, the four linear chambers inside each segment were positioned to be parallel to the external contour and, hence, tilt towards each other along the tapering of the segment (see Figure 2a). This ensures a homogeneous expansion of the chambers when pressurized. Moreover, in order to eliminate any bending dead zone throughout the arm, each segment is designed to overlap with its neighboring segments by 10 mm (see insert in Figure 2a). Finally, two sets of rigid links, 3d-printed in PLA material, are embedded in the arm, serving a dual function: they are primarily used to constrain the expansion of the chambers to the desired direction and also to minimize the interference between neighboring chambers. More specifically, a set of inner rings is installed on the outer surface of the assembly of the linear chambers (Figure 2b) and separates them from the helical chambers, while a second set of outer rings is installed over the outer edge of the arm (Figure 2d) and provides increased stiffness in the radial direction. In this sense, the rigid rings antagonize the linear chambers, similarly to the way that the transverse muscles antagonize the longitudinal ones in the octopus arm. A section view of the mid-segment, shown in Figure 3, further illustrates the layered arrangement of the elements comprising the arm.

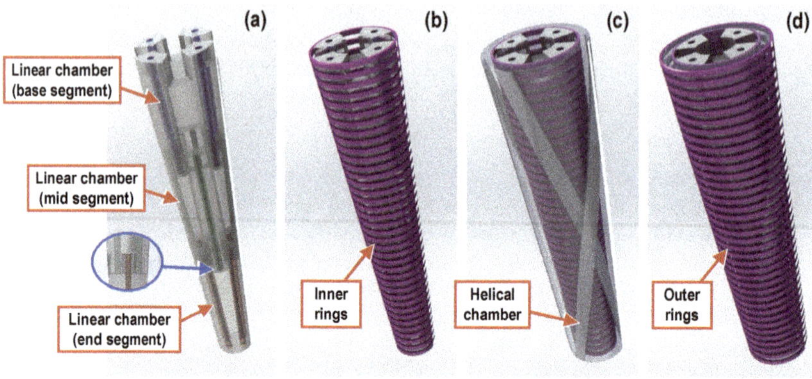

Figure 2. Three-segment soft manipulator arm design elements: (**a**) soft linear chamber assemblies of the three segments (insert shows overlap between consecutive segments), (**b**) set of rigid inner rings constraining the soft linear chambers, (**c**) soft helical chambers, and (**d**) set of rigid outer rings.

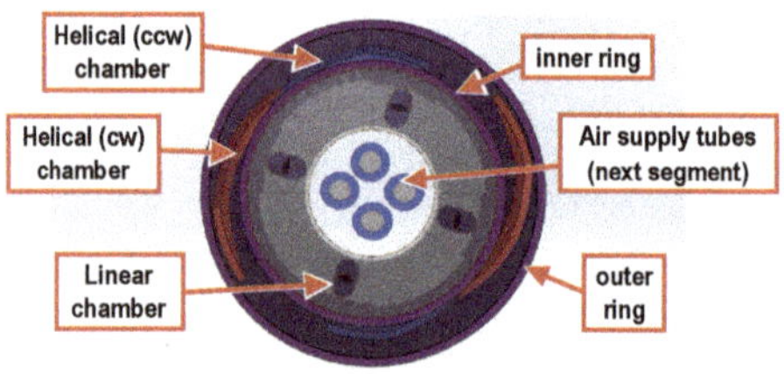

Figure 3. Section view of the second (mid) segment, as seen when looking from the base of the arm towards its tip, showing the arrangement of the different arm design elements.

3.2. Fabrication of the Soft Arm Prototypes

The material used for the fabrication of both prototype arms is a two-component silicone (Ecoflex 00-10 Smooth-On). The molds (see Figure 4), which were used for casting into shape the silicon, were fabricated with PLA material in a 3d printer for each arm design. A room-temperature casting procedure is followed, whereby the liquid silicone mixture is slowly poured from the top of the mold to avoid entrapment of air bubbles that would cause inconsistency in the final part. The pot life of the mixture is 30 min, and it cures over a 4 h period into a 00-10 shore hardness silicone rubber, with a 100% modulus of 8 psi, a tensile strength of 120 psi, and 800% of elongation at break. An alternative option for silicone would be the Dragon Skin 10, with a shore hardness of 10 A, a 100% modulus of 22 psi, a tensile strength of 475 psi, and 1000% of elongation at break. However, the higher shore hardness would require a significantly higher air pressure in order to actuate the arm (from 1 to 2 bar, as operated, to 5 to 6 bar). The selected silicone's properties, and especially its flexibility, make this silicone one of the most appropriate commercially available options.

To fabricate the three-segment arm, due to the difference in segment dimensions, multiple molds had to be created, one for each segment; hence, multiple castings were necessary in order to bring all segments together in one single arm.

First, for each of the three segments, the core element (comprising the assembly of the four linear chambers) was cast separately, and silicone tubes were attached to them (see Figure 5a,b) and secured with thin copper wire. All three segments were then joined, each one's tubes passing through the center of the preceding segments, and the set of inner rings were installed (Figure 5c). The segments were permanently joined together by exploiting the ability of silicone to bond on parts made by the same silicon. Using the molds and the helix-shaped metal inserts shown in Figure 4b, the helical chambers were then cast to surround the three joined segments (Figure 5d). Finally, the outer rings were inserted over the arm and molded into place, creating the complete arm, shown in Figure 6. The overall length of the final prototype is 150 mm, while the diameters of the base and the tip are 49 mm and 35 mm, respectively. The overall mass of the arm is 830 g.

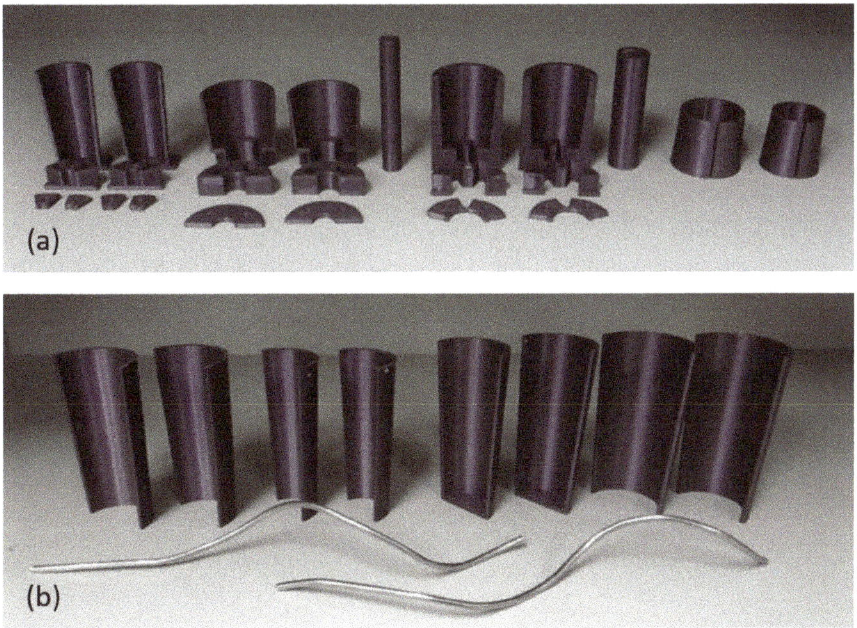

Figure 4. The 3d printed molds employed in the fabrication of the soft manipulator arm: (**a**) linear chamber molds, (**b**) helical chambers molds.

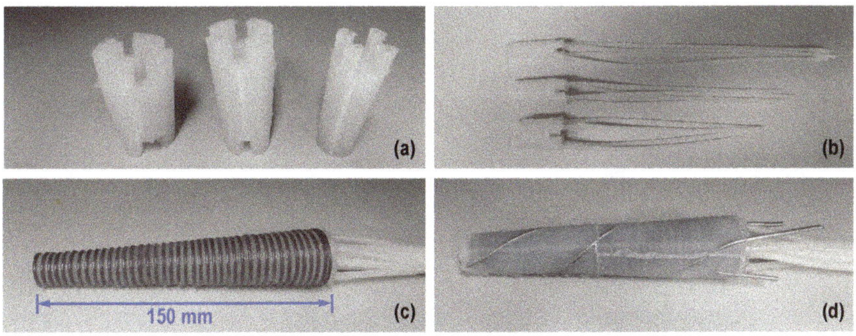

Figure 5. Different stages of the soft multi-segment arm fabrication: (**a**) the three segment cores, (**b**) attachment of the tubes to the linear chambers of the three segments, (**c**) fitting of the inner rings, and (**d**) casting of the helical chambers (the helix-shaped metal inserts are subsequently removed).

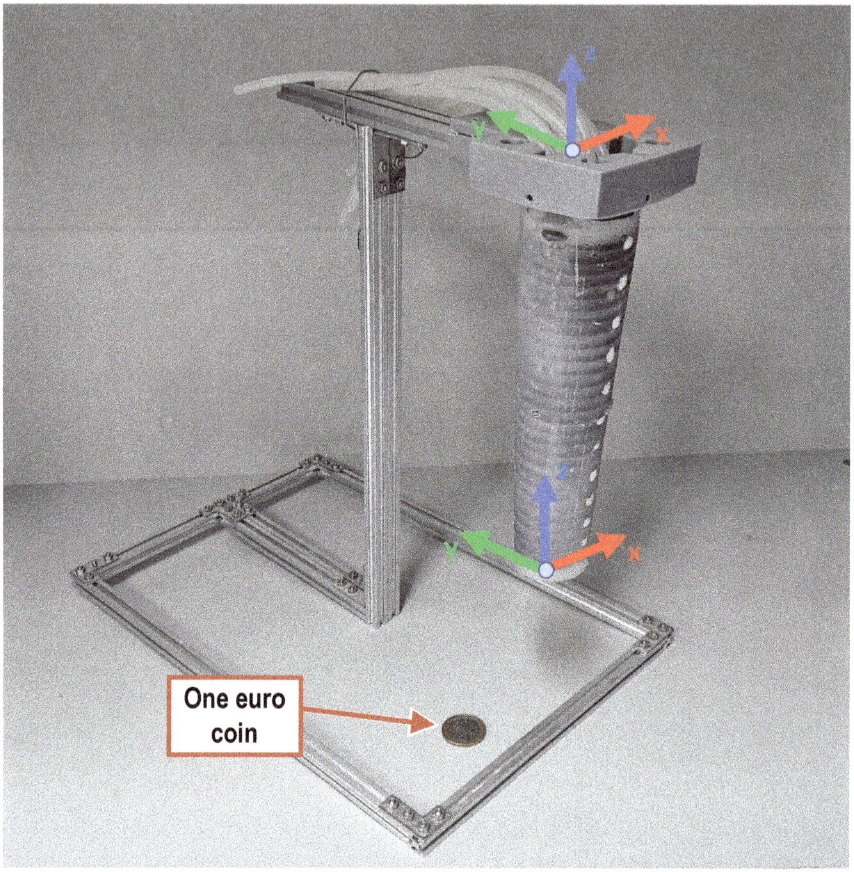

Figure 6. The three-segment soft arm prototype, shown mounted on the stand used during the experiments. The world frame XYZ is located at the base of the arm, while the local xyz frame is attached at the tip of the arm. For size reference, a one-Euro coin (⌀23.25 mm) has been included in the photograph.

3.3. *Pneumatic Actuation and Control Unit*

A schematic overview of the actuation and control system for the soft arm prototypes is provided in Figure 7, while Figure 8 shows the custom-developed control unit. The latter incorporates a total of 28 miniature solenoid valves (zhv-0519, Zonhen Electric Appliances, Shenzhen, China) that actuate the pneumatic chambers, along with their drive electronics and a 32-bit microcontroller board (Teensy 3.6) that controls their operation. The latter also handles communication via a USB interface, with a host computer that runs the arms control software. These components are housed inside a custom-designed, 3d-printed enclosure that includes fixtures for securing the solenoid valves, a top tray for mounting the main control board, as well as the housing for a small fan providing active cooling to the solenoid valves. The assembled control unit measures 146 mm × 131 mm × 100 mm (length × width × height).

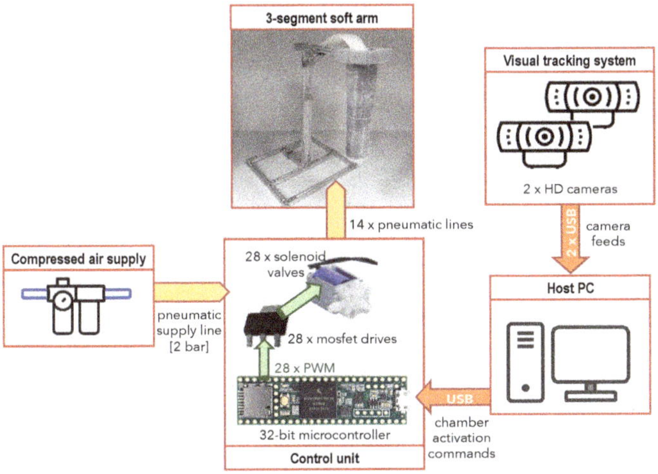

Figure 7. Diagram of the soft arm actuation and control system.

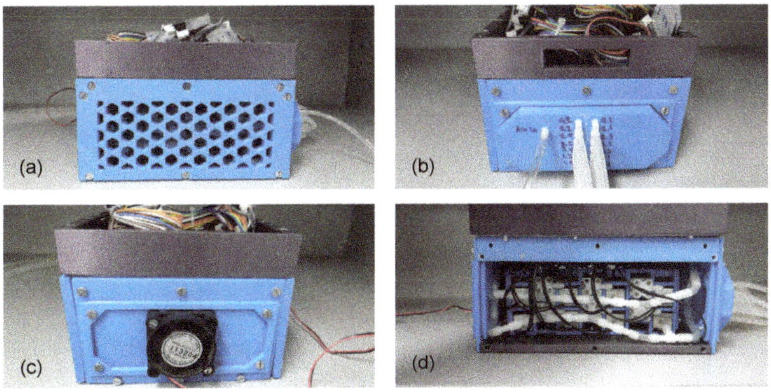

Figure 8. Control unit: (**a**) Side view, (**b**) Front view, (**c**) Back view, and (**d**) Internal view.

Each pneumatic chamber is actuated by a pair of three-port solenoid valves, arranged to achieve, by proper activation, the three necessary operations, namely inserting air into the chamber, retaining the inserted air, and expelling the air from the chamber. This is illustrated in Figure 9, where "on" indicates that a valve is at an energized state that seals the inlet (port A in Figure 9d) and air from the main outlet (port B) can move to the auxiliary outlet (port C), while "off" indicates that air entering the intake is routed to the main outlet. The valve pair is configured so that the inlet port of the first valve is connected to the air supply, while its main outlet is connected to the chamber, and its auxiliary outlet is connected to the inlet port of the second valve. In the case of the helicoidal chambers, the two clockwise chambers are interconnected and thus need only one pair of solenoid valves for their operation, likewise for the counterclockwise helicoidal chambers.

The valves are mounted on two custom 3d-printed racks, each housing 14 valves inside the control unit (see Figure 10). Compressed air is provided to the control unit by an external source. The housing has vents in its side wall (see Figure 8a) to allow the expelled air from the valves to escape. In addition, a 12 V fan, used to cool the solenoid valves, is mounted on the housing's rear end (Figure 8c).

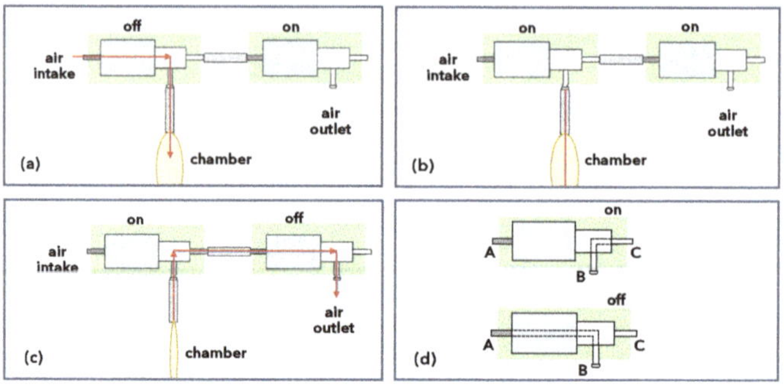

Figure 9. Activation scheme for the pair of solenoid valves controlling air flow to the pneumatic chambers, for (**a**) inflating the chamber, (**b**) maintaining pressurized air in the chamber, and (**c**) deflating the chamber. (**d**) Three-port solenoid valve operation.

Each solenoid valve is powered with 6 V, and is operated through a dedicated MOSFET (STN4NF03L by STMicroelectronics) circuit that is driven by a 20 Hz PWM signal generated by the microcontroller. By appropriately varying the duty cycle (hence, the duration of the pulse) of the PWM signals driving the valve pair of a particular segment, the rate of inflating/deflating the segment can be varied, thus controlling the speed of arm movements. Due to bandwidth limitations of the solenoid valves' mechanics, the minimum pulse duration to which the valves can respond is 10 ms (equivalent to a 20% duty cycle of the 20 Hz PWM signal). The control electronics, along with the power, data, and control connectors, are mounted on a custom-made PCB, shown in Figure 11.

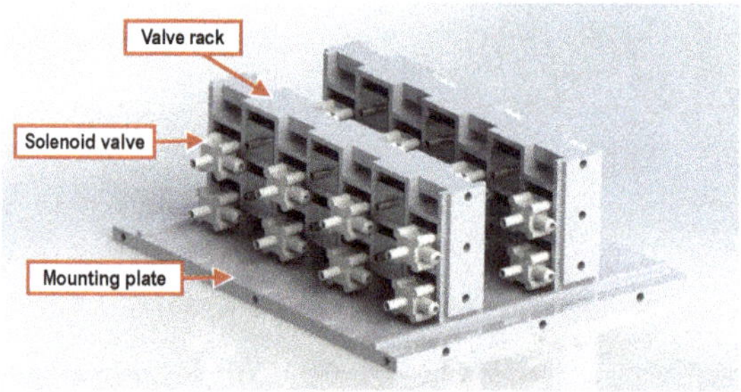

Figure 10. Layout of the solenoid valve assembly, housed inside the control unit.

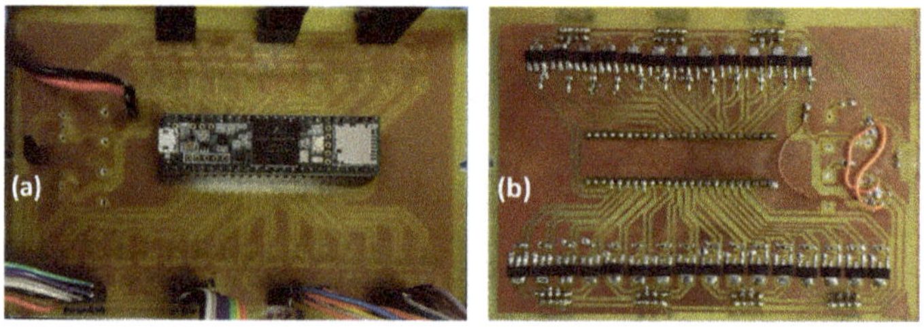

Figure 11. The printed circuit board hosting the arm control electronics: (**a**) top side with the Teensy 3.6 microcontroller board installed, and (**b**) bottom side, with the surface-mounted MOSFET drive circuits (28 in total) of the solenoid valves.

3.4. Arm Motion Control

The soft arm can move and change its shape through appropriate activation of the linear and helical pneumatic chambers, taking into consideration their topology inside the arm. With regard to the linear chambers, since each segment is rotated by 45° with respect to the previous one, different chambers on each segment must be activated to achieve bending of both segments in the same direction. For example, in order to bend the whole arm in the positive X direction, the chambers identified as 1.1, 2.1, 2.4, and 3.1 in Figure 12 must be inflated. Hence, in this case, for the base and the end segment, only one chamber is inflated, while for the middle segment, two chambers are activated (note that these two chambers must be inflated by an equal amount of air). Similarly, to achieve whole-arm bending in the plane rotated around the Z-axis by 45° with respect to the XZ-plane, chambers 1.3, 1.4, 2.3, 3.3, and 3.4 are activated. For arbitrary such rotations of the desired motion plane around the z-axis, the chambers that are involved from each segment are inflated to a different degree. Table 1 summarizes the linear chamber activation combinations for attaining the basic shape configurations of the three-segment arm.

For twisting motions, the control is simpler, since the whole arm can be twisted either clockwise or counterclockwise by activation of the respective pair of helical chambers. Twisting can be combined with other motions induced by the activation of linear chambers.

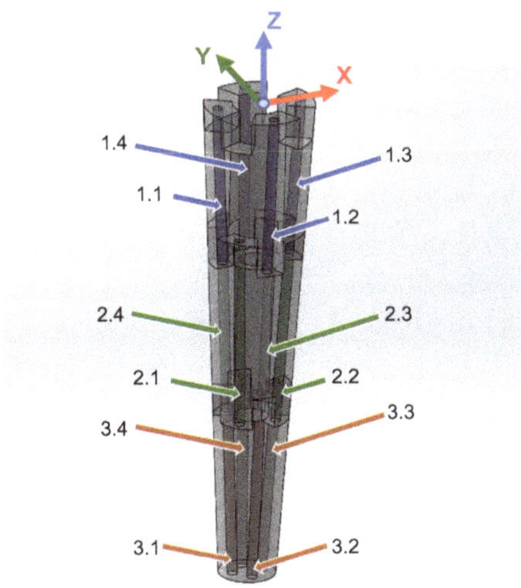

Figure 12. Location and numbering scheme of the linear chambers in the three-segment arm.

Table 1. Examples of linear chamber activation combinations for attaining different shape configurations of the three-segment arm. Filled circles denote the inflated chambers. Chambers are identified according to Figure 12.

	Linear Chamber											
Direction	**3.1**	**3.2**	**3.3**	**3.4**	**2.1**	**2.2**	**2.3**	**2.4**	**1.1**	**1.2**	**1.3**	**1.4**
whole-arm bending												
$+X$	●	○	○	○	●	○	○	●	●	○	○	○
$-X$	○	○	●	○	○	●	●	○	○	○	●	○
$+Y$	○	●	○	○	●	●	○	○	○	●	○	○
$-Y$	○	○	○	●	○	○	●	●	○	○	○	●
whole-arm longitudinal extension												
$-Z$	●	●	●	●	●	●	●	●	●	●	●	●
composite configurations												
S-shape	●	○	○	○	○	●	●	○	●	○	○	○
Ƨ-shape	○	○	●	○	●	○	○	●	○	○	●	○

3.5. Control Software

Low-level control of the soft arm is implemented through the custom firmware, developed in C language, that is executed on the Teensy 3.6 microcontroller. The firmware is mainly responsible for generating 28 PWM signals driving, through MOSFET gates, the solenoid valves that actuate each one of the pneumatic chambers, according to activation commands received via USB from a host PC at a 100 Hz update rate.

High-level control of the soft arm is provided through a graphical user interface (GUI) developed in MATLAB and running on the host PC that communicates with the microcontroller. The host PC also receives and processes visual data from the cameras used to track the arm's motion (see Section 4.2) and to provide feedback for the closed-loop schemes for motion control of the arm (see Section 5.5). The GUI provides a series of panels (see Figure 13) for setting up the communication with the control unit, configuring the cameras, issuing commands to manually set the state of the arm's chambers, as well as for configuring and executing the closed-loop control schemes. Exporting data, including chamber activation signals, sensor readings, video recordings, etc., to log files for post-processing is also readily supported by the GUI.

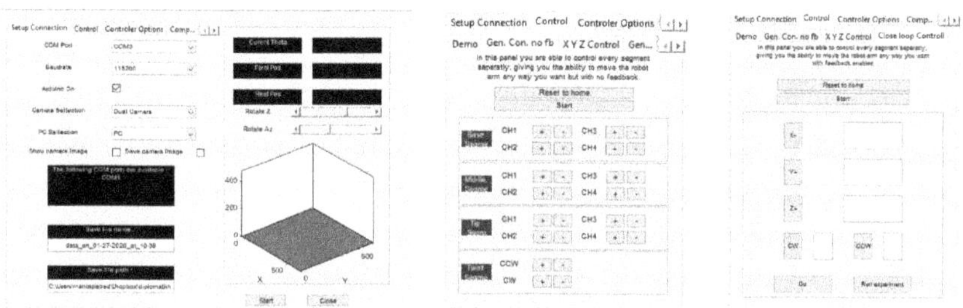

(a) (b) (c)

Figure 13. Elements of the Graphical User Interface (GUI) developed for controlling the three-segment arm: (**a**) Startup screen with initialization setup and 3d core pose estimation representation, (**b**) GUI panel for open-loop control of individual chambers, (**c**) GUI panel for closed-loop control.

4. Experimental Setup

Both prototype arms were tested while rigidly mounted on a custom stand constructed using aluminum extrusions (stand footprint: 220 mm × 200 mm stand height: 320 mm), which holds the arm in a vertical position. Figure 6 illustrates the experimental setup along with the global reference frame XYZ (defined to be located on top of the base segment of the arm and oriented so that the X- and Y-axes are aligned with the stand's base) and the moving frame xyz (positioned at the tip of the arm so that it has the same orientation as the XYZ frame and is displaced along the Z-axis when the arm is at rest).

4.1. Force Measurement

A high-precision digital dynamometer (FMI-210A5, Alluris GmbH & Co, Freiburg, Germany) was used to measure, at a 1 kHz rate, the force exerted by the arm, as shown in Figure 14. The dynamometer is placed underneath and in contact with the resting arm. The arm is used in extension mode, where all linear chambers are filled equally with air, and as the chambers expand, the arm presses upon the dynamometer.

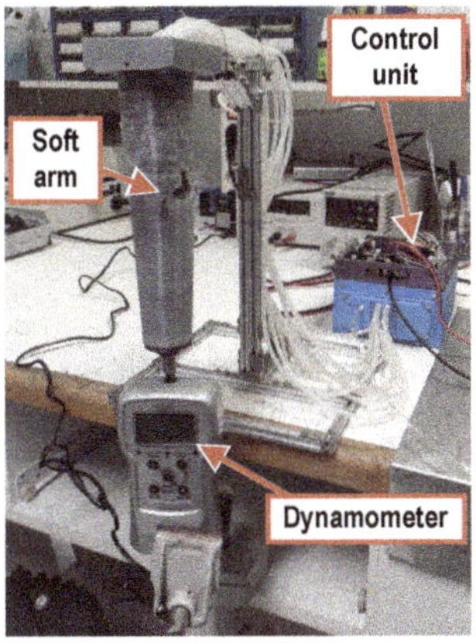

Figure 14. Test setup for force measurement experiments.

4.2. Vision-Based Arm Motion Tracking

The position and shape configuration of the arm were tracked using a pair of high-definition cameras (HD Pro C920, Logitech International S.A., Lausanne, Switzerland) placed 30 cm away from the arm and at a 90° angle with respect to each other, as shown in Figure 15. Each camera was calibrated with a checkerboard of known dimensions to determine its extrinsic and intrinsic parameters prior to setting up [48], using the Camera Calibration App of MATLAB [49]. The pair is oriented so that each camera captures the movement of the arm in one plane of movement with regard to the XYZ frame (specifically, Camera 1 for the XZ-plane and Camera 2 for the YZ-plane), using computer vision methods that track two series of painted dot-markers on the arm. Each series consists of 11 such markers positioned at known intervals along a straight line that is parallel to the Z-axis when the arm is at its resting state. Combining the images captured by the pair of cameras, a representation of the arm's configuration in 3d space may be obtained. In particular, calculating the intersections of corresponding markers from the two series yields the 3d coordinates of 11 points forming the center-line of the arm. Real-time processing of the video feeds from the two cameras is performed in MATLAB at a 30 fps rate and is integrated with the GUI described in Section 3.5.

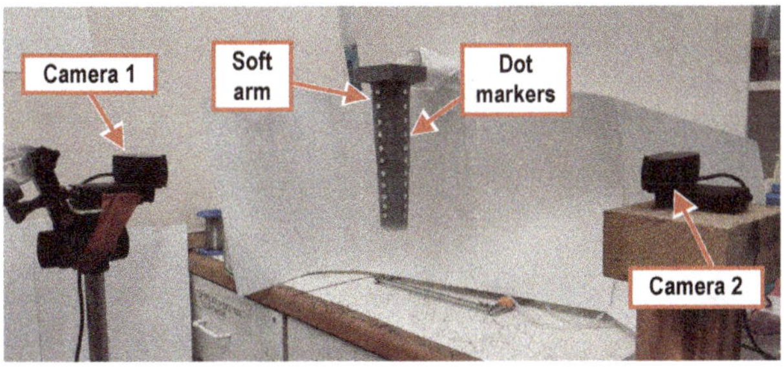

Figure 15. The setup employed for visual tracking of the arm motion.

5. Experimental Results

Utilizing the experimental setups described in Section 4, tests were conducted to characterize the three-segment arm with respect to the forces it is capable of exerting and the configurations it can achieve through activation of the linear chambers, while the single-segment arm was tested for its longitudinal extension and twisting capabilities.

5.1. Arm Force Experiments

The force generated by the three-segment arm in the direction of the Z-axis was assessed using the experimental setup of Figure 14. Pressurized air was incrementally supplied to all the arm's linear chambers by providing a sequence of 5 discrete 10 ms activation pulses, where 7 s were allowed between consecutive pulses. The exerted force was measured 6 s after each pulse to ensure the absence of any transients. This test was repeated for different air pressures provided to the system, ranging from 0.1 to 0.5 bar.

The average values at steady-state of the thus measured forces are summarized in Figure 16a. It can be seen that the force exerted by the arm increases with the number of air input pulses, as well as with the pressure of the air provided to the system, reaching a maximum of approximately 10 N. Moreover, we observe that the exerted force magnitude is not linearly proportional to the number of input pulses; this is an anticipated result, since the pressure provided by the external system remains constant, while the pressure inside the chambers increases, resulting in a lower pressure difference between them and thus the pressure increase in a chamber for the constant time pulses drops.

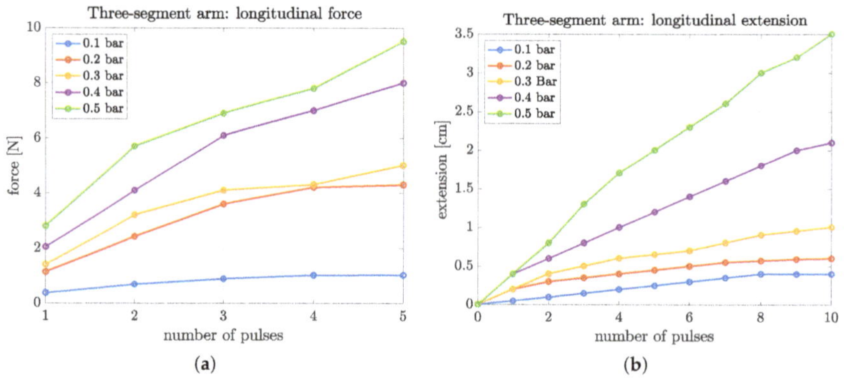

Figure 16. Measurements of (**a**) the exerted force, and (**b**) the extension of the three-segment arm during experiments involving the simultaneous activation of all the arm's linear chambers.

5.2. Arm Extension Experiments

The vision-based arm tracking setup in Figure 15 was used to measure the elongation characteristics of the two arm prototypes. For the three-segment arm, the arm extension along the Z-axis was measured with respect to the pressure of the air intake (ranging from 0.1 to 0.5 bar) and the number of pulses (1 to 10) provided to the linear chambers. The pulsing inputs were applied following an activation sequence like the one used in the force measurement tests. The results are summarized in Figure 16b. It can be seen that increasing the supply pressure affords a significant increase in arm elongation (up to approximately 24% compared to its non-inflated state).

The longitudinal extension of the single-segment arm was assessed by a different method, namely using a syringe to manually insert known volumes of air in the segment's four linear segments. Results, shown in Figure 17a, suggest a quadratic relationship between the volume of air inserted in the arm's chambers and its elongation. It can be seen that the arm is elongated by 2.5 cm (i.e., by 50%) through the injection of a total of 25 mL of air in its four linear chambers. Compared to the experimental results shown in Figure 16b, where, instead of inserting a known volume of air, pressurized air is inserted in the chamber in pulses, we can deduce that the slower increase of displacement based on the number of pulses is due to the lowering pressure difference between the pressure of the air in the chamber and the pressure provided to the control unit, which results in a decrease of the volume of air inserted in each successive pulse.

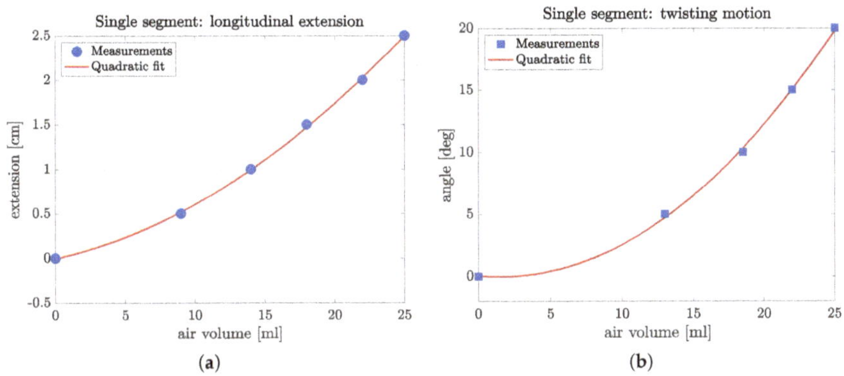

Figure 17. Measurements of (**a**) the longitudinal extension, and (**b**) the twist angle of the single-segment arm prototype as a function of the input air volume.

5.3. Arm Twist Experiments

To determine the arm's twisting capability, tests were conducted with the single-segment prototype mounted on the stand shown in Figure 15. During this experiment, the air was inserted with a syringe in the helical chambers, and measurements were taken of the arm's twist angle (rotation around the Z-axis) as a function of the volume of air inserted.

Inserting air in the helical chambers inflates them, and, due to the restriction of expansion in the lateral direction, the chambers expand by increasing their length. This expansion results in a twisting motion. The results, shown in Figure 17b, indicate a quadratic relation of the twist angle with respect to the volume of air inserted, similar to the arm extension experiments.

5.4. Composite Arm Movements

In this set of experiments, the motion tracking setup of Figure 15 was used to capture and analyze the movement of the arm for various open-loop activation patterns of the linear chambers, with the pressure of the air supply set to 1.0 bar.

Results from an indicative such experiment are provided in Figure 18a. Solid lines correspond to the displacement of the arm's tip along the X, Y, and Z axes, resulting from the sequence of activation, indicated with the dotted line, of a single linear chamber of the end segment, namely the segment identified as 3.3 in Figure 12 (all the other chambers are deflated throughout the experiment). Specifically, from 0.5 s to 1.7 s, the valve pair is activated so that compressed air is inserted into the chamber (Figure 9a) via continuous pulsing. This results in a bending motion of the arm's lower part, occurring, as anticipated, predominantly on the XZ-plane (there is also some limited displacement along the Y-axis). At 1.7 s, the valve pair is activated to seal the chamber (Figure 9b), thus stabilizing the tip. Finally, at 2.8 s, the valve pair is activated to deflate the chamber (Figure 9c), and the arm returns, within about 2.5 s, to its original, non-actuated configuration. An analogous response is recorded in the results shown in Figure 18b, taken from a similar experiment that involved activation of a different chamber of the end segment (namely the chamber identified as 3.4 in Figure 12), resulting in a bending motion on the YZ-plane.

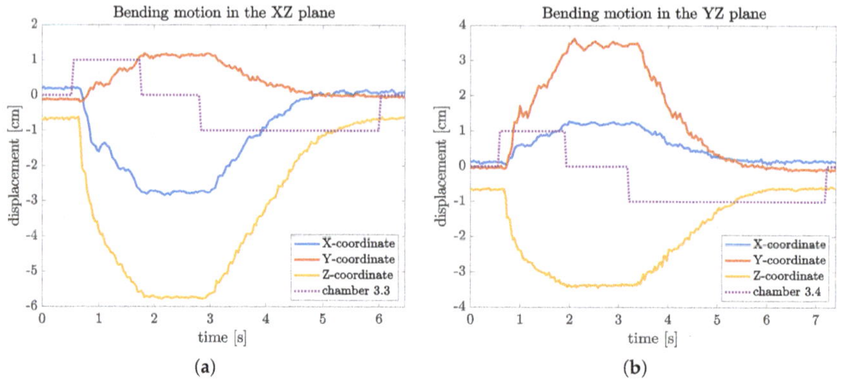

Figure 18. Experimental results for bending motion under activation of a single chamber of the end segment: (**a**) bending motion in the XZ-plane (activation of chamber 3.3), and (**b**) bending motion in the YZ-plane (activation of chamber 3.4).

Figure 19 shows the data from an experiment involving the longitudinal extension of the arm through the activation of all 12 of the linear chambers with the same pattern (indicated with the dotted line). It can be seen that the tip of the arm is displaced along the Z-axis by as much as 11 cm (i.e., the arm is elongated by approximately 72%) in about 1.8 s. During this extension, there is very limited displacement along the X- and Y-axes. When the valves are commanded, at 4.5 s, to release the air from the chambers, the arm returns to its resting, non-actuated position within 2.6 s, exhibiting a certain number of oscillations. These can be attributed to the fact that when the chambers are depressurized, the arm is at its lowest stiffness.

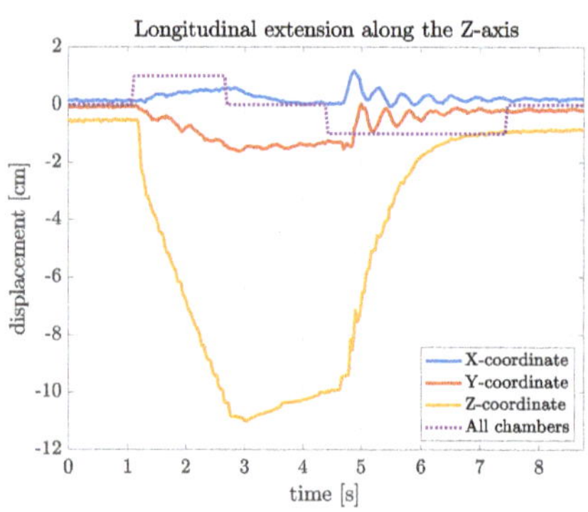

Figure 19. Results for the motion of the arm's tip during an experiment involving the longitudinal extension of the soft robot, with the pressure of the air supply set to 1.0 bar.

Finally, Figure 20 shows three indicative sequences of frames from various tests with the three-segment arm. The last sequence (Figure 20c) is taken from an experiment demonstrating the capability of the soft arm for attaining S-shaped configurations through the bending of consecutive segments on opposite sides by the activation pattern registered in the last row of Table 1.

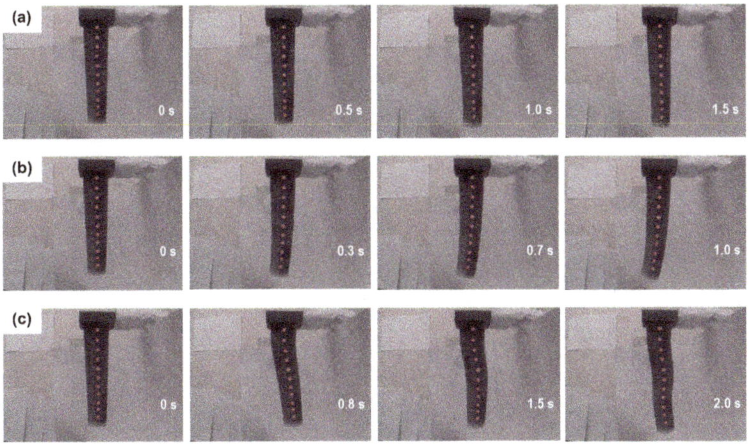

Figure 20. Series of frames from experiments involving (**a**) longitudinal extension, (**b**) whole-arm bending, and (**c**) S-shaped bending, of the three-segment soft arm.

5.5. Closed-Loop Arm Control

Experiments were also conducted where the external stereo vision system was employed to provide a three-dimensional representation of the arm in real time. The latter can be used as feedback for closed-loop control of, e.g., the position of the arm's tip, the twisting angle, or the motion plane and extent of bending.

Here, we present results from experiments for closed-loop control of the bending angle θ (see inset in Figure 21a), defined as the angle between the Z-axis of the global frame (arm base) and the z-axis of the local frame (arm tip), during whole-arm bending motion on the YZ-plane (see Table 1).

Two types of controllers were considered, namely a proportional-only (P-control) and a proportional-derivative (PD-control) one, using as input the error between the setpoint and the vision-estimated value of θ. The output of the controller is used to specify the duty cycle of the PWM signals activating the solenoid valve pairs of the appropriate chambers, namely chambers {1.2, 2.1, 2.2, 3.2} for $\theta > 0$ and chambers {1.4, 2.3, 2.4, 3.4} for $\theta < 0$.

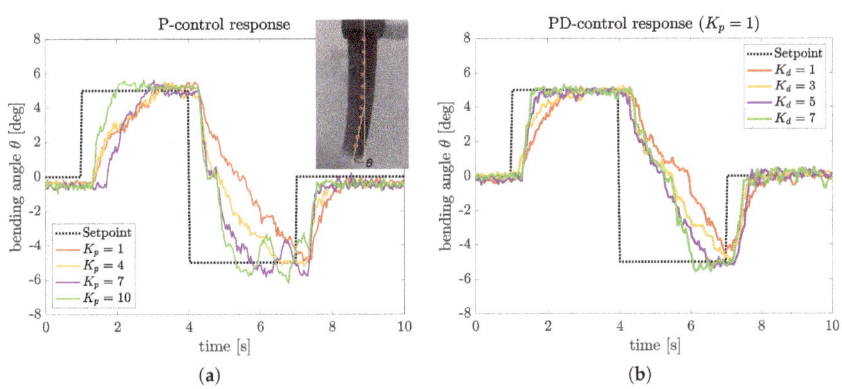

Figure 21. Bending angle response for the three-segment arm under closed-loop control: (**a**) P-controller for different values of the proportional gain K_p, (**b**) PD-controller for fixed K_p and different values of the derivative gain K_d.

Initial experiments focused on the effect of the P-controller gain K_p on the obtained response, involving both positive and negative values for the bending angle setpoint. The results, summarized in Figure 21a, indicate that the P-controller can regulate the bending angle to the specified setpoint with good accuracy. As anticipated, increasing the control gain results in a faster overall response. However, for large K_p values ($K_p = 7$ and $K_p = 10$ in Figure 21a) there are undesirable oscillations, as well as a small dead zone near the zero degrees mark when the system moves from a positive to a negative setpoint.

As shown in Figure 21b, the PD-controller yields marked improvements, where the introduction of the error-derivative control action, with appropriate gain settings (e.g., for $K_p = 1$, $K_d = 5$), allows for efficient damping of the oscillations while maintaining a fast and accurate response. Given, in particular, the minimal steady-state error thus attained, we opted against including an integral-error term to the controller to avoid degradation of the transient response.

6. Discussion and Conclusions

We have presented the development of an octopus-inspired soft arm prototype that closely emulates the musculature of an octopus, utilizing pneumatically actuated chambers to achieve comparable motion capabilities. Linear chambers mimic the longitudinal muscles, enabling bending and elongation, while helical chambers replicate the oblique muscles, allowing for twisting motions—a novel feature rare among similar prototypes. This bi-directional twisting ability enhances the arm's versatility and is achieved through both clockwise and counterclockwise helical chambers. Rigid 3d-printed rings function similarly to the transverse muscles by constraining radial expansion, ensuring precise, directed movement. The three-segment arm incorporates a conical, overlapping design to avoid bending dead zones, with appropriately rotated and tapered chambers for uniform actuation. Compressed air is delivered through internal silicone tubes, making the design robust and capable of complex, multi-segment motion akin to the natural dexterity of an octopus arm, as demonstrated through experimental validation studies. Moreover, we have demonstrated the efficacy of simple, model-free closed-loop controllers to position the arm tip with speed and accuracy.

A noteworthy finding to emerge from our studies is that thanks to the durability of the silicone employed, a carefully fabricated mechanism without air leaks allows the arm to hold any given shape indefinitely, provided external forces do not change. Even while resisting varying external forces, this enables the arm to perform consistently in any orientation and maintain its functionality.

It is noted that the conventional casting methods used to fabricate our prototypes are quite involved and are prone to inconsistencies, primarily due to the multi-step process required to accommodate the arm's complex morphology. To address this issue, one promising alternative is the use of multi-material 3d printing technologies, which could enable the creation of the entire arm in a single, streamlined stage. This would not only eliminate inconsistencies in the arm's soft materials but also allow for the seamless integration of pneumatic circuits and other functional components directly into the structure. Such advancements could significantly enhance the arm's precision, durability, and overall controllability while also reducing production time and effort.

While the foundation for the creation of a functional pneumatic arm has been established in this study, there are still considerable challenges that need to be addressed, opening up new and interesting research directions. One priority is the integration of embedded sensors to measure the arm's movements directly. For instance, flex sensors could be incorporated into the inner and outer surfaces of the arm, enabling precise deformation measurements at known locations. Such data would allow the development of real-time pose estimation models, eliminating the need for external tracking systems and greatly improving the control, precision, and application scope of the arm.

Exploring alternative chamber shapes and placements is another promising avenue to expand the variety of achievable motion patterns. Adjustments to the configuration of linear and helical chambers could enhance dexterity and efficiency. Additionally, experimenting with different actuation fluids, such as water or oil, could reveal how these fluids affect the arm's performance, particularly in terms of force output, responsiveness, and environmental suitability.

Miniaturization of the arm holds substantial potential, particularly for medical applications. A smaller arm would reduce resource demands while maintaining functionality. Scaling down components, including solenoid valves and air compressors, could enable the creation of a soft robotic arm small enough for laparoscopic procedures. Its compliance and soft materials would provide a safer alternative for interacting with delicate tissues, reducing the risk of damage during surgical operations.

Conversely, a scaled-up version of the arm could be explored for applications requiring safe, dexterous manipulation of larger objects, e.g., in manufacturing, where it could handle delicate assembly tasks; in agriculture, for gentle crop harvesting; or in challenging environments like disaster zones, where its flexibility could aid in search-and-rescue missions. Moreover, the arm's compliance and adaptability could make it valuable for space exploration, where soft robotic systems are particularly advantageous for interacting with fragile equipment or unpredictable environments.

Author Contributions: Conceptualization, E.P., D.P.T. and M.S.; methodology, E.P., D.P.T. and M.S.; software, E.P.; validation, E.P.; investigation, E.P.; writing—original draft preparation, E.P., D.P.T. and M.S.; writing—review and editing, E.P., D.P.T. and M.S.; visualization, E.P. and M.S.; supervision, D.P.T. and M.S. All authors have read and agreed to the published version of the manuscript.

Funding: This research received no external funding.

Institutional Review Board Statement: Not applicable.

Data Availability Statement: Data are contained within the article.

Conflicts of Interest: The authors declare no conflict of interest.

References

1. Walker, I.D.; Dawson, D.M.; Flash, T.; Grasso, F.W.; Hanlon, R.T.; Hochner, B.; Kier, W.M.; Pagano, C.C.; Rahn, C.D.; Zhang, Q.M. Continuum robot arms inspired by cephalopods. In Proceedings of the Unmanned Ground Vehicle Technology VII. International Society for Optics and Photonics, Orlando, FL, USA, 29–31 March 2005; Volume 5804, pp. 303–314. [CrossRef]
2. Laschi, C.; Mazzolai, B.; Cianchetti, M. Soft robot arm inspired by the octopus. *Adv. Robot.* **2012**, *26*, 709–727. [CrossRef]
3. Sfakiotakis, M.; Kazakidi, A.; Pateromichelakis, N.; Ekaterinaris, J.A.; Tsakiris, D.P. Robotic underwater propulsion inspired by the octopus multi-arm swimming. In Proceedings of the 2012 IEEE International Conference on Robotics and Automation, St. Paul, MN, USA, 14–18 May 2012; pp. 3833–3839. [CrossRef]
4. Sfakiotakis, M.; Kazakidi, A.; Tsakiris, D.P. Octopus-inspired multi-arm robotic swimming. *Bioinspir. Biomim.* **2015**, *10*, 035005. [CrossRef] [PubMed]
5. Sfakiotakis, M.; Kazakidi, A.; Chatzidaki, A.; Evdaimon, T.; Tsakiris, D.P. Multi-arm robotic swimming with octopus-inspired compliant web. In Proceedings of the 2014 IEEE/RSJ International Conference on Intelligent Robots and Systems, Chicago, IL, USA, 14–18 September 2014; pp. 302–308. [CrossRef]
6. Calisti, M.; Giorelli, M.; Levy, G.; Mazzolai, B.; Hochner, B.; Laschi, C.; Dario, P. An octopus-bioinspired solution to movement and manipulation for soft robots. *Bioinspir. Biomim.* **2011**, *6*, 036002. [CrossRef]
7. Wehner, M.; Truby, R.L.; Fitzgerald, D.J.; Mosadegh, B.; Whitesides, G.M.; Lewis, J.A.; Wood, R.J. An integrated design and fabrication strategy for entirely soft, autonomous robots. *Nature* **2016**, *536*, 451–455. [CrossRef] [PubMed]
8. Rus, D.; Tolley, M.T. Design, fabrication and control of soft robots. *Nature* **2015**, *521*, 467–475. [CrossRef] [PubMed]
9. Marchese, A.D.; Onal, C.D.; Rus, D. Autonomous soft robotic fish capable of escape maneuvers using fluidic elastomer actuators. *Soft Robot.* **2014**, *1*, 75–87. [CrossRef]
10. Xie, Z.; Domel, A.G.; An, N.; Green, C.; Gong, Z.; Wang, T.; Knubben, E.M.; Weaver, J.C.; Bertoldi, K.; Wen, L. Octopus arm-inspired tapered soft actuators with suckers for improved grasping. *Soft Robot.* **2020**, *7*, 639–648. [CrossRef] [PubMed]
11. Kim, S.; Laschi, C.; Trimmer, B. Soft robotics: A bioinspired evolution in robotics. *Trends Biotechnol.* **2018**, *36*, 478–494. [CrossRef] [PubMed]
12. Thuruthel, T.G.; Shih, B.; Laschi, C.; Tolley, M.T. Soft robot perception using embedded soft sensors and recurrent neural networks. *Sci. Robot.* **2019**, *4* . [CrossRef]
13. Wang, X.; Xu, C.; Aabloo, A.; Medina, F.; Xie, M.; Punning, A.; Chen, X. Soft robots in healthcare: A concise review. *Robot. Biomim.* **2018**, *5*, 11. [CrossRef]
14. Antonelli, M.G.; Beomonte Zobel, P.; Sarwar, M.A.; Stampone, N. Seahorse-tail-inspired soft pneumatic actuator: Development and experimental characterization. *Biomimetics* **2024**, *9*, 264. [CrossRef]

15. Deshpande, S.; Almubarak, Y. Octopus-inspired robotic arm powered by Shape Memory Alloys (SMA). *Actuators* **2023**, *12*, 377. [CrossRef]
16. Falco, I.D.; Cianchetti, M.; Menciassi, A. A soft multi-module manipulator with variable stiffness for minimally invasive surgery. *Bioinspir. Biomim.* **2017**, *12*, 056008. [CrossRef] [PubMed]
17. Shiva, A.; Stilli, A.; Noh, Y.; Faragasso, A.; Falco, I.D.; Gerboni, G.; Cianchetti, M.; Menciassi, A.; Althoefer, K.; Wurdemann, H.A. Tendon-Based Stiffening for a Pneumatically Actuated Soft Manipulator. *IEEE Robot. Autom. Lett.* **2016**, *1*, 632–637. [CrossRef]
18. Qi, X.; Mei, Y.; Chen, D.; Li, Z.; Tan, X. Design and nonlinear modeling of a modular cable-driven soft robotic arm. *IEEE/ASME Trans. Mechatron.* **2024**, *29*, 3083–3091. [CrossRef]
19. Mazzolai, B.; Margheri, L.; Cianchetti, M. Bioinspired Soft Robotics: Material, Actuation, and Design. *Front. Robot. AI* **2019**, *6*, 84. [CrossRef]
20. Li, S.; Wang, K.W. Bioinspired self-powered soft robotic arm with controllable motions based on bistable origami and liquid crystal elastomers. *Soft Robot.* **2019**, *6*, 745–758. [CrossRef]
21. Coyle, S.; Majidi, C.; LeDuc, P.; Hsia, K.J. Bio-inspired soft robotics: Material selection, actuation, and design. *Extrem. Mech. Lett.* **2018**, *22*, 51–59. [CrossRef]
22. Ma, J.; Han, Z.; Liu, Z.; Li, G.; He, W.; Ge, S. Modeling and control of an octopus inspired soft arm under prescribed spatial motion constraints. *J. Intell. Robot. Syst.* **2023**, *109*, 94. [CrossRef]
23. Wang, T.; Halder, U.; Gribkova, E.; Gillette, R.; Gazzola, M.; Mehta, P.G. A sensory feedback control law for octopus arm movements. In Proceedings of the IEEE Conference on Decision and Control (CDC), Cancun, Mexico, 6–9 December 2022; pp. 1059–1066. [CrossRef]
24. Cianchetti, M.; Calisti, M.; Margheri, L.; Kuba, M.; Laschi, C. Bioinspired soft actuation system using Shape Memory Alloys. *Actuators* **2018**, *7*, 40. [CrossRef]
25. Majidi, C. Soft-matter engineering for soft robotics. *Adv. Mater. Technol.* **2019**, *4*, 1800477. [CrossRef]
26. Hu, W.; Lum, G.; Mastrangeli, M.; Sitti, M. Small-scale soft-bodied robot with multimodal locomotion. *Nature* **2018**, *554*, 81–85. [CrossRef]
27. Rupert, L.; Saunders, B.O.; Killpack, M.D. Performance metrics for fluidic soft robot rotational actuators. *Front. Robot. AI* **2021**, *8*, 632835. [CrossRef]
28. Doi, T.; Wakimoto, S.; Suzumori, K.; Mori, K. Proposal of flexible robotic arm with thin McKibben actuators mimicking octopus arm structure. In Proceedings of the 2016 IEEE/RSJ International Conference on Intelligent Robots and Systems (IROS), Daejeon, Republic of Korea, 9–14 October 2016; pp. 5503–5508.
29. Yamamoto, Y.; Wakimoto, S.; Kanda, T.; Yamaguchi, D. A soft robot arm with flexible sensors for master–slave operation. *Eng. Proc.* **2021**, *10*, 34. [CrossRef]
30. Chen, G.; Lin, T.; Ding, S.; Chen, S.; Ji, A.; Lodewijks, G. Design and test of an active pneumatic soft Wrist for soft grippers. *Actuators* **2022**, *11*, 311. [CrossRef]
31. Wu, S.; Ze, Q.; Dai, J.; Udipi, N.; Paulino, G.H.; Zhao, R. Stretchable origami robotic arm with omnidirectional bending and twisting. *Proc. Natl. Acad. Sci. USA* **2021**, *118*, e2110023118. [CrossRef]
32. Flash, T.; Zullo, L. Biomechanics, motor control and dynamic models of the soft limbs of the octopus and other cephalopods. *J. Exp. Biol.* **2023**, *226*, jeb245295. [CrossRef]
33. Kier, W.M.; Smith, K.K. Tongues, tentacles and trunks: The biomechanics of movement in muscular-hydrostats. *Zool. J. Linn. Soc.* **1985**, *83*, 307–324. [CrossRef]
34. Yekutieli, Y.; Sumbre, G.; Flash, T.; Hochner, B. How to move with no rigid skeleton? The octopus has the answers. *Biologist* **2003**, *49*, 250–254.
35. Kier, W.; Stella, M. The arrangement and function of octopus arm musculature and connective tissue. *J. Morphol.* **2007**, *268*, 831–843. [CrossRef] [PubMed]
36. Smith, K.K.; Kier, W.M. Trunks, tongues, and tentacles: Moving with skeletons of muscle. *Am. Sci.* **1989**, *77*, 28–35.
37. Mather, J.A. How do octopuses use their arms? *J. Comp. Psychol.* **1998**, *112*, 306. [CrossRef] [PubMed]
38. Gutfreund, Y.; Flash, T.; Yarom, Y.; Fiorito, G.; Segev, I.; Hochner, B. Organization of octopus arm movements: A model system for studying the control of flexible arms. *J. Neurosci.* **1996**, *16*, 7297–7307. [CrossRef]
39. Sumbre, G.; Gutfreund, Y.; Fiorito, G.; Flash, T.; Hochner, B. Control of octopus arm extension by a peripheral motor program. *Science* **2001**, *293*, 1845–1848. [CrossRef]
40. Yekutieli, Y.; Zohar, R.; Hochner, B.; Flash, T. Dynamic model of the octopus arm. II. Control of reaching movements. *J. Neurophysiol.* **2005**, *94*, 1459–1468. [CrossRef]
41. Yekutieli, Y.; Sagiv-Zohar, R.; Aharonov, R.; Engel, Y.; Hochner, B.; Flash, T. Dynamic model of the octopus arm. I. Biomechanics of the octopus reaching movement. *J. Neurophysiol.* **2005**, *94*, 1443–1458. [CrossRef] [PubMed]
42. Sumbre, G.; Fiorito, G.; Flash, T.; Hochner, B. Octopuses Use a human-like strategy to control precise point-to-point arm movements. *Curr. Biol.* **2006**, *16*, 767–772. [CrossRef]
43. Sumbre, G.; Fiorito, G.; Flash, T.; Hochner, B. Neurobiology: Motor control of flexible octopus arms. *Nature* **2005**, *433*, 595–596. [CrossRef] [PubMed]

44. Kang, R.; Kazakidi, A.; Guglielmino, E.; Branson, D.T.; Tsakiris, D.P.; Ekaterinaris, J.A.; Caldwell, D.G. Dynamic model of a hyper-redundant, octopus-like manipulator for underwater applications. In Proceedings of the 2011 IEEE/RSJ International Conference on Intelligent Robots and Systems, San Francisco, CA, USA, 25–30 September 2011; pp. 4054–4059. [CrossRef]
45. Vavourakis, V.; Kazakidi, A.; Tsakiris, D.; Ekaterinaris, J. A nonlinear dynamic finite element approach for simulating muscular hydrostats. *Comput. Methods Biomech. Biomed. Eng.* **2014**, *17*, 917–931. [CrossRef] [PubMed]
46. Kazakidi, A.; Tsakiris, D.P.; Angelidis, D.; Sotiropoulos, F.; Ekaterinaris, J.A. CFD study of aquatic thrust generation by an octopus-like arm under intense prescribed deformations. *Comput. Fluids* **2015**, *115*, 54–65. [CrossRef]
47. Chang, H.S.; Halder, U.; Shih, C.H.; Naughton, N.; Gazzola, M.; Mehta, P. Energy-shaping control of a muscular octopus arm moving in three dimensions. *Proc. R. Soc. A Math. Phys. Eng. Sci.* **2023**, *479*, 20220593. [CrossRef]
48. Corke, P. *Robotics, Vision and Control: Fundamental Algorithms in MATLAB*; Springer: Berlin/Heidelberg, Germany, 2011. [CrossRef]
49. The MathWorks Inc. *Computer Vision Toolbox (R2022b)*; The MathWorks Inc.: Natick, MA, USA, 2022.

biomimetics

Article

Neural Network-Based Shape Analysis and Control of Continuum Objects

Yuqiao Dai [1], Shilin Zhang [1], Wei Cheng [2,*] and Peng Li [1,*]

[1] School of Mechanical Engineering and Automation, Harbin Institute of Technology Shenzhen, Shenzhen 518055, China
[2] Devol Advanced Automation, Inc., Jiaan Science Park, Liuxian 1st Road, Bao an District, Shenzhen 518101, China
* Correspondence: davidcheng@daa-inc.com (W.C.); peng.li@hit.edu.cn (P.L.)

Abstract: Soft robots are gaining increasing attention in current robotics research due to their continuum structure. However, accurately recognizing and reproducing the shape of such continuum robots remains a challenge. In this paper, we propose a novel approach that combines contour extraction with camera reconstruction to obtain shape features. Neural networks are employed to model the relationship between motor inputs and the resulting shape output. A simulation environment is established to verify the shape estimation and shape control of the flexible continuum. The outcomes demonstrate that this approach effectively predicts and reproduces the shape of flexible continuum robots, providing a promising solution for continuum shape control.

Keywords: continuum object; shape analysis; shape control; neural networks

1. Introduction

Small tubular surgical instruments are widely used in minimally invasive surgery. The surgeon manipulates these instruments using images provided by the endoscope to perform the procedure. Different from conventional rigid robots, flexible instruments introduce redundant degrees of freedom, making it more difficult to calculate their position. This complicates real-time shape perception and accurate three-dimensional (3D) modeling. Precise shape perception of the robot is critical in surgery, as it provides the surgeon with information that cannot be obtained visually.

Various flexible continuum robots have been developed to perform complex tasks in robot-assisted minimally invasive procedures [1]. According to [1], these continuum robots can be categorized into several types, including cable-driven robots [2,3], concentric tubular robots [4,5], catheters [6,7], soft robots [8,9], flexible needles [10,11], fluid-operated robots [12,13], and shape memory alloy manipulators [14,15]. Creating a kinematic model of these flexible structures in 3D space enables real-time shape perception of these instruments. Arata et al. presented a four-degree-of-freedom flexible robotic arm in [16]. This robotic arm demonstrates high accuracy and repeatability, distinguishing it from previous designs. The authors successfully modeled its kinematics and verified its performance through elastodynamic methods. Subsequently, in [17], Camarillo and Loewke introduced a vision-based method for real-time 3D shape perception of flexible robots. Their method uses sampling point measurements to improve the precision of 3D shape estimation, contrasting with sensor-based methods commonly used in traditional rigid robotic arms. The correctness of their results was verified through experiments under open-loop control. Loutfi et al. evaluated the performance of four different learning models for the forward kinematics of multi-segment continuous manipulators within the framework of forward kinematics, drawing conclusions for various scenarios in [18].

Recently, a neural network-based method has been presented to solve the forward and inverse kinematics issue, accounting for non-negligible mass and elasticity in [19].

Simulation experiments were performed in a redundant constraint plane and a minimum constraint space to compare their method with conventional approaches. In [20], Godage et al. developed a kinematic model for a multi-segment continuous arm based on mode shape functions (MSF). They successfully applied modal approximation techniques to create a new kinematic model for a generalized variable-expansion, multi-segment continuum robot arm. It illustrates the model's capability to simulate both spatial bending and straight-arm motions, introducing inverse kinematics for multi-segment continuum systems. An approach for reconstructing the 3D shape of minimally invasive surgical devices has been introduced to enhance the navigation of interventional tools in [21]. Their method integrates physics-based simulation with a nonlinear Bayesian filter, where the physics-based model predicts the shape, and the filter uses external image features to refine the navigation model.

In [22], an actor-critic design has been presented to address tracking control challenges in jointed robots with uncertainties. Neural networks were utilized to handle uncertainty, allowing for the approximation of motion control in these elastic systems. In [23], an approach for tracking and shape reconstruction of wire-driven flexible robots has been designed. Their approach utilizes a third-order Bessel curve-based algorithm, which reconstructs the robot's shape using position and orientation data acquired from an electromagnetic sensor. The method demonstrated excellent tracking performance and high accuracy in shape reconstruction. Zou et al. proposed an algorithm for 3D shape reconstruction of soft robotic arms based on neural networks in [24]. Their method employs a deep neural network (DNN) to correct the continuum robot's kinematics through precise visual estimation. The performance of the trained DNN was evaluated on both test sets and real-time bending deformation experiments. The results demonstrated that the DNN can accurately and stably learn the 3D shape of the robotic arm. Recently, an approach utilizing soft e-textile resistive sensors was proposed in [25], offering seamless integration with the robot's structure. A deep Convolutional Neural Network (CNN) is employed to decode sensor signals, enabling precise shape estimation and control. Unlike the notable results achieved with e-textile sensors in [25], this work focuses on control based on visual feedback.

In this paper, we adopt a neural network-based kinematic modeling approach for shape estimation. First, the flexible continuum's centerline is extracted from the camera using a contour extraction algorithm. The 3D coordinates of this centerline are then reconstructed using a binocular camera system. These coordinates provide a dataset that describes essential shape information, which is used to train two neural networks. Finally, these networks are trained to accurately estimate and control the shape of the robotic arm.

The remainder of this paper is organized as follows. Section 2 introduces the edge centerline extraction (ECE) algorithm. In Section 3, neural networks are applied to model the input-output relationship of the instrument. Simulation and experimental results are presented in Section 4. Finally, Section 5 offers conclusions based on the findings.

2. ECE Algorithm

2.1. Pre-Process Image

First, color thresholding segmentation is performed in the HSV color space, employing color information to eliminate irrelevant objects quickly. Next, a joint bilateral filter is applied to eliminate residual noise, ensuring the retention of critical contour edge details. Following this, the Canny operator is employed to extract image contours, producing a binarized contour map and identifying all contour point coordinates.

2.2. Self-Organizing Map Algorithm for Contour Centerline Extraction

Self-organizing map (SOM) is a neural network model based on unsupervised learning. It is particularly useful for generating a low-dimensional, discrete map by learning from input data, making it a form of dimensionality reduction. Unlike traditional neural networks that rely on backpropagation to minimize loss functions, SOMs employ competitive learning, where neurons compete to adjust their weights. This approach helps maintain the topology of the input space by using a nearest-neighbor function, as illustrated in Algorithm 1.

Algorithm 1 ECE Algorithm

```
Function:   Centerline Extraction (Image, Contour Point)
Initialize SOM
For k = 0,1,2,...,T
      Select a contour point D(t) randomly
      Identify the distance in the SOM D(t) nearest neuron v. Denote it by u
    For all neurons in the SOM
        Update weights using (1)
    End
End
Extract neuron weights W as centroids
Output W.
```

The SOM architecture consists of two layers: the input layer and the output layer. The number of neurons in the input layer is determined by the dimensionality of the input vector. In this case, because the input features consist of pixel coordinates (x, y), the input layer is represented in two dimensions. The output layer, or competitive layer, can be either one-dimensional or two-dimensional. A one-dimensional structure forms a line, while a two-dimensional structure creates a grid. For this task, where the goal is to extract a centerline contour (a line), we adopt a one-dimensional structure with 10 neurons to fit the centerline [26,27]:

$$W_u(k+1) = W_u(k) + \theta(u, v, k)\alpha(k)(D(t) - W_u(k)) \tag{1}$$

where k is the step index, v is the index of the node. The weights of the node at position u are represented as $W_u(k+1)$ at the next training step and $W_u(k)$ at the current training step. $\theta(u, v, k)$ is the neighborhood function, with a bubble function applied in this case. The learning rate $\alpha(k)$ is given by:

$$\alpha(k) = \frac{lr}{1 + \frac{k}{T}}. \tag{2}$$

Here, lr is the learning rate and T is half of the total training time. $D(t)$ is a target input data vector.

2.3. *Curve Fitting*

After training the network, keypoints are extracted to aid in reconstructing the centerline shape. These keypoints are used to generate a smooth curve through interpolation or fitting techniques. In control theory, the curves generated by PID controllers tend to be smoother. By adjusting the damping ratios, we can achieve different curve effects, which we applied to the spline curves. A two-dimensional PID controller is employed, where the coordinates of the keypoints serve as the control objects, and the positions of the pixel points act as the inputs. A time axis is introduced to describe the states of the points over time. The spline curve is then constructed by projecting the points across all time steps onto the image.

Starting from the initial state defined by the first keypoint, a reference input is generated by interpolating between the first and second keypoints using time-based weighting. As the control point moves closer to the second keypoint or reaches the threshold time, interpolation shifts to the second and third keypoints to produce the next reference input. This process repeats iteratively until the control point approaches the final keypoint or the threshold time limit is reached. By mapping the control points at each time step onto the image, the resulting spline curve is obtained. The results of centerline extraction and spline curve fitting for six different bending states are presented in Figure 1.

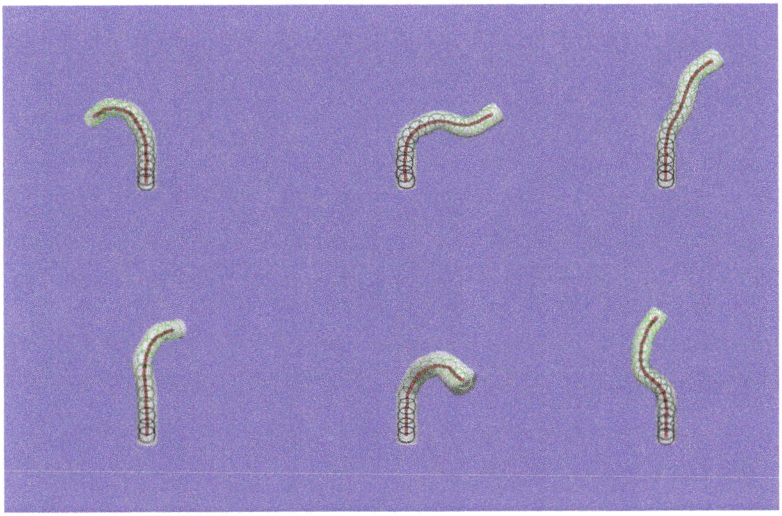

Figure 1. Centerline extraction and spline curve fitting results. The green circles are the extracted center points of the robot.

2.4. 3D Reconstruction Using a Binocular Camera

Since depth information is not directly available, we utilize binocular camera matching to determine the 3D coordinates. This is achieved by applying the ray intersection algorithm, using the intrinsic camera parameters and the corresponding centroids identified in the two images. The centroid's world coordinates are calculated through the following transformation:

$$T_{world} = T_{cam} + Q_{cam} \times P_{cam}. \quad (3)$$

Here, T_{world} represents the centroid's coordinates in the world coordinate system (WCS). T_{cam} denotes the position of the camera focal point's position in WCS, Q_{cam} is the matrix that represents the camera's orientation in WCS, and P_{cam} represents the coordinates of the centroid in the camera coordinate system. Mapping the irradiance of robots in space onto the corresponding pixels on the screen, using the perspective transformation matrix, we can project the vertex positions of each triangle in the model onto their corresponding positions on the camera screen, simulating the process of a binocular camera capturing an image.

3. Neural Networks

Using contour centerline extraction and 3D reconstruction with binocular cameras, we obtain spatial data on the instrument's tip shape under various motor inputs. To achieve a functional mapping between the complete set of motor inputs and the corresponding instrument shapes, we develop a model. Given the intricate nature of this input-output relationship, influenced by numerous factors, traditional modeling approaches prove insufficient. Neural networks, known for their strong generalization capabilities and ability to fit complex functions, provide a more effective solution [28–30]. In this paper, we train a neural network on a data subset to effectively model the surgical instrument's input-output dynamics.

3.1. The Selection of Neural Networks

We utilize two neural network models for different tasks: one to interpret the instrument's shape based on the current motor input and another to regulate the shape by modifying the motor input to achieve a desired shape.

The first network, referred to as the encoder, takes the motor input as its input and outputs a mathematical representation of the instrument's shape. This network consists of 6 layers with 32 neurons in each hidden layer. The hidden layers use the ELU activation function, and linear activation function is applied to the output layer. Key hyperparameters include a seed value of 1234, a batch size of 4, a learning rate of 0.001, and a total of 500,000 training iterations. The second network, referred to as the decoder, takes as input a mathematical representation of the instrument's shape and outputs the motor input. This network consists of six layers, with a hidden layer size of 96 neurons. The ELU activation function is used in the hidden layers, and linear activation function is applied to the output layer. The model is trained using a seed value of 1234, a batch size of 4, a learning rate of 0.001, and for 500,000 iterations.

3.2. Process Dataset

An essential factor in training a neural network is the quality of the dataset provided. A well-prepared dataset is important for reducing data coupling, preserving key features, and optimizing the network's training process.

Data Preprocessing. Once the shape recognition algorithm identifies the centerline of the instrument, appropriate mathematical transformations can be applied, followed by feature extraction from the raw data. To decouple the data as much as possible, algorithms should be designed to reduce its dimensionality, enhancing the overall efficiency of the data. The relative positions between joints are inherently coupled, meaning that a shift in a parent joint affects its child. However, the relative postures between joints remain independent of one another. To capture this information, we use quaternions to extract and store relative posture data for each joint. We first connect the origin of the coordinate system to the root node to establish the initial vector, and then iteratively connect the center points from the root node to the end node, forming the remaining $N-1$ vectors. Using a forward kinematics algorithm, we can derive the relationship between the equipment's posture and its relative position and posture, leading to the following equation:

$$T_{child} - T_{parent} = Q_{parent} P_{offset} \tag{4}$$

$$R_{child} = Q_{parent}^{-1} Q_{child} \tag{5}$$

where T_{child} is the world coordinate of the child joint, T_{parent} is the world coordinate of the parent joint, Q_{parent} is the absolute pose of the parent joint, Q_{child} is the absolute pose of the child joint, P_{offset} is the difference between the initial world coordinates of the child joint and the parent joint, and R_{child} is the relative pose of the child joint and the parent joint.

Physically, this relationship represents the rotation of the child joint around the parent joint, encompassing three degrees of freedom. This can be further broken down into two components: a two-degree-of-freedom rotation, which transitions the current direction vector to the target direction vector (referred to as rotational) [31], and a one-degree-of-freedom spin around the final direction vector (referred to as orientational). While rotation influences both the position and orientation of the child joint, orientation affects only the joint's posture.

After obtaining the center point coordinates, we acquire the position data for all joints. The joint and its associated posture in three different bending states are shown in Figure 2. However, we still lack the pose information necessary to determine the relative orientations of these joints accurately. Since each joint's movement is constrained, the three rotational degrees of freedom are effectively reduced to two, focusing solely on rotational components. As shown in (4), there is a defined mathematical relationship between the current coordinate difference (i.e., vector) between parent and child joints, the initial coordinate difference, and the absolute pose of the parent joint. Given that only rotational components are involved, the pose can be treated as a rotational transformation. The rotation axis is derived from the cross product of the two vectors, while the dot product yields the rotation angle. We then use the axis-angle relationship to convert this information into a quaternion pose representation. From (5), we can derive the relative orientations of each joint from their absolute orientations, which are represented and computed using quaternion algorithms.

Figure 2. Joint and its associated posture in three different bending states.

Data normalization. Prior to inputting data into the network, it should be normalized by adjusting its mean to 0 and standard deviation to 1. This is done by calculating the overall mean and standard deviation of the dataset, then subtracting the mean from each data point and dividing by the standard deviation.

Data regularization. To mitigate overfitting, we introduce noise into the data, with its range determined by the standard deviation of the input data. While this added noise enhances the network's generalization capability, excessive noise can hinder convergence.

Once the network is trained, the output data are scaled back to its original form by multiplying by the standard deviation and adding the mean.

4. Simulation Environment Construction and Experiments

Before conducting physical tests, we verify the algorithm's feasibility in a simulation environment. This approach, implemented on a computer, minimizes reliance on physical setups and streamlines the testing process. The simulation environment is a simplified representation of the physical world, reducing noise and creating near-ideal conditions. Once the algorithm performs successfully in this virtual setting, it can then be considered for implementation in the physical environment.

4.1. Skeletal Animation Environment

In computer graphics, geometric models are typically depicted as surfaces composed of multiple triangles. These triangles store essential information, such as vertex coordinates, normals, and other attributes that define the geometry's surface characteristics.

The model's motion is defined by the movement of its vertices, which are bound to a rigid skeleton structure. The root of the skeleton has properties that include spatial position and orientation, while the rest of the skeleton is defined by the relative position and rotation of each segment compared to its parent segment. The relative positions are described by three-dimensional vectors, while orientations and rotations are expressed using quaternions. The motion of the skeleton is governed by forward kinematics, allowing for the iterative calculation of the spatial position and orientation of each skeletal segment. Vertices are connected to bones and move according to the bones' transformations. If a vertex is influenced by multiple bones, its motion is determined through weighted interpolation of the transformations from each bone.

To begin, a cylinder model is constructed, with multiple sequentially bound bones. Each node on the cylinder is assigned a weight according to their influence by the bones, and their behavior is tested across various angles. For smooth deformation of the flexible structure, weights should be assigned to each vertex appropriately. However, manual weight adjustment becomes impractical due to the large number of vertices. Therefore, we develop an automated algorithm specifically for weight assignment in the flexible body model.

1. Compute the distances from each vertex to all skeletal root nodes and arrange them in ascending order based on proximity.
2. For each vertex, identify the four closest skeletal root nodes and identify their corresponding parent joints. Compute the projection of the vector from the parent joint to the vertex onto the vector from the parent joint to its child. Then, find the ratio of this projection to the relative distance between the parent and child nodes.
3. Derive the influence weight of each vertex based on the calculated vector ratio with respect to the parent joint.
4. Scale the four weights simultaneously to ensure that their total equals 1.

$$f(x) = \begin{cases} e^{-\frac{(x-0.5)^2}{1.5}}\,(x > 0.5) \\ e^{-\frac{(x-0.5)^2}{0.25}}\,(x < 0.5). \end{cases} \tag{6}$$

4.2. Image Acquisition

The model of the instrument is loaded into the rendering environment, where camera settings, lighting, and other parameters are configured before initiating the rendering process. The obtained image is shown in Figure 3.

We simulate motor inputs using a controller's two joysticks, with each joystick's X and Y axes representing the inputs for separate motors. This setup enables control of four motors, simulating the deformation of the flexible continuum. The inputs from the controller influence the posture of multiple joints through pre-defined functional relationships, accurately modeling the continuum's behavior.

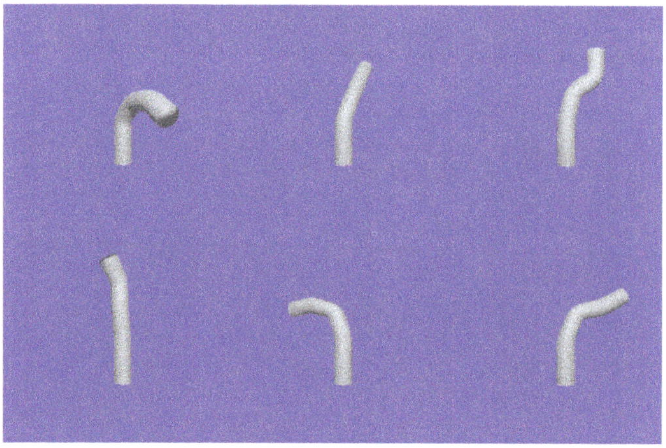

Figure 3. Rendered image of the instrument model.

Once extraction is complete, a binocular 3D reconstruction algorithm is applied to the center points, yielding the 3D spatial coordinates. After generating substantial raw data using the shape recognition algorithm within the renderer, the datasets are processed according to the method described in Section 3.2 and stored as binary files. These datasets are subsequently used to train both the encoder and decoder networks, following the network parameters outlined in Section 3.1.

4.3. Network Training Variance

We performed multiple rounds of network training on the dataset, adjusting the level of noise introduced in each round to develop a stable encoder-decoder model. The results of the training process are presented in Figure 4, where the blue line represents the actual values and the yellow line indicates the fitted values. The figure demonstrates that the neural network effectively captures and fits the complex structures within the data.

Before verifying the model in the simulation environment, we evaluate its performance by calculating the mean square error (MSE) using the verification set. The network's behavior under various noise inputs is summarized in Table 1.

Table 1. MSE and actual variance of neural networks across various noise levels during training.

Noise	Network Name	MSE	Variance
0.1	encoder	0.28037	0.28059
0.33	encoder	0.30727	0.29850
0.5	encoder	0.32699	0.30925
1.0	encoder	0.35881	0.34083
0.1	decoder	0.35031	0.49787
0.33	decoder	0.41572	0.52816
0.5	decoder	0.45248	0.52723
1.0	decoder	0.42029	0.52670

For a comprehensive evaluation, we selected an encoder with a noise level of 0.25 and a decoder with a noise level of 1.0 for participation in the following experiments.

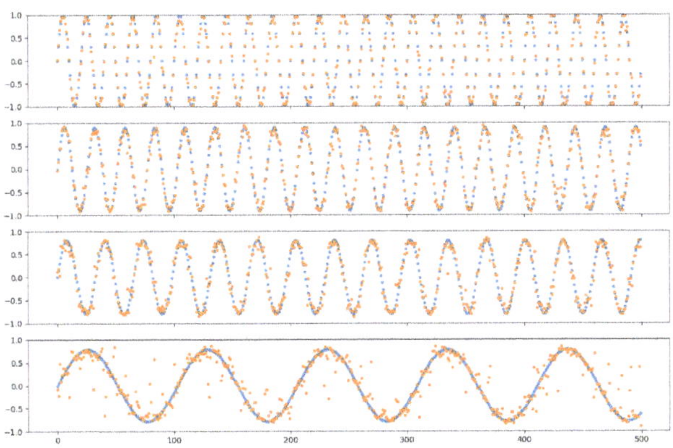

(a)

Figure 4. *Cont.*

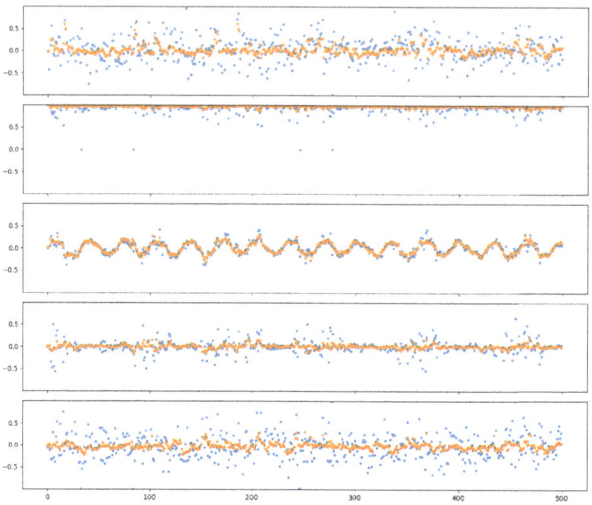

(b)

Figure 4. Example of Neural Network Fitting Results: (**a**) Motor Input Fitting: The horizontal axis represents sampling time, while the vertical axis represents input values. The four motor inputs are shown from top to bottom, with yellow indicating actual values and blue indicating fitted values. (**b**) Joint Pose Fitting (Quaternion): The horizontal axis denotes sampling time, and the vertical axis indicates component size. From top to bottom, the components of the quaternion (x, y, z, w) are displayed, with yellow representing actual values and blue representing fitted values.

4.4. Result Analysis

After completing the network training, we integrated it into the simulation model for verification. To verify the performance of the shape recognition algorithm and assess the quality of the network training, we separately evaluated the encoder and decoder networks.

First, we validated the encoder network by confirming that it can directly compute the corresponding shape output from motor inputs, effectively achieving shape estimation analogous to solving forward kinematics. By manipulating the control handle, we influenced the relative positions of the joints based on the functional relationships defined in the simulation model, inducing deformations in the flexible continuum and obtaining the actual shape output. Simultaneously, the encoder network processed the input signal to calculate the joints' relative positions, which were then mapped onto the bone joints to generate the simulation output. As illustrated in Figure 5, the left side represents the shape output from the simulation model, while the right side shows the shape predicted by the encoder network. Overall, the predicted shape closely approximates the actual shape, though certain errors are observed in the bending of certain joints. Notably, the prediction error increases when the instrument shape features a high degree of curvature.

To rigorously assess the prediction accuracy, we calculated the average joint deviation between the predicted results and the actual results in the validation dataset. As illustrated in Figure 6, the horizontal axis represents the sampling time, while the vertical axis shows the average joint deviation at each time step. The results demonstrate a strong correlation between the predicted outcome and the input signal, although the model's performance degrades when predicting highly bent shapes.

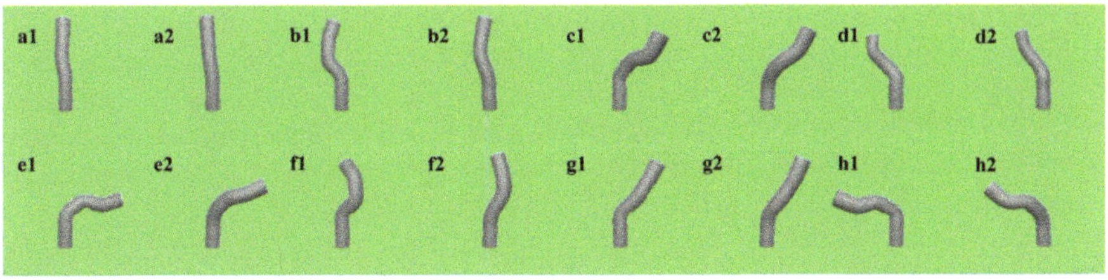

Figure 5. Shape Prediction Results of 8 Bending States: The left side (**a1–h1**) displays the actual shape, while the right side (**a2–h2**) presents the shape predicted by the network.

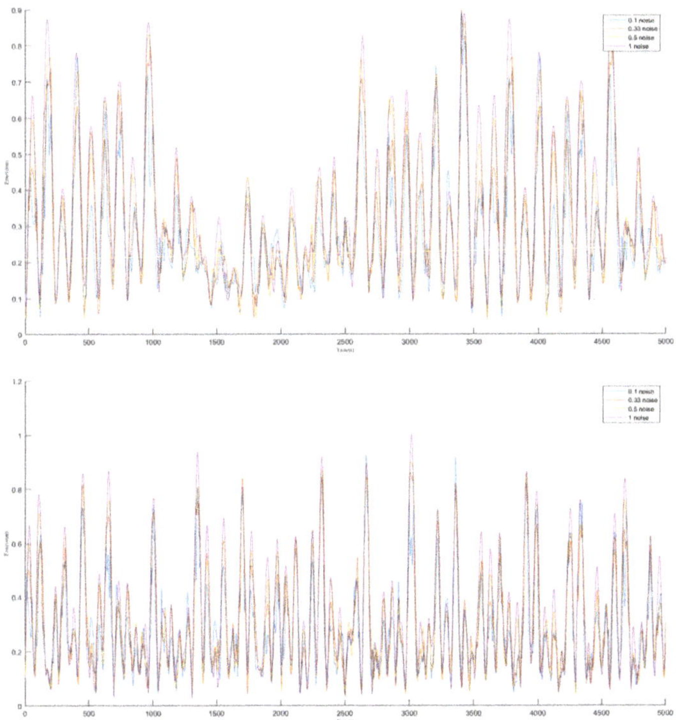

Figure 6. The average joint deviation for Validation Dataset 1 and Validation Dataset 2. The horizontal axis represents the sampling time points, while the vertical axis indicates the corresponding average joint deviation. The lines in different colors represent the network results under various noise training conditions.

Next, we validate the decoder network to confirm its ability to calculate appropriate motor inputs from shape information, effectively addressing an inverse kinematics problem. Given a specific target shape, the decoder generates corresponding handle input data, which is applied to a mathematical model to simulate the instrument's shape. The simulated shape is then compared to the target shape to evaluate accuracy. The results are illustrated in Figure 7, where the left panel displays the target shape and the right panel shows the shape produced using motor inputs generated by the decoder and controlled via the simulation model. To validate the effectiveness of shape control, we calculate the mean joint position error over a certain sampling time, and the results are shown in Figure 8.

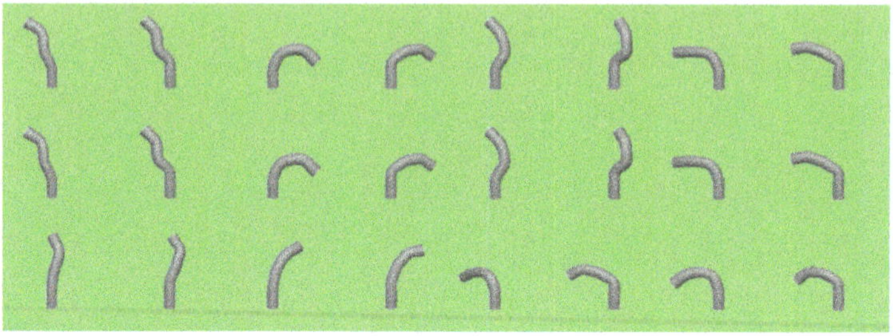

Figure 7. Shape control results with the actual shape displayed on the left and the reconstructed shape control results shown on the right.

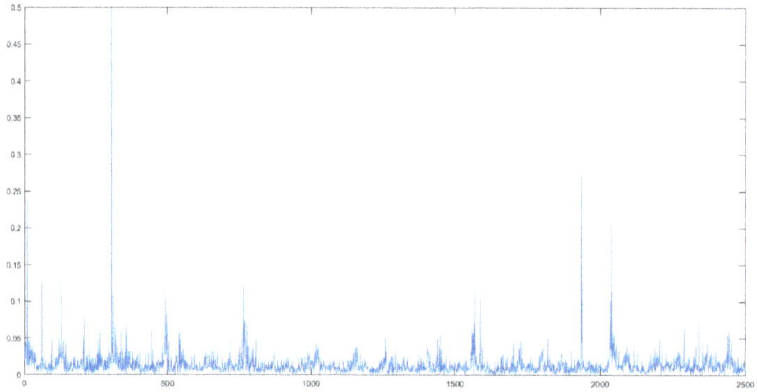

Figure 8. The shape control joint deviation curve.

During testing, we observed that the shape control algorithm performs more effectively under non-extreme bending conditions. To quantify this, we calculated the average joint position error over a specified sampling period, as shown in Figure 9. The horizontal axis represents the sampling time, and the vertical axis indicates the average joint deviation between the two shapes at each time point.

From Figures 7 and 9, we infer that the decoder network efficiently translates shape data into appropriate motor inputs for devices with geometries similar to those in the dataset, enabling accurate shape control.

To further evaluate the effectiveness of this shape control, we conducted two sets of experiments. In the first set, the method in [31] was used to directly manipulate the joints of flexible objects, guiding them to the desired positions in a simulation environment. The shape and orientation of these objects were then fed into the network, which generated motor inputs to control the flexible object. We compared the joint positions obtained using the method in [31] with those achieved by the network, as shown in Figure 10.

In the second set of experiments, we focused on reconstructing flexible objects in a real-world environment. Utilizing a binocular camera, we captured the shape of the flexible object and reconstructed the 3D coordinates of key points based on the camera's position and orientation. These coordinates were then used to compute the object's shape, which was input into the decoder network to generate the corresponding analog handle inputs. The reconstruction was finalized through a simulation-based mathematical model.

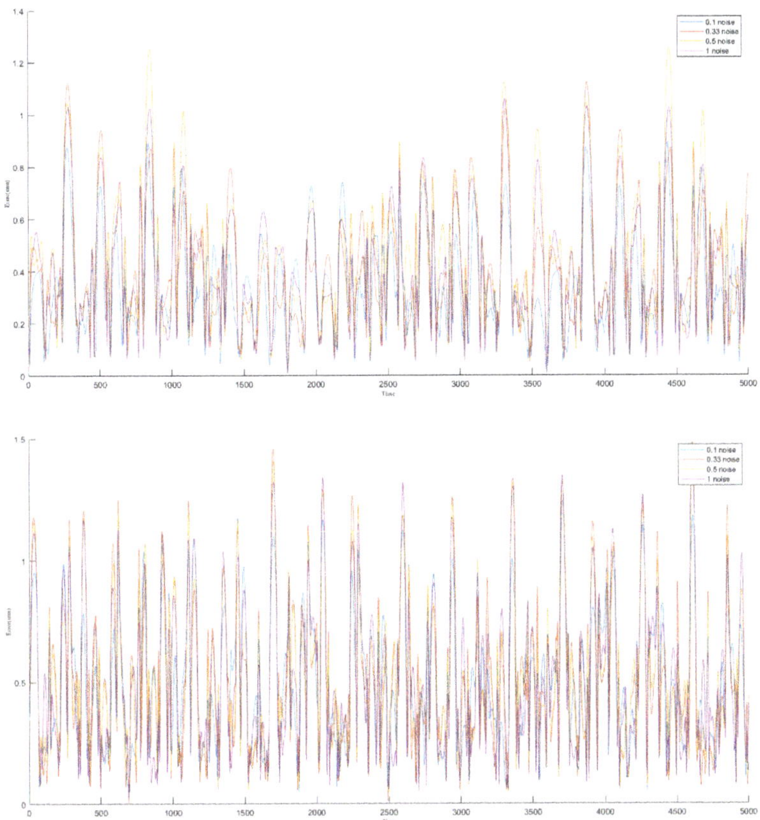

Figure 9. The average joint deviation for Validation Dataset 1 and Validation Dataset 2. The horizontal axis represents the sampling time points, while the vertical axis indicates the corresponding average joint deviation. The lines in different colors represent the network results under various noise training conditions.

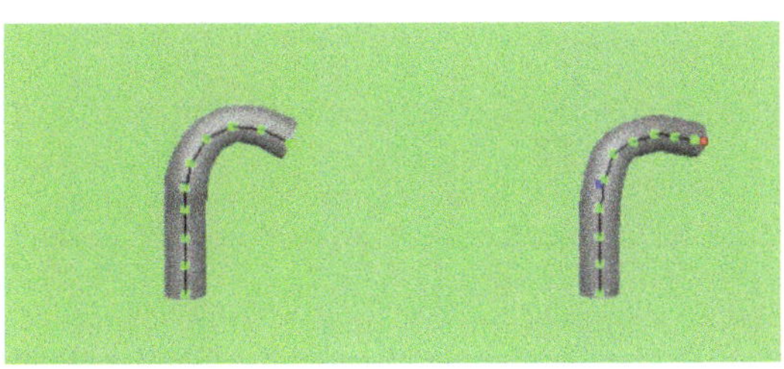

Figure 10. *Cont.*

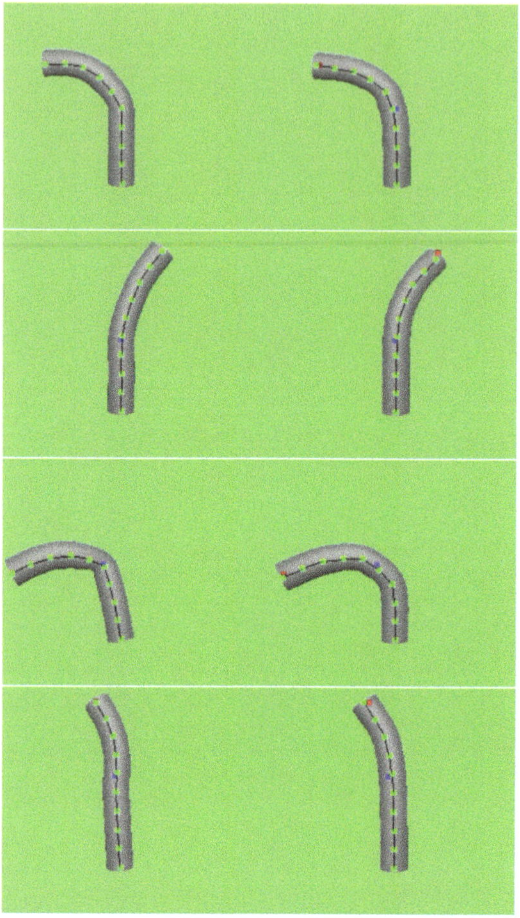

Figure 10. The results from the method in [31] are compared with our reconstruction. The red square represents the expected position of the object's end joint, while the blue square represents the expected position of the object's middle joint. On the left, we show the outcome using the method from [31] to control the joints of the flexible object, and on the right, we present the result after the network resolves the left posture.

As shown in Figure 11, the shape recognition and control algorithm successfully met its initial objectives. For a given instrument shape, we decoded the necessary motor inputs through the decoder network. By using these inputs, we can control the flexible surgical instrument to achieve the desired shape. In this manner, the decoder network enabled precise shape control of the flexible continuum.

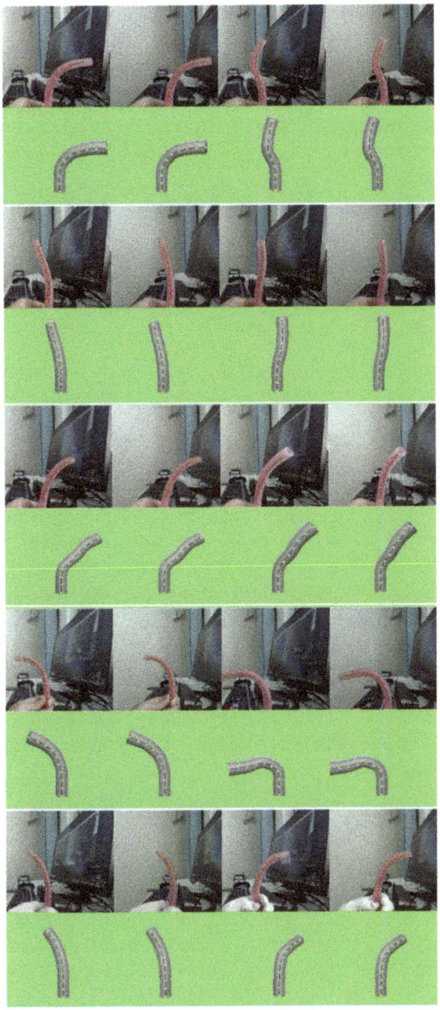

Figure 11. Real-world Shape Control Results: The shape of the flexible device in the real-world environment is replicated in a simulated environment, with its form controlled by a network-driven system.

5. Conclusions

This study introduces a neural network designed to tackle the challenge of end shape control for flexible surgical instruments, focusing on two key aspects: perception and control. The training data are derived from a specially developed simulation environment, with significant features identified through a shape recognition algorithm. To assess its effectiveness, the neural network is tested in practical applications, showcasing the potential of learning-based approaches for managing the shape of flexible surgical tools.

Our shape recognition algorithm performs effectively in cases involving slight to moderate instrument bending and is also capable of addressing more severe bending situations. By utilizing a trained encoder network, motor inputs are translated into shape outputs, ensuring precise shape estimation within the range of the training dataset. In addition, the network's predictions provide valuable insights for input data beyond the training set. The decoder network, in turn, derives the motor inputs from the shape output, enabling precise control over the instrument's shape. When the instrument's shape deviates only slightly from the training data, the network successfully controls the device to achieve the desired shape.

In extreme bending scenarios, shape recognition becomes particularly challenging due to the significant deformations objects undergo. Utilizing topological invariants and curvature analysis may help preserve recognition accuracy, though this approach lies beyond the scope of the current paper and represents an avenue for future research. In addition, integrating data from multiple sensors—such as visual, haptic, or 3D scanning technologies—can improve shape recognition and will be a key focus of our future work.

Author Contributions: Conceptualization, P.L.; validation, Y.D. and S.Z.; formal analysis, Y.D.; investigation, Y.D. and S.Z.; writing—original draft preparation, Y.D.; writing—review and editing, W.C. and P.L.; supervision, W.C. and P.L.; funding acquisition, W.C. and P.L. All authors have read and agreed to the published version of the manuscript.

Funding: This research was partly funded by the Natural Science Foundation of China under Grant 62073099 and Shenzhen Science and Technology Program JSGG20211108092806010 and JSGG20220831-104403007.

Institutional Review Board Statement: Not applicable.

Data Availability Statement: Data are contained within the article.

Conflicts of Interest: Author Wei Cheng was employed by the company Devol Advanced Automation, Inc. The remaining authors declare that the research was conducted in the absence of any commercial or financial relationships that could be construed as a potential conflict of interest.

References

1. Sahu, S.K.; Sozer, C.; Rosa, B.; Tamadon, I.; Renaud, P.; Menciassi, A. Shape reconstruction processes for interventional application devices: State of the art, progress, and future directions. *Front. Robot. AI* **2021**, *8*, 758411. [CrossRef] [PubMed]
2. Tamadon, I.; Soldani, G.; Dario, P.; Menciassi, A. Novel robotic approach for minimally invasive aortic heart valve surgery. In Proceedings of the 2018 40th Annual International Conference of the IEEE Engineering in Medicine and Biology Society (EMBC), Honolulu, HI, USA, 18–21 July 2018 ; IEEE: Piscataway, NJ, USA, 2018; pp. 3656–3659.
3. Abdelaziz, S.; Esteveny, L.; Barbé, L.; Renaud, P.; Bayle, B.; de Mathelin, M. Development of a MR-compatible cable-driven manipulator: Design and technological issues. In Proceedings of the 2012 IEEE International Conference on Robotics and Automation, St Paul, MN, USA, 14–18 May 2012; IEEE: Piscataway, NJ, USA, 2012; pp. 1488–1494.
4. Gafford, J.B.; Webster, S.; Dillon, N.; Blum, E.; Hendrick, R.; Maldonado, F.; Gillaspie, E.A.; Rickman, O.B.; Herrell, S.D.; Webster, R.J. A concentric tube robot system for rigid bronchoscopy: A feasibility study on central airway obstruction removal. *Ann. Biomed. Eng.* **2020**, *48*, 181–191. [CrossRef] [PubMed]
5. Girerd, C.; Rabenorosoa, K.; Rougeot, P.; Renaud, P. Towards optical biopsy of olfactory cells using concentric tube robots with follow-the-leader deployment. In Proceedings of the 2017 IEEE/RSJ international conference on intelligent robots and systems (IROS), Vancouver, BC, Canada, 24–28 September 2017; IEEE: Piscataway, NJ, USA, 2017; pp. 5661–5887.
6. Murai, E.H.; Homer-Vanniasinkam, S.; Silveira, P.G.; Dai, J.S.; Martins, D.; Wurdemann, H.A. Towards a modular suturing catheter for minimally invasive vascular surgery. In Proceedings of the 2018 IEEE International Conference on Robotics and Automation (ICRA), Brisbane, Australia, 21–25 May 2018; IEEE: Piscataway, NJ, USA, 2018; pp. 44–49.
7. Iacovacci, V.; Ricotti, L.; Sinibaldi, E.; Signore, G.; Vistoli, F.; Menciassi, A. An intravascular magnetic catheter enables the retrieval of nanoagents from the bloodstream. *Adv. Sci.* **2018**, *5*, 1800807. [CrossRef] [PubMed]
8. De Falco, I.; Cianchetti, M.; Menciassi, A. A soft multi-module manipulator with variable stiffness for minimally invasive surgery. *Bioinspiration Biomim.* **2017**, *12*, 056008. [CrossRef] [PubMed]
9. Ranzani, T.; Gerboni, G.; Cianchetti, M.; Menciassi, A. A bioinspired soft manipulator for minimally invasive surgery. *Bioinspiration Biomim.* **2015**, *10*, 035008. [CrossRef] [PubMed]
10. Culmone, C.; De Falco, I.; Menciassi, A.; Dankelman, J.; van den Dobbelsteen, J. A variable stiffness mechanism for minimally invasive surgical needles. In Proceedings of the The Hamlyn Symposium on Medical Robotics, London, UK, 25–28 June 2017; Volume 7.
11. Engh, J.A.; Podnar, G.; Khoo, S.Y.; Riviere, C. Flexible needle steering system for percutaneous access to deep zones of the brain. In Proceedings of the IEEE 32nd Annual Northeast Bioengineering Conference, Easton, PA, USA, 1–2 April 2006; IEEE: Piscataway, NJ, USA, 2006; pp. 103–104.
12. Shiva, A.; Stilli, A.; Noh, Y.; Faragasso, A.; De Falco, I.; Gerboni, G.; Cianchetti, M.; Menciassi, A.; Althoefer, K.; Wurdemann, H.A. Tendon-based stiffening for a pneumatically actuated soft manipulator. *IEEE Robot. Autom. Lett.* **2016**, *1*, 632–637. [CrossRef]

13. Sozer, C.; Paternò, L.; Tortora, G.; Menciassi, A. Pressure-driven manipulator with variable stiffness structure. In Proceedings of the 2020 IEEE International Conference on Robotics and Automation (ICRA), Paris, France, 31 May–31 August 2020; IEEE: Piscataway, NJ, USA, 2020; pp. 696–702.
14. Shi, Z.Y.; Liu, D.; Wang, T.M. A shape memory alloy-actuated surgical instrument with compact volume. *Int. J. Med Robot. Comput. Assist. Surg.* **2014**, *10*, 474–481. [CrossRef] [PubMed]
15. Kadir, M.R.A.; Dewi, D.E.O.; Jamaludin, M.N.; Nafea, M.; Ali, M.S.M. A multi-segmented shape memory alloy-based actuator system for endoscopic applications. *Sensors Actuators A Phys.* **2019**, *296*, 92–100. [CrossRef]
16. Arata, J.; Fujisawa, Y.; Nakadate, R.; Kiguchi, K.; Harada, K.; Mitsuishi, M.; Hashizume, M. Compliant four degree-of-freedom manipulator with locally deformable elastic elements for minimally invasive surgery. In Proceedings of the 2019 International Conference on Robotics and Automation (ICRA), Montreal, QC, Canada, 20–24 May 2019; IEEE: Piscataway, NJ, USA, 2019; pp. 2663–2669.
17. Camarillo, D.B.; Loewke, K.E.; Carlson, C.R.; Salisbury, J.K. Vision based 3-D shape sensing of flexible manipulators. In Proceedings of the 2008 IEEE International Conference on Robotics and Automation, Pasadena, CA, USA, 19–23 May 2008; IEEE: Piscataway, NJ, USA, 2008; pp. 2940–2947.
18. Loutfi, I.M.; Boutchouang, A.B.; Melingui, A.; Lakhal, O.; Motto, F.B.; Merzouki, R. Learning-based approaches for forward kinematic modeling of continuum manipulators. *IFAC-PapersOnLine* **2020**, *53*, 9899–9904. [CrossRef]
19. Chawla, I.; Pathak, P.M.; Notash, L.; Samantaray, A.K.; Li, Q.; Sharma, U.K. Neural network-based inverse kineto-static analysis of cable-driven parallel robot considering cable mass and elasticity. In Proceedings of the International Conference on Cable-Driven Parallel Robots, Nantes, France, 7–9 July 2021; Springer: Berlin/Heidelberg, Germany, 2021; pp. 50–62.
20. Godage, I.S.; Guglielmino, E.; Branson, D.T.; Medrano-Cerda, G.A.; Caldwell, D.G. Novel modal approach for kinematics of multisection continuum arms. In Proceedings of the 2011 IEEE/RSJ International Conference on Intelligent Robots and Systems, San Francisco, CA, USA, 25–30 September 2011; IEEE: Piscataway, NJ, USA, 2011; pp. 1093–1098.
21. Trivisonne, R.; Kerrien, E.; Cotin, S. Constrained stochastic state estimation of deformable 1D objects: Application to single-view 3D reconstruction of catheters with radio-opaque markers. *Comput. Med. Imaging Graph.* **2020**, *81*, 101702. [CrossRef] [PubMed]
22. Ouyang, Y.; Dong, L.; Wei, Y.; Sun, C. Neural network based tracking control for an elastic joint robot with input constraint via actor-critic design. *Neurocomputing* **2020**, *409*, 286–295. [CrossRef]
23. Song, S.; Li, Z.; Yu, H.; Ren, H. Shape reconstruction for wire-driven flexible robots based on Bézier curve and electromagnetic positioning. *Mechatronics* **2015**, *29*, 28–35. [CrossRef]
24. Zou, S.; Lyu, Y.; Qi, J.; Ma, G.; Guo, Y. A deep neural network approach for accurate 3D shape estimation of soft manipulator with vision correction. *Sens. Actuators A Phys.* **2022**, *344*, 113692. [CrossRef]
25. Galeta, E.V.; Nada, A.A.; Hameed, I.; El-Hussieny, H. Curvature Sensing and Control of Soft Continuum Robots Using e-Textile Sensors. *Appl. Syst. Innov.* **2024**, *7*, 84. [CrossRef]
26. Kohonen, T. The self-organizing map. *Proc. IEEE* **1990**, *78*, 1464–1480. [CrossRef]
27. Kohonen, T. Essentials of the self-organizing map. *Neural Netw.* **2013**, *37*, 52–65. [CrossRef] [PubMed]
28. Gurney, K. *An Introduction to Neural Networks*; CRC Press: Boca Raton, FL, USA, 2018.
29. Wang, G.; Wang, C.; Li, L.; Du, Q. Distributed adaptive consensus tracking control of higher-order nonlinear strict-feedback multi-agent systems using neural networks. *Neurocomputing* **2016**, *214*, 269–279. [CrossRef]
30. Wang, G. Distributed control of higher-order nonlinear multi-agent systems with unknown non-identical control directions under general directed graphs. *Automatica* **2019**, *110*, 108559. [CrossRef]
31. Aristidou, A.; Lasenby, J. FABRIK: A fast, iterative solver for the Inverse Kinematics problem. *Graph. Model.* **2011**, *73*, 243–260. [CrossRef]

biomimetics

MDPI

Article

Magnetically Driven Quadruped Soft Robot with Multimodal Motion for Targeted Drug Delivery

Huibin Liu [1], Xiangyu Teng [1], Zezheng Qiao [1], Wenguang Yang [1,*] and Bentao Zou [2,*]

[1] School of Electromechanical and Automotive Engineering, Yantai University, Yantai 264005, China; cactusil@163.com (H.L.); 15166828596@163.com (X.T.); qiaojingbiao6656@163.com (Z.Q.)
[2] Engineering Training Center, Yantai University, Yantai 264005, China
* Correspondence: yangwen-guang@ytu.edu.cn (W.Y.); zou53168@ytu.edu.cn (B.Z.)

Abstract: Untethered magnetic soft robots show great potential for biomedical and small-scale micromanipulation applications due to their high flexibility and ability to cause minimal damage. However, most current research on these robots focuses on marine and reptilian biomimicry, which limits their ability to move in unstructured environments. In this work, we design a quadruped soft robot with a magnetic top cover and a specific magnetization angle, drawing inspiration from the common locomotion patterns of quadrupeds in nature and integrating our unique actuation principle. It can crawl and tumble and, by adjusting the magnetic field parameters, it adapts its locomotion to environmental conditions, enabling it to cross obstacles and perform remote transportation and release of cargo.

Keywords: quadruped soft robot; magnetically driven; multimodal motion; targeted drug delivery

1. Introduction

As medical standards advance and technology progresses, biomedicine is increasingly moving towards standardized procedures and targeted drug delivery [1,2]. In this context, soft robots exhibit significant potential due to their compact size, flexible manipulation, and programmable operation [3,4]. Because of their reduced size, soft robots are usually powered by external energy sources, such as magnetic, ultrasonic, electric, optical, or chemical fields [5–9]. Magnetic soft robots have been widely studied because the low-intensity magnetic fields used to drive them are harmless to biological cells and tissues and they offer programmable manipulation capabilities [2,10].

In terms of a single structure, current magnetic soft robots are mainly divided into two categories: continuum and untethered soft robots. Compared to continuum structures, untethered magnetic soft robots show higher flexibility in terms of maneuvering and workspace. Moreover, the untethered design allows for truly noninvasive treatments to a certain extent [2,11]. Zhao et al. proposed a magnetic robot kinematics mechanism based on magnetic moment-induced asymmetric friction effects, which enables a bipedal robot to maneuver flexibly at a speed of 25.33 BL/s fine in advance and to achieve a loading capacity of up to four times its own weight [12]. Du et al. designed a magnetic fish-shaped microrobot with a multifunctional tail, which is highly adaptable to avoid obstacles, as well as pass through a 45 μm wide channel, and morph and change colors in a complex environment with a height of 2 mm [13]. Hu et al. designed millimeter-scale magnetic microrobots with multiple motion modes that can roll, jump, swim, crawl, and transport goods two times their own weight under the manipulation of an externally programmed magnetic field [14]. Soon et al. designed a double-layer soft robot with programmable deformation inspired by the pangolin, which was able to achieve >70 °C heating at a distance of >5 cm in <30 s. It proved to be effective in cargo release, in situ demagnetization, thermotherapy, and hemostasis experiments on isolated tissues [15]. Although the aforementioned footless or biped designs present excellent multimodal locomotion in unstructured environments, developing a magnetically driven quadruped soft robot that walks like quadrupeds still faces many challenges [16,17]. In microrobotics, quadruped structures offer more stable support and locomotion for soft robots, enhancing both their operational accuracy and their ability to avoid obstacles and adapt to various environments [18,19]. Therefore, it is important to study the design of magnetic quadruped soft robots with strong locomotion and operational capabilities for their potential biomedical applications.

In this work, we present a bionic magnetic quadruped soft robot for intestinal-targeted drug delivery (Figure 1). This robot consists of a magnetic top cover and four flexible feet, each with specific magnetization directions. Its magnetic responsiveness is derived from N52 NdFeB magnetic particles, and it possesses two modes of motion, crawling and tumbling, under the manipulation of a programmed magnetic field. Moreover, the combination of these two motion modes enables the transportation and release of cargo to a specified location. In addition, we analyzed the motion principle and motion model of the quadruped structure and verified the motion characteristics of its two motion modes, its ability to cross obstacles, and its ability to transport and release cargo through experiments. The theoretical analysis and experimental results show that our magnetically driven quadruped soft robot has multi-modal motion capability, which provides a typical case for the design and research of magnetic quadruped microrobots. Moreover, the excellent motion characteristics and manipulation operation capability are expected to be applied to targeted drug delivery.

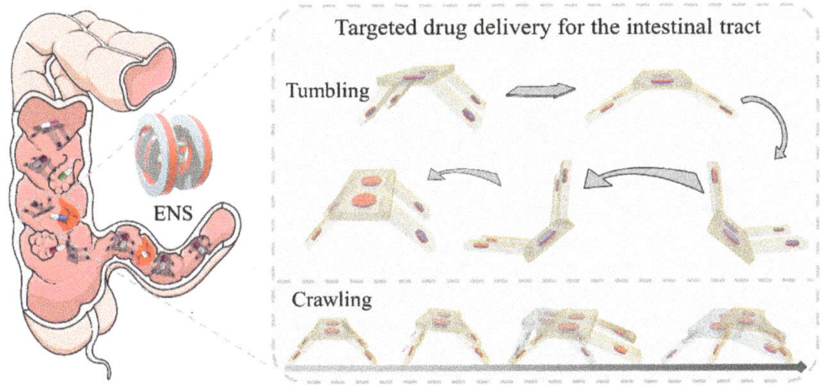

Figure 1. Schematic representation of the two motion modes and targeted drug delivery of a magnetically driven quadruped soft robot.

2. Material and Methods

2.1. Design and Fabrication of Magnetically Driven Quadruped Soft Robot

Inspired by the way crabs move in nature, we designed a magnetically driven quadruped soft robot that mimics the movement of crabs. This robot consists of a magnetic top cover and four feet, each with specific magnetization directions. Both the top cover and the feet are composed of polydimethylsiloxane (PDMS) and N52 NdFeB magnetic particles. First, the PDMS colloid was formulated according to the ratio (A:B: 10:1), and the molds of the top cover and the feet were prepared using a high-precision 3D printer (Figure 2A). The A and B formulations of PDMS before mixing and curing are clear fluids. After homogeneous mixing, the PDMS prepolymer remains a clear solution with a viscosity of 4000 at 25 °C room temperature. Next, the N52 magnetic particles were placed in the designated positions in the mold and poured into the formulated PDMS colloid for heating and curing. Modally stable PDMS flakes were obtained after curing at 80 °C ambient for 1 h or at 100 °C ambient for 30 min. The position height of the magnetic pellet was slightly adjusted during the curing process so that it was completely encapsulated by the PDMS. At this stage, the sheet exhibited a hardness of 50 Shore A, a tensile strength of 7 MPa, and an elongation at break of 140%.

After curing and cooling to room temperature, the flexible hydrogel of the top cover and foot were carefully detached from the mold. According to the designed dimensions of the mold, the magnetic top cover was prepared with a length of 18 mm, a width of 8 mm, and a thickness of 0.9 mm. The four magnetic feet were prepared with the same dimensions, specifically with a length of 11 mm, a width of 4 mm, and a thickness of 0.9 mm. In addition, the N52 magnetic particles for the top cover were cylindrical, with a diameter of 4 mm and a thickness of 0.5 mm. For the feet, the N52 magnetic particles were also cylindrical but with a diameter of 2 mm and a thickness of 0.5 mm. As shown in Figure 2B, the four flexible feet were arranged with a specific magnetization direction, ensuring that each foot was positioned at a 45° angle relative to the top cover and the magnetic particles were oriented in different directions. The feet were then bonded to the top cover using PDMS.

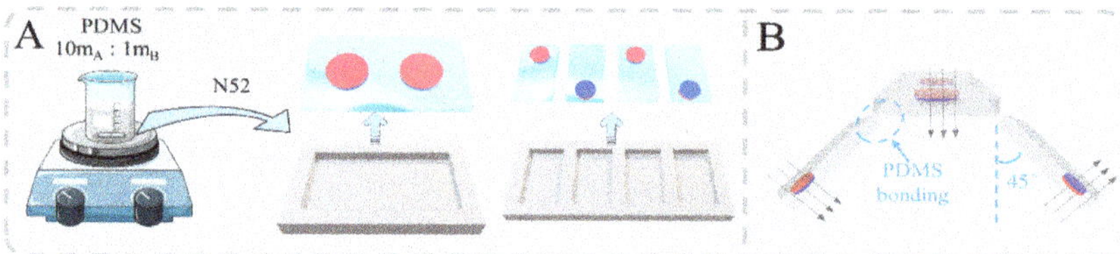

Figure 2. Preparation and assembly of magnetic quadruped soft robot. (**A**) Preparation of magnetic quadruped soft robot. (**B**) Assembly of magnetic quadruped soft robot.

2.2. Magnetic Field Properties and Magnetic Drive Bending Model

Compared to NdFeB and Fe_3O_4 magnetic powders, N52 NdFeB magnetic particles have a stronger magnetic field response. The magnetic particles used in this study were purchased from Shenzhen Lala Magnetic Materials Development Co., Ltd., Shenzhen, China. They were measured using a vibrating sample magnetometer (VSM) and the obtained material data were processed, as shown in Figure 3A. The N52 grains we used have a remanent magnetization (Mr) of 182 emu/g and a coercivity (Ms) of 209 emu/g. The grains have excellent remanent magnetization and coercivity, enabling them to generate a strong driving torque in response to manipulation by a uniformly strong magnetic field. In addition, we used a programmable and highly maneuverable 3D Helmholtz coil electromagnetic field as an electromagnetic navigation system (ENS). In the uniformly strong field generated by the electromagnetic coil, the deformation of the magnetic top cover and foot was driven by torque to achieve ordered motion. Figure 3B,C illustrate the magnetic field distribution during the operation of the ENS. The simulations across multiple sections and work planes reveal that the magnetic field is uniformly distributed in the three-dimensional operation space along a certain direction. Moreover, by adjusting the current signal of the coil, it is possible to establish a uniform magnetic field in any direction in space.

The specifically assembled robot exhibits different deformations of the joints in each magnetization direction in response to manipulation by the ENS. However, since the two legs on the same side have the same magnetization direction, their deformations when facing the same homogeneous magnetic field are also the same. As shown in Figure 3D, when the ENS is off, the assembled robot relies on the support of the quadruped to maintain a smooth attitude. When the ENS is turned on and provides a uniformly strong field to the left, the top cover, the left two feet, and the right two feet deform as shown in Figure 3D. This deformation results from the coupling of the magnetic moment and the deformation between the constituent structures. Based on such ordered deformation characteristics, we can program the direction and magnitude of the magnetic field in the ENS to manipulate the motion of the robot.

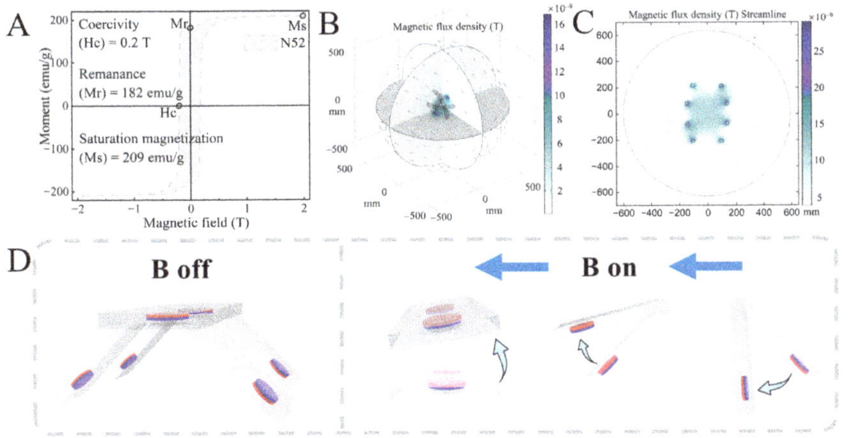

Figure 3. Material parameters and magnetic field properties and magnetically actuated deformation of magnetic quadruped soft robot. (**A**) Material parameters of N52 NdFeB magnetic particles. (**B**) Simulation of ENS in COMSOL with multi-cut magnetic field distribution. (**C**) Simulation of ENS in COMSOL with magnetic field distribution in the work plane. (**D**) Deformation effect of the robot in response to ENS actuation.

2.3. Two Motion Modes for Magnetically Driven Quadruped Soft Robot

Here, we manipulated the robot using two different signals of the drive field to realize two motion modes, tumbling and crawling. As shown in Figure 4A(a), the current signal for manipulating the robot's tumbling motion consists of triangular waves on the *X*-axis and *Z*-axis. Through superposition, a uniform magnetic field of varying size and direction is generated in the ENS operation space. The tumbling motion occurring in the robot driven by this uniform field can be decomposed into five deformation steps as described in ①–⑤. The deformation steps correspond to the signal diagrams. First, in the ①–② process, the magnetic field in the XZ plane, angled diagonally downward, causes the left and right quadrupeds to extend outward, and the robot's overall attitude is downward. Then, in the ②–③ process, the magnetic field direction is reversed and gradually intensified, causing the robot to undergo tumbling deformation. Next, in the ③–④ process, the magnetic moment gradually decreases and, in the coupling with gravity, the top cover of the robot gradually makes contact with the support plane. Finally, in the ④–⑤ process, the magnetic field direction is again reversed and gradually intensified, completing the tumbling deformation and restoring the initial attitude under the dual response of the feet and the top cover. Figure 4B illustrates the deformation process of the robot, showing how it undergoes tumbling motion as manipulated by this signal in the experiment. Based on different motion postures, the manipulated deformation of the tumbling motion is realized by adjusting the size and direction of the magnetic field in the operation space. Moreover, the observed physical deformations are consistent with our theoretical analysis that the tumbling process of magnetic quadruped microrobots includes four steps of attitude downward pressure, lateral tumbling, and back to the correct attitude. In addition, subsequent experiments verified that its motion characteristics can be adjusted by changing the magnetic field strength and frequency. In addition to tumbling motion, the robot can also be manipulated to perform a crawling motion by adjusting the drive signal. As shown in Figure 4A(b), a single *X*-axis drive signal drives the robot to mimic the crawling motion of a crab. As with the tumbling motion, the top cover and quadruped are magnetically bent and deformed during crawling. It should be noted that the magnetic field strength is reduced by adjusting the size of the input current, so that the top cover responds with less deformation, which is insufficient to induce a tumbling motion. At this time, the top cover exhibits the deformation characteristic of lifting up at a small angle under the drive of the ENS. Additionally, coupled with the bending deformation of the four feet, different deformations of the left and right feet are realized by decreasing or increasing the support force of the feet. By combining these motion features and programming the control signals, a crab-like crawling motion is realized. First, the left side of the top cover is lifted to promote the contraction of the left two feet while inhibiting the movement of the right two feet. Then, the magnetic field is reversed and the right side of the top cover is lifted to promote the extension of the right two feet and inhibit the movement of the left two feet. By repeating the above deformation processes and combining them, an effective movement similar to crab crawling is realized. Again, we verified this analysis in our experiments. As shown in Figure 4C, the deformation effects of the feet and the top cover during the crawling motion in the experiment are basically consistent with our analysis. Based on different motion postures, the manipulation of the crawling motion is realized by adjusting the size and direction of the magnetic field in the operation space. And the observed physical deformation is consistent with our theoretical analysis that the magnetic quadruped microrobots' crawling process includes four steps of posture down, alternating unilateral warping, and returning to the correct posture. We also explored how variations in magnetic field strength and frequency affect the crawling motion characteristics in subsequent experiments.

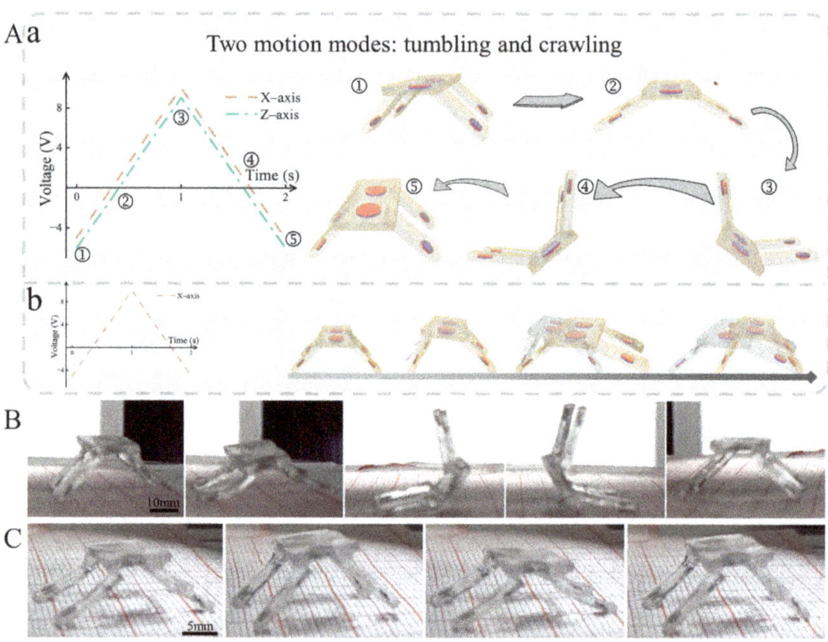

Figure 4. Manipulation of two motion modes. (**A**) Manipulation signal and motion decomposition diagrams for two motion modes, (a) tumbling and (b) crawling. (**B**) Experimental screenshot of tumbling motion. (**C**) Experimental screenshot of crawling motion.

3. Results and Discussion

3.1. Deformation Response of the Magnetically Driven Quadruped Soft Robot

The effect of magnetic field strength on the bending deformation of the robot was investigated. As shown in Figure 5A, the strength of the external DC magnetic field gradually increases while the robot is flipped and fixed on a flat surface. We also observed and counted the bending deformation angle of the foot. The deformation angle of the foot in the fixed attitude gradually increases with the gradual increase in the external magnetic field strength. The quadruped deformation angle of the robot gradually increases from the initial 45° to 55°. Moreover, the robot quadruped deformation angle is positively correlated with the magnetic field strength. It should be noted that the robot in the natural attitude may change its attitude due to the coupled magnetic response of the magnetic top cover, which, in turn, affects the deformation angle of the foot. In addition, to investigate the effect of magnetic field strength on the attitude of the robot, we applied a gradually increasing magnetic field to the robot in its natural position. We then observed and counted the robot's attitude as well as the deformation angle of the top cover at each magnetic field strength. As shown in Figure 5B, the extension inclination angle of the quadruped gradually increases with higher magnetic field strength, resulting in a downward compression of its attitude. Moreover, the magnetic top cover responds to the magnetic field drive and deforms under the action of torque to produce a certain inclination angle. With the increase in the inclination angle, the tumbling motion characteristic eventually appears. Furthermore, due to the coupling effect of the robot's gravity and magnetic moment, the magnetic top cover establishes a specific angular relationship with the support plane. As the magnetic moment increases, the coupled attitude aligns more closely with the direction of the magnetic field. In general, the motion attitude of magnetic quadruped microrobots is positively correlated as a function of magnetic field strength. And the magnetic quadruped microrobots' motion characteristics and attitude can be controlled by adjusting the magnetic field strength and direction. The magnetic field of the ENS is generated by an electromagnetic coil, and its strength can be controlled by adjusting the input current. The relationship between the input current and the magnetic field strength of the 3D Helmholtz coil triaxial is illustrated in Figure 5C. By adjusting the size and direction of the triaxial magnetic field in space, a uniform magnetic field of any desired direction and size can be synthesized. In addition, the statistical data on how the magnetic field strength affects the robot were organized and are presented in Figure 5D. From the intersection of the data in the figure, it can be seen that the robot used in the experiment tumbled and deformed, i.e., changed its attitude, at a magnetic field strength of 11.5 mT. At this strength, the magnetic field's impact on the quadruped's deformation also changes. Moreover, after the experimental test, it can be obtained that, under the action of the field strength at this transition point, 9 motion modal transitions occurred in 10 motion controls, and the stability is as high as 90%. Recognizing this will aid in analyzing the control and transformation between the robot's two motion modes: crawling and tumbling. This will also help us further analyze how the magnetic field frequency affects the motion speed of the two motion modes in the corresponding magnetic field strength interval.

Since the motion of magnetic quadruped microrobots is mainly obtained by deforming the PDMS sheet wrapped with N52 magnetic particles, it is necessary to examine the deformation effect of the PDMS sheet under various ambient temperatures. Accordingly, only if the PDMS sheet maintains good deformation performance, the designed programmed motion can be realized. The material specification shows that the cured PDMS maintains stable mechanical properties in the temperature interval of −50 to 200 °C. In addition, its thermal conductivity is 0.17 $W/(m \cdot K)$, which has a certain temperature resistance and can better meet the operational requirements. Combined with the characteristics of the operating environment, the PDMS sheet still maintains good elasticity and deformation ability under the environment of −4 °C after experimental testing. Moreover, since the PDMS sheet is cured by heating, it maintains stable mechanical properties even under a high-temperature environment. Therefore, the PDMS sheet for responding to magnetic torque deformation has good operational stability at various ambient temperatures.

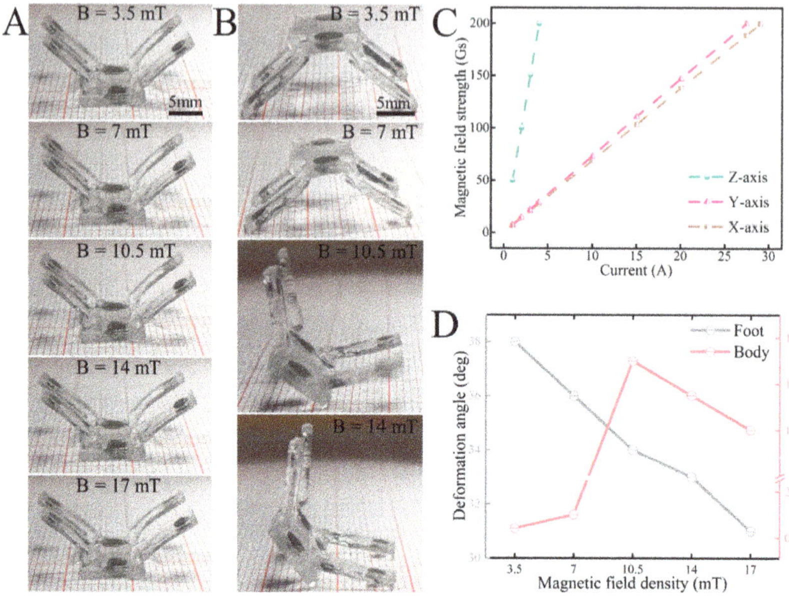

Figure 5. Deformation response of the magnetically driven quadruped soft robot. (**A**) Bending response of the robot's feet. (**B**) Response of the robot to tumbling deformation. (**C**) Conversion of magnetic field input current of solenoid coil versus magnetic field strength. (**D**) Driving effect of magnetic field strength on foot bending and top cover deformation.

3.2. Kinematic Characterization of the Magnetically Driven Quadruped Soft Robot

Previous analyses and experiments indicate that the magnetic quadruped soft robot may experience tumbling attitude changes under a magnetic field strength of 11 mT. Therefore, in this section, we set the maximum magnetic field strength for crawling at 9 mT and the minimum for tumbling at 11 mT. We then further investigated the motion characteristics of these two motion modes by adjusting the magnetic field frequency and strength within the intervals. As shown in Figure 6A, we statistically analyzed how the magnetic field frequency affects the motion speed of the magnetic quadruped soft robot within a magnetic field strength range of 4 mT to 9 mT. The figure shows that the magnetic quadruped soft robot exhibits lower motion velocity at all frequencies when the magnetic field strength is 9 mT. Observations during the experimental process reveal that this is due to the increased deformation angle of the magnetic top cover at larger magnetic field strengths. This greater deformation significantly reduces the stability and effectiveness of the robot's motion and can even cause unexpected deformation during tumbling. The robot also moves at the same lower speed at all frequencies when subjected to a 4 mT magnetic field strength. However, the difference is that the smaller magnetic field strength results in a smaller deformation angle of the magnetic top cover, which limits the bending deformation of the foot and thus leads to a lower movement speed. The motion characteristics of the robot at each frequency under a magnetic field strength of 5 mT to 8 mT are also affected by the deformation angle of the magnetic top cover and the degree of foot bending deformation. Overall, the robot exhibits better motion speeds when subjected to 7 mT and 8 mT magnetic field strengths. However, at 8 mT, the robot may tumble and deform when driven in a lower-frequency magnetic field, which can prevent it from completing the crawling motion. Through observation and analysis of the experimental phenomena, we have identified a compelling reason: at 8 mT, the magnetic top cover exhibits a larger deformation angle, increasing the likelihood of tumbling deformation. Under high-frequency magnetic fields, the timely reversal of moments prevents tumbling deformation caused by inertia. Conversely, at low magnetic field frequencies, tumbling deformation occurs due to inertia. Therefore, for the purpose of crawling motion drive in this study, a magnetic field strength of 7 mT is more appropriate.

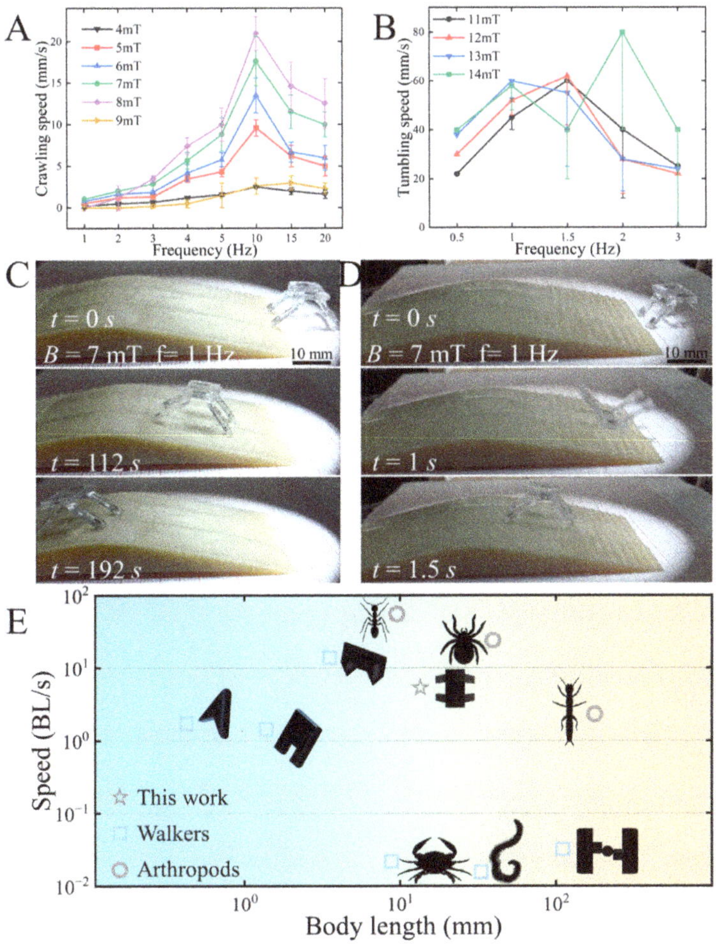

Figure 6. Kinematic characteristics of magnetically driven quadruped soft robot and its ability to traverse obstacles. (**A**) Effect of magnetic field strength and frequency on the robot's crawling kinematic speed. (**B**) Effect of magnetic field strength and frequency on the robot's tumbling kinematic speed. (**C**) The robot crawling through an obstacle. (**D**) The robot tumbling through an obstacle. (**E**) Performance comparison with reported robots.

In general, as the magnetic field frequency increases, the motion cycle time decreases, resulting in faster speeds. However, observing the robot speed versus frequency curve in the figure reveals that, as the frequency increases, the motion speed first increases and then appears to decrease. Observations during the experiment indicate that, when the frequency increases to a certain degree, the deformation and recovery speed of the robot cannot keep pace with changes in the magnetic field frequency. Therefore, the stability and effectiveness of the motion decrease, leading to a natural reduction in speed. For this experimental phenomenon, we adopt the definition from other studies and refer to it as the cutoff frequency, i.e., the maximum frequency at which the robot can match changes in the magnetic field. Clearly, the cutoff frequency for the robot's crawling motion in our study is 10 Hz. Through experimental analysis, we can further optimize the motion effect by choosing the appropriate magnetic field strength and frequency within the parameter interval of the crawling motion mode according to the operational requirements.

In addition, we conducted a statistical analysis of how the magnetic field frequency affects the robot's tumbling motion mode within varying intervals of magnetic field strength. As shown in Figure 6B, we tested the tumbling motion speed of the robot at each magnetic field frequency within the magnetic field strength varying from 11 mT to 14 mT. It can be seen that the motion speed of the robot increases gradually with the increase in the magnetic field strength. However, when the magnetic field strength reaches a certain degree, the motion speed of the robot shows little increase and even decreases at certain frequencies. Based on the experimental process, we found that the deformation of the robot's feet also increases as the magnetic field strength increases. However, as the bending of the foot increases, the bending moment increases dramatically, leading to a dramatic increase in the required magnetic moment, which, in turn, requires a stronger magnetic field strength. At this stage, the regular incremental increase in magnetic field strength is insufficient to provide the required magnetic moment.

The robot's speed is also affected by the magnetic field frequency, and we tested its movement speed across a frequency range of 0.5 to 3 Hz. As illustrated in the figure, the tumbling motion speed gradually increases as the magnetic field frequency increases. However, as the magnetic field frequency continues to increase, the speed of the tumbling motion starts to decrease. During the tumbling motion experiment, we observed a cutoff frequency phenomenon similar to that seen in the crawling motion. Additionally, due to the larger deformation involved in the tumbling motion, it is more sensitive to changes in the magnetic field frequency. According to the experimental data, the tumbling motions all showed a certain speed decrease at a magnetic field frequency of 1.5 Hz, with the speed continuing to decrease as the frequency increased. However, we also noticed that the robot reached a tumbling motion speed of 80 mm/s at 2 Hz under a magnetic field strength of 14 mT. Several experiments have confirmed that the robot does not exhibit a low tumbling motion speed. We found that, while higher magnetic field strength can cause instability in the tumbling motion, increasing the magnetic field frequency makes the acting time of the magnetic moment shorter. Instead of optimizing this problem, this results in an abnormal increase in speed.

Through an in-depth investigation of the motion characteristics of the robot, we identified and summarized the characteristic parameters of effective motion. There is a situation where different parameters can achieve the same control effect for these motion control parameters. For this situation, we take improving the stability of motion manipulation and reducing the accidental damage to human organs and tissues as the guideline and choose as much as possible a low magnetic field frequency (improve the stability of the motion) and a small magnetic field strength (avoid the unknown damage under the high field strength). Moreover, according to the operational requirements, we flexibly adjust the parameters of the control variables so as to achieve the best control effect. Building on these parameters, we conducted an experimental test to investigate how adjusting the robot's manipulation parameters according to its motion characteristics can enhance its ability to navigate through obstacles. The complete experimental procedure is detailed in Video S1, and Figure 6C,D provide representative screenshots of the experiments. As shown in Figure 6C, the robot can pass through a bridge-shaped environment with a certain slope using a crawling motion. When facing a sloped environment, its motion capability can be enhanced by adjusting the inclination angle of its crawling motion, based on the crawling motion characteristics analyzed earlier, so as to navigate the slope smoothly. However, while it can navigate a slope through crawling, it takes a considerable amount of time. Therefore, further exploration and optimization are needed to maintain a stable attitude and movement speed. On the other hand, the robot can traverse a ramp with a cross-grain in 1.5 s and return to its initial upright attitude (Figure 6D). In this case, the robot's locomotion capability and speed are greatly improved. However, the tumbling process implies a drastic change in attitude, which can negatively impact cargo stability during transport. Therefore, it is important to find a balance between maintaining cargo stability and achieving movement speed in the tumbling motion mode. As the magnetic quadruped microrobots can flexibly change the two motion modes of crawling and tumbling, therefore, the motion characteristics can be flexibly adjusted according to the operational requirements during the actual operation so as to meet the positioning accuracy and speed requirements.

After experimental testing and analytical verification, our designed magnetic quadruped microrobots have two freely switchable motion modes and excellent motion characteristics. In order to further highlight the advantages of the motion characteristics, we made a comparison with some typical studies that have been reported. We selected some typical walking untethered magnetic microrobots as well as the locomotion characteristics of arthropods in nature for comparison. The comparisons were made in terms of body size and locomotion speed, respectively, to visualize the advantages and characteristics of each magnetically driven microrobot design. For the data to be more informative, we use the body length and the number of moving body lengths per second as comparison parameters. As can be seen from the data in Figure 6E, the magnetic microrobots that currently exist include multiple orders of magnitude sizes ranging from 0.1 mm to 100 mm. Moreover, they have different motion characteristics based on design features, for example, triangular [20], portal [21], biconical [12], multipedal [22], crab [23], and bionic crab shapes [24] with walking ability. Among them, the magnetic quadruped bionic microrobots we designed have excellent locomotion characteristics by achieving the first gradient of locomotion speed while maintaining the more common body size. On the other hand, comparing with typical arthropods in nature, such as ants, spiders, web spinners, Hediste diadroma, etc., the magnetic quadrupedal microrobots still have comparable locomotion speeds and locomotion characteristics.

3.3. *Transportation and Release of Cargo by the Magnetically Driven Quadruped Soft Robot*

In the previous section, we explored and verified the kinematic properties and the obstacle-crossing ability of the designed robot. Based on this research and experimental observations, we designed in vitro simulated cargo transportation experiments. Considering the characteristics of the two motion modes, crawling and tumbling can be combined to effectively navigate a complex operating environment. First, the magnetic quadruped microrobots are maneuvered to the target location. Then, the magnetic quadrupedal microrobots' tumbling motion characteristics are utilized to adjust the control signal so that they tumble rapidly and continuously at the targeted drug delivery position to achieve drug release. After realizing the targeted drug delivery operation, the magnetic quadruped microrobots' tumbling motion control signal is restored. The magnetic quadruped microrobots are made to complete the tumbling motion, recover the attitude, and leave the operation area. As shown in Figure 7A, by adjusting the control signal of the electromagnetic coil and the magnetic field strength, the characteristic attitude of the tumbling motion can be changed, thus realizing the half-tumbling motion with left and right swaying. Once the cargo has been released, restoring the magnetic field parameters of the tumbling motion will enable the soft robot to recover its attitude. Based on this control strategy, we designed and carried out experiments for cargo transportation and release. The complete experimental procedure can be seen in Video S2, with representative screenshots provided in Figure 7B. As shown in Figure 7B, after crawling and tumbling to reach the target position, the robot performs a half-tumbling motion in response to the control signal of the tumbling motion (Target position). Then, the magnetic field strength and frequency are varied to perform a left–right swinging motion to release the transported cargo (Oscillation release). Finally, the tumbling motion is resumed to control the magnetic field, thus adjusting the robot's attitude to upright (Targeted delivery). This process combines the stability and the ability of the crawling motion to handle small displacements with the ability of the tumbling motion to manage large displacements and release cargo through oscillation. Based on this mode of operation, the ability to achieve work accuracy and movement speed is greatly enhanced. From the experimental observations, we successfully applied the magnetic quadruped soft robot for cargo transportation and release following the designed experimental steps. However, several areas require improvement, such as enhancing the stability during cargo transportation and addressing challenges related to cargo release. In the future, our goals include further reducing the robot's size, enhancing its movement ability based on its kinematic characteristics, and optimizing the methods for cargo transportation and release.

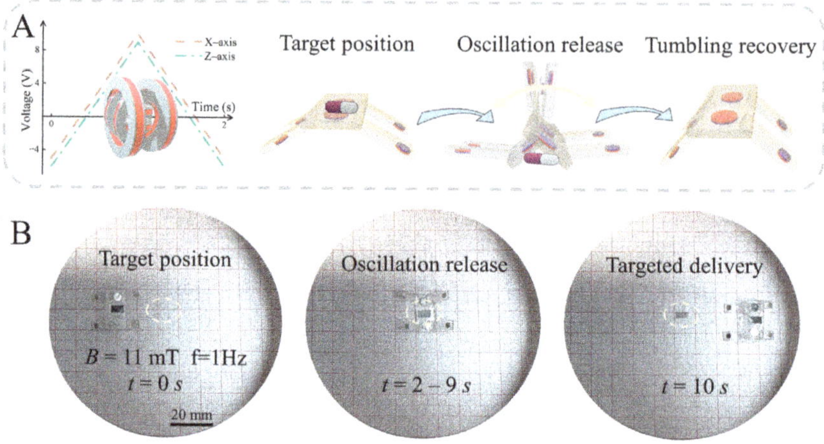

Figure 7. The magnetic quadruped soft robot transporting and releasing cargo. (**A**) Schematic diagram of the robot transporting and releasing cargo using tumbling and swinging motions. (**B**) Screenshots of experiments of the robot transporting and releasing cargo using tumbling and swinging motions.

4. Conclusions

In this work, we have introduced a bionic magnetic quadruped soft robot featuring a magnetic top cover and four magnetic flexible feet with specific magnetization angles. We introduced and analyzed the material parameters and magnetic field properties of the robot, including the hysteresis loop of magnetic particles and the deformation of magnetic flexible feet under magnetic field drive. After experimental testing, the overall structure maintains stable mechanical deformation properties in environments ranging from −4 °C to 80 °C. Then, we analyzed and verified through experiments two motion modes of the robot, crawling and tumbling, controlled by magnetic field parameters. In addition, we analyzed the effect of magnetic field strength on the motion modes of crawling and tumbling, considering their respective attitude characteristics. By adjusting the parameters of the control variables, the designed magnetic quadruped microrobots are able to achieve a motion speed of 80 mm/s. According to the experimentally obtained magnetic field strength interval, the stability of modal transition is as high as 90% and we further explore the motion characteristics of the two motion modes. Finally, we experimentally verified the robot's robust ability to traverse obstacles and transport and release cargo in a targeted manner and also demonstrated its effectiveness and promising application for tissue-targeted drug delivery in tissues such as the intestinal tract. Considering the operational requirements and issues identified in this study, our next step will be to further reduce the size of the robot, e.g., by preparing a magnetic microrobot body using an optical 3D printing system based on digital micromirror devices (DMD). Ferromagnetic compound particles are used instead of N52 magnetic particles, resulting in more flexible setting of the magnetization direction as well as shrinking the volume of the magnetic response layer. In addition, the structure design is further optimized to enhance the motion characteristics. Finally, according to the operational requirements, we try to increase the operational module child to optimize the operational process and enhance the application potential.

Supplementary Materials: The following are available online at https://www.mdpi.com/article/10.3390/biomimetics9090559/s1, Video S1: The robot crawling and tumbling through an obstacle. Video S2: The robot transporting and releasing cargo using tumbling and swinging motions.

Author Contributions: Conceptualization, H.L. and W.Y.; methodology, H.L.; software, X.T.; validation, X.T., Z.Q. and B.Z.; formal analysis, H.L.; investigation, H.L.; data curation, H.L.; writing—original draft preparation, H.L.; writing—review and editing, W.Y. and B.Z.; supervision, W.Y.; funding acquisition, W.Y. All authors have read and agreed to the published version of the manuscript.

Funding: This research was funded by the National Natural Science Foundation of China (Project No. 62273289), The Youth Innovation Science and Technology Support Program of Shandong Province (Project No. 2022KJ274), and Graduate Innovation Foundation of Yantai University, GIFYTU.

Institutional Review Board Statement: Not applicable.

Data Availability Statement: The data that support the findings of this study are available on request from the corresponding author.

Conflicts of Interest: The authors declare no conflicts of interest.

References

1. Go, G.; Yoo, A.; Nguyen, K.T.; Nan, M.; Darmawan, B.A.; Zheng, S.; Kang, B.; Kim, C.-S.; Bang, D.; Lee, S.; et al. Multifunctional microrobot with real-time visualization and magnetic resonance imaging for chemoembolization therapy of liver cancer. *Sci. Adv.* **2022**, *8*, eabq8545. [CrossRef] [PubMed]
2. Nauber, R.; Goudu, S.R.; Goeckenjan, M.; Bornhäuser, M.; Ribeiro, C.; Medina-Sánchez, M. Medical microrobots in reproductive medicine from the bench to the clinic. *Nat. Commun.* **2023**, *14*, 728. [CrossRef] [PubMed]
3. Iacovacci, V.; Diller, E.; Ahmed, D.; Menciassi, A. Medical Microrobots. *Annu. Rev. Biomed. Eng.* **2024**, *26*, 561–591. [CrossRef] [PubMed]
4. Liu, D.; Liu, X.; Chen, Z.; Zuo, Z.; Tang, X.; Huang, Q.; Arai, T. Magnetically Driven Soft Continuum Microrobot for Intravascular Operations in Microscale. *Cyborg Bionic Syst.* **2022**, *2022*, 9850832. [CrossRef] [PubMed]
5. Gardi, G.; Ceron, S.; Wang, W.; Petersen, K.; Sitti, M. Microrobot collectives with reconfigurable morphologies, behaviors, and functions. *Nat. Commun.* **2022**, *13*, 2239. [CrossRef]
6. Liang, X.; Chen, Z.; Deng, Y.; Liu, D.; Liu, X.; Huang, Q.; Arai, T. Field-Controlled Microrobots Fabricated by Photopolymerization. *Cyborg Bionic Syst.* **2023**, *4*, 0009. [CrossRef]
7. Abbasi, S.A.; Ahmed, A.; Noh, S.; Gharamaleki, N.L.; Kim, S.; Chowdhury, A.M.M.B.; Kim, J.-Y.; Pané, S.; Nelson, B.J.; Choi, H. Autonomous 3D positional control of a magnetic microrobot using reinforcement learning. *Nat. Mach. Intell.* **2024**, *6*, 92–105. [CrossRef]
8. Wang, M.; Wu, T.; Liu, R.; Zhang, Z.; Liu, J. Selective and Independent Control of Microrobots in a Magnetic Field: A Review. *Engineering* **2023**, *24*, 21–38. [CrossRef]
9. Hou, Y.; Wang, H.; Fu, R.; Wang, X.; Yu, J.; Zhang, S.; Huang, Q.; Sun, Y.; Fukuda, T. A review on microrobots driven by optical and magnetic fields. *Lab A Chip* **2023**, *23*, 848–868. [CrossRef]
10. Li, N.; Fei, P.; Tous, C.; Rezaei Adariani, M.; Hautot, M.-L.; Ouedraogo, I.; Hadjadj, A.; Dimov, I.P.; Zhang, Q.; Lessard, S.; et al. Human-scale navigation of magnetic microrobots in hepatic arteries. *Sci. Robot.* **2024**, *9*, eadh8702. [CrossRef]
11. Gao, Y.; Guo, Y.; Yang, Y.; Tang, Y.; Wang, B.; Yan, Q.; Chen, X.; Cai, J.; Fang, L.; Xiong, Z.; et al. Magnetically Manipulated Optoelectronic Hybrid Microrobots for Optically Targeted Non-Genetic Neuromodulation. *Adv. Mater.* **2024**, *36*, 2305632. [CrossRef] [PubMed]
12. Zhao, J.; Xin, C.; Zhu, J.; Xia, N.; Hao, B.; Liu, X.; Tan, Y.; Yang, S.; Wang, X.; Xue, J.; et al. Insect-Scale Biped Robots Based on Asymmetrical Friction Effect Induced by Magnetic Torque. *Adv. Mater.* **2024**, *36*, 2312655. [CrossRef] [PubMed]
13. Du, X.; Cui, H.; Xu, T.; Huang, C.; Wang, Y.; Zhao, Q.; Xu, Y.; Wu, X. Reconfiguration, Camouflage, and Color-Shifting for Bioinspired Adaptive Hydrogel-Based Millirobots. *Adv. Funct. Mater.* **2020**, *30*, 1909202. [CrossRef]
14. Hu, W.; Lum, G.Z.; Mastrangeli, M.; Sitti, M. Small-scale soft-bodied robot with multimodal locomotion. *Nature* **2018**, *554*, 81–85. [CrossRef]
15. Soon, R.H.; Yin, Z.; Dogan, M.A.; Dogan, N.O.; Tiryaki, M.E.; Karacakol, A.C.; Aydin, A.; Esmaeili-Dokht, P.; Sitti, M. Pangolin-inspired untethered magnetic robot for on-demand biomedical heating applications. *Nat. Commun.* **2023**, *14*, 3320. [CrossRef]
16. Huang, C.; Lai, Z.; Wu, X.; Xu, T. Multimodal Locomotion and Cargo Transportation of Magnetically Actuated Quadruped Soft Microrobots. *Cyborg Bionic Syst.* **2022**, *2022*, 0004. [CrossRef]
17. Hong, S.; Um, Y.; Park, J.; Park, H.-W. Agile and versatile climbing on ferromagnetic surfaces with a quadrupedal robot. *Sci. Robot.* **2022**, *7*, eadd1017. [CrossRef]
18. Li, Z.; Chen, L.; Ma, Y.; Weng, D.; Wang, Z.; Zhang, X.; Wang, J. Multi-physics coupling simulation and design of magnetic field-driven soft microrobots in liquid environments. *Int. J. Mech. Sci.* **2024**, *272*, 109136. [CrossRef]

19. Wang, Y.; Wang, B.; Zhang, Y.; Wei, L.; Yu, C.; Wang, Z.; Yang, Z. T-phage inspired piezoelectric microrobot. *Int. J. Mech. Sci.* **2022**, *231*, 107596. [CrossRef]
20. He, Y.; Dong, S.; Wang, L.; Rong, W.; Sun, L. Bipedal microwalkers actuated by oscillating magnetic fields. *Soft Matter* **2020**, *16*, 7927–7934. [CrossRef]
21. Li, J.; Wang, H.; Shi, Q.; Zheng, Z.; Cui, J.; Sun, T.; Ferraro, P.; Huang, Q.; Fukuda, T. Biped Walking of Magnetic Microrobot in Oscillating Field for Indirect Manipulation of Non-Magnetic Objects. *IEEE Trans. Nanotechnol.* **2020**, *19*, 21–24. [CrossRef]
22. Sun, D.; Zhang, J.; Fang, Q.; Xiang, P.; Xue, Y.; Wang, Y.; Xiong, R.; Lu, H. Analysis and control for a bioinspired multi-legged soft robot. *Biomim. Intell. Robot.* **2022**, *2*, 100030. [CrossRef]
23. Han, B.; Ma, Z.-C.; Zhang, Y.-L.; Zhu, L.; Fan, H.; Bai, B.; Chen, Q.-D.; Yang, G.-Z.; Sun, H.-B. Reprogrammable Soft Robot Actuation by Synergistic Magnetic and Light Fields. *Adv. Funct. Mater.* **2022**, *32*, 2110997. [CrossRef]
24. Li, Q.; Chen, T.; Chen, Y.; Wang, Z. An underwater bionic crab soft robot with multidirectional controllable motion ability. *Ocean Eng.* **2023**, *278*, 114412. [CrossRef]

biomimetics

MDPI

Article

Finite-Time Line-of-Sight Guidance-Based Path-Following Control for a Wire-Driven Robot Fish

Yuyang Mo, Weiheng Su, Zicun Hong , Yunquan Li * and Yong Zhong *

Shien-Ming Wu School of Intelligent Engineering, South China University of Technology, Guangzhou 511442, China
* Correspondence: yunquanli@scut.edu.cn (Y.L.); zhongyong@scut.edu.cn (Y.Z.)

Abstract: This paper presents an adaptive line-of-sight (LOS) guidance method, incorporating a finite-time sideslip angle observer to achieve precise planar path tracking of a bionic robotic fish driven by LOS. First, an adaptive LOS guidance method based on real-time cross-track error is presented. To mitigate the adverse effects of the sideslip angle on tracking performance, a finite-time observer (FTO) based on finite-time convergence theory is employed to observe the time-varying sideslip angle and correct the target yaw. Subsequently, classical proportional–integral–derivative (PID) controllers are utilized to achieve yaw tracking, followed by static and dynamic yaw angle experiments for evaluation. Finally, the yaw-tracking-based path-tracking control strategy is applied to the robotic fish, whose motion is generated by an improved central pattern generator (CPG) and equipped with a six-axis inertial measurement unit for real-time swimming direction. Quantitative comparisons in tank experiments validate the effectiveness of the proposed method.

Keywords: biomimetic robotic fish; path following; line-of-sight guidance law; directional tracking

1. Introduction

As the demand for marine exploration and development continues to grow, underwater robots have emerged as the preferred choice to replace human intervention to ensure safer and more stable underwater operations. Traditionally, underwater robots have relied primarily on propellers for propulsion. However, propeller-based propulsion systems have several limitations, such as high noise, low efficiency, poor yaw stability [1], and limited maneuverability, which make them unsuitable for certain mission areas. Fish, having evolved over millions of years, demonstrate remarkable underwater locomotion capabilities [2]. Leveraging advances in disciplines such as biology and robotics, researchers have developed bio-inspired robotic fish that mimic the propulsion mechanisms of real fish. Chen et al. introduce a tensioning integral robotic fish and provide a detailed description of its overall structure, stiffness, and mechatronics. The results indicate that the use of tensioning integral nodes can improve swimming performance [3]. A bionic robotic fish driven by ionic polymer–metal composite (IPMC) actuators is presented, and the effect of passive fins on swimming performance is investigated [4]. Michael S et al. [5] were the first to mimic the propulsion mechanism of fish in the presentation of an efficient bionic robotic fish. Du et al. [6] designed a novel tuna-inspired robot with rapid swimming and exceptional maneuverability. Dong et al. designed an innovative bionic gliding robotic fish, which has excellent gliding performance [7]. Liao et al. [8] propose a robotic fish consisting of multiple pulling wires to simulate muscles, as well as an elastic element to simulate the spine of the fish, achieving a swimming speed of 0.54 m/s. Zhong et al. presented a robotic fish featuring an active line-driven body combined with a soft compliant tail for efficient wave swimming, closely resembling real fish behavior, reaching a maximum speed of 2.15 body length/s [9]. These advancements in robotic fish technology demonstrate the potential to overcome limitations associated with traditional underwater propulsion methods.

Early research in the field of robotic fish has primarily focused on the structural design and implementation of locomotion. To further advance the practical applications of bio-inspired robotic fish, current efforts are directed toward enhancing their autonomy and intelligence, with the ultimate goal of improving their operational capabilities and cognitive functions. These endeavors encompass a range of tasks, including water exploration [10], autonomous obstacle avoidance [11,12], kinematic optimization [13], target tracking [14], path planning [15], and path following [16]. These studies are dedicated to developing control strategies that empower robotic fish to effectively handle diverse tasks in underwater environments. In practical applications, robotic fish can be applied to ocean monitoring, undersea operations, aquatic life observation, pollution search, etc. [17], showcasing its immense potential value. Among the various features, path following plays a critical role in enhancing the autonomy and intelligence of bio-inspired robotic fish. It enables them to efficiently perform tasks, such as patrolling and sampling, in underwater environments.

According to existing research, a typical path-following controller consists of two main components, namely a guidance subsystem and an inner-loop controller. The guidance subsystem commonly employs a line-of-sight (LOS) guidance law that aligns the fish's heading with the line of sight to achieve the desired position. This approach is straightforward and practical and has been widely applied in the control of unmanned surface vessels. Conversely, the inner-loop controller employs various control methods for angle control, such as sliding mode control, PID control, and model predictive control. For example, Liu et al. [18] used a fixed-radius LOS and an improved deep deterministic policy gradients (DDPG) algorithm to control a bio-inspired dolphin to follow a three-dimensional path. Chen et al. [16] employed traditional LOS to achieve real-time desired yaw and pitch angles and used a fuzzy PID controller for tail-joint control and a PID controller for pectoral fin attack-angle control to achieve bio-inspired three-dimensional spiral path tracking. However, the simulation results showed less than ideal tracking performance.

In practical applications, LOS, which relies solely on position error as feedback, may demonstrate reduced tracking performance in the presence of external disturbances. Furthermore, when the lateral tracking error of the robotic fish is zero, regardless of any errors along the path, the desired angle provided by the guidance system will align with the path angle, resulting in deviation from the intended path and consequent degradation of tracking performance. Fossen et al. [19] introduced an integral LOS (ILOS) to compensate for the sideslip angle caused by ocean currents. They utilized an adaptive approach to estimate and compensate for the sideslip angle and demonstrated the effectiveness of the proposed ILOS using the Lyapunov stability theory. In order to overcome the mutual constraint between lateral tracking error and along-track error in traditional LOS, Qu et al. [20] introduced an iterative mechanism to find the optimal path point within a single control cycle to ensure zero along-track error. Through theoretical analysis, an extended-state LOS (ELOS) was derived based on the variable distance to adaptively compensate for the current yaw-tracking error and sideslip angle. Experimental results demonstrated that ELOS surpasses traditional LOS and ILOS in terms of tracking performance. However, its convergence rate is slow, and its observer is complex. Dai et al. [21] proposed a novel path factor update law utilizing a logarithmic-type boundary-layer function (BLF) to limit the along-track error within a constrained range during a single control cycle, effectively addressing the issue of along-track error impeding the convergence of lateral tracking error. These innovative approaches are aimed at enhancing the tracking performance of bio-inspired robotic fish by compensating for the limitations of traditional LOS and mitigating the effects of external disturbances as well as mutual constraints between different tracking errors. However, it should be noted that, in certain cases, automatic compensation for errors was not achieved.

In summary, LOS shows great potential for path-tracking applications. However, traditional LOS methods for path tracking are limited by slow convergence, inadequate tracking accuracy, and significant interference from the sideslip angle of robotic fish. We work on solving the problems of traditional LOS for robotic fish path-tracking applications and the perturbation problems caused by the side-slip angle of robotic fish. Therefore, this study introduces an adaptive LOS (ALOS) control strategy that incorporates a compensation mechanism for the sideslip angle using a finite-time observer (FTO). The results of experiments conducted in a water tank validate the effectiveness of the proposed guidance strategy, showing faster convergence speed and lower steady-state error when compared to traditional guidance methods. The bio-inspired robotic fish achieves precise and stable path tracking, thus confirming the effectiveness of this closed-loop control strategy.

The paper makes two significant contributions:

(1) A dynamic look-ahead distance adjustment mechanism was adopted to improve the convergence speed and reduce the overshoot of the traditional LOS;
(2) An FTO is proposed to estimate and correct the sideslip angle, effectively solving the path deviation problem.

The remainder of this paper is organized as follows. Section 2 describes the mechanical structure and electrical system of the wire-driven robotic fish. Section 3 presents the kinematic model based on the CPG model, as well as the proposed FTO-based ALOS (FTALOS) as a path-following control guidance subsystem for the robotic fish and stability analysis. Pool experimental validation and result analysis are performed in Section 4. Finally, conclusions and future work are summarized in Section 5.

2. Design of Robotic Fish

2.1. Mechanical Design

Fish in their natural environment exhibit a remarkable level of body flexibility. During locomotion, the fish's body adopts a sinusoidal pattern, forming an "S" shape, which has been scientifically proven to optimize the transmission of swimming waves and enhance movement stability [2]. Biologists have utilized electromyography to confirm that the majority of the fish's swimming energy is produced by the muscles in the anterior and middle sections of the body, with the muscles in the caudal fin serving as transmitters. In other words, the anterior and middle sections act as an engine for oscillatory motion, propelling the posterior half like a propeller [22,23]. During tail swinging, the fish exhibits lateral deviation of its posterior body to one side, while the tail and caudal fin bend towards the opposite side as a result of fluid resistance in the water, leading to a distinctive "S" shape. This observation implies that fish do not exert intentional control over their caudal fin to counter water resistance. Instead, the fin flexes autonomously in response to inertia, contributing to increased efficiency.

Drawing inspiration from the physiological structure and propulsion mechanism of real fish, we have proposed a segmented robotic fish design (Figure 1a,b) consisting of a rigid head, a movable driving component, and a flexible tail fin. The active driving component includes a primary servo motor located in the head, along with a movable body section. In contrast, as shown in Figure 1c, the flexible component constitutes a passive flexible tail fin. The amalgamation of two components gives rise to an active compliant propulsion mechanism that closely mimics the physiological structure of actual fish.

2.2. Electronics

The electronic components used in the robotic fish are shown in Figure 1a and Table 1. We utilize the STM32F103 as the microcontroller for the robotic fish, and the E62-433T20D as the communication module, which operates at a frequency of 433 MHz with a data transmission baud rate of 115,200 bits/s, facilitating effective underwater communication. Additionally, the power source driving the caudal fin's movement is the SAVOX SW-1210SG servo, which is capable of providing an ultra-high torque of 3.2 N·m and oscillating at a frequency of 2 Hz under load, meeting the requirements for the normal swimming behavior of the robotic fish. The robotic fish is also equipped with a six-axis sensor MPU6050, primarily used to measure the real-time yaw angle of the fish. All electronics are powered by a 7.4 V DC lithium battery.

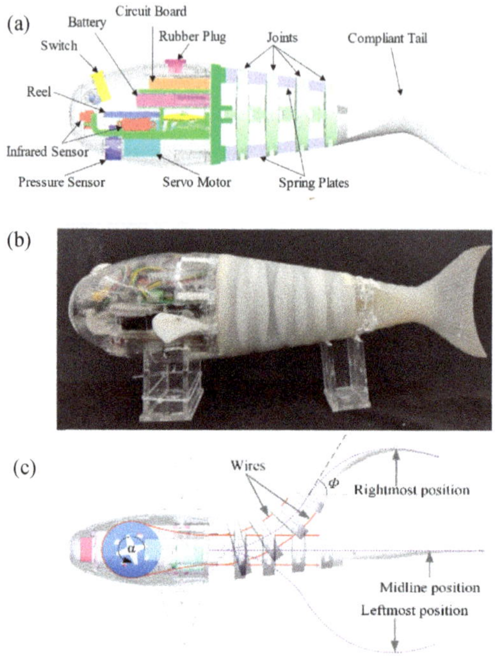

Figure 1. Developed wire-driven active body and compliant tail biomimetic robotic fish. (**a**) Overall CAD structure; (**b**) prototype; and (**c**) illustration of the swinging motion of the robotic fish.

Table 1. The electronic components used in the robotic fish.

Items	Characteristics
Microcontroller	STM32F103
Servo motor	SAVOX SW-1210SG
Communication module	E62-433T20D
Inertial Measurement Unit (IMU)	MPU6050
Power Supply	DC 7.4 V

3. CPG-Based Path Following Control Design

3.1. CPG-Base Model

3.1.1. CPG Model

The rhythmic tail-swinging motion observed in natural fish is driven by a CPG and dynamically modulated by sensory inputs, such as water flow, temperature, pressure, and olfactory and visual information. This movement aids fish in adapting to changes in water flow and aquatic environments. Based on meticulous observations of actual fish movement, we propose an enhanced CPG model that is suitable for the proposed wire-driven biomimetic robotic fish. The model is derived from Ijspeert's successful CPG model applied to gait generation control in biomimetic salamanders [24,25]. Manduca et al. [26] used a CPG model to control the torque on the caudal fin. Korkmaz et al. [27] employed the CPG model, fuzzy logic, and a sensory feedback mechanism to achieve three-dimensional motion. Given that our robotic fish has only one motor drive, we simplify the CPG model by removing complex coupling and introducing the concept of temporal ratio to describe varying durations in recovery and flapping phases during turning. The final expression of the CPG model is as follows:

$$\begin{cases} \ddot{b} = k_b\left(0.25k_b(B-b) - \dot{b}\right) \\ \ddot{m} = k_m(0.25k_m(M-m) - \dot{m}) \\ \dot{\phi} = \left[\frac{(1+R)^2}{4R} - \frac{R^2-1}{4R}sign(sin(\phi))\right]\omega \\ \alpha = b + mcos(\phi) \\ \beta = b + msin(\phi) \\ sign(\lambda) = \begin{cases} 1, & if\ \lambda > 0 \\ 0, & if\ \lambda = 0 \\ -1, & if\ \lambda < 0 \end{cases} \end{cases} \tag{1}$$

where, b, m, and ϕ represent the real-time values of the caudal fin beating bias, amplitude, and phase, while (M, B, ω, R) represent the beating amplitude, bias, frequency, and time ratio taken for one complete cycle in the CPG high-level control commands. Additionally, k_b and k_m are design parameters used to control the sensitivity of the CPG commands. The time ratio is defined as the ratio between the restore phase (t_r) and the beat phase (t_b). This parameter can be utilized to control the turning rates. By adjusting the parameters in the high-level control commands of CPG, different swimming modes can be generated, and smooth transitions between different modes can be achieved without sudden changes in the control variables.

The first and second differential equations represent second-order linear differential equations, indicating that the current amplitude m converges to the CPG input M. The rate of convergence is contingent upon the coefficient k_m, while the offset b exhibits similar behavior. When given inputs $M > 0$ and $\omega > 0$, an oscillatory gait is initiated, at which point a non-zero input b begins to polarize.

3.1.2. Kinematic Model

The kinematic model diagram is shown in Figure 2. Due to the fact that robotic fish typically do not engage in long-term depth adjustments, and adjustments to weight distribution are made during design to ensure stability in the roll direction during swimming, depth variations can be disregarded, and the roll angle and pitch angle can be set to zero, resulting in the simplified two-dimensional planar kinematic equation as follows:

$$\begin{cases} \dot{X}_f = ucos\psi - vsin\psi \\ \dot{Y}_f = usin\psi + vcos\psi \\ \dot{\psi} = r \end{cases} \tag{2}$$

where $X_W O_W Y_W$ stands for the world coordinate system which is stationary, and X_f and Y_f represent the position of the robotic fish in the world coordinate system. xoy denotes the coordinate system bound to the motion of the robotic fish, whose direction is shown in Figure 2, and the velocity of the robotic fish in the x-axis and y-axis are u and v, respectively. Meanwhile, r denotes the angular velocity of rotation around the z-axis. ψ represents the yaw angle of the robotic fish, i.e., the angle between the x-axis of the robotic fish's motion coordinate system and the X_W-axis of the world coordinate system.

3.1.3. Combined Motion

This subsection describes how the CPG model drives the robotic fish to perform motion. The CPG model, as presented in Equation 1, is capable of generating stable swimming vibration gaits of the caudal fin. Theoretically, the swinging caudal fin will generate the force to be applied to the dynamics model, and the generalized acceleration generated by the dynamics model will be integrated to get the generalized velocity, which will be applied to the kinematic model, thus driving the robotic fish to move. However, due to the complex structure of the highly propulsive robotic fish designed in this paper and the complexity of fluid mechanics, it is difficult to establish an accurate dynamics model. So, this paper does not establish a dynamics model but assumes that the acceleration process of the robotic fish is very short and that the robotic fish can soon reach equilibrium. That is, the total propulsive force is equal to the drag force. Therefore, the dynamics model can be ignored, and the mapping relationship from the CPG input to the motion state of the robot fish, i.e., the kinematic model, can be established. When the CPG input parameters are $M, w > 0,\ b = 0,$ and $R = 1$, the caudal fin of the robotic fish oscillates symmetrically along the central axis, which produces forward thrust, i.e., $u > 0$ in Equation 2, and the average angular velocity r is zero. Therefore, the robotic fish swims forward, as shown in Figure 3a. While the input is $b \neq 0$, the caudal fin of the robotic fish is polarized, thus generating a lateral force, and the angular velocity $r \neq 0$, as shown in (b) and (c) of Figure 3. It can be seen that the purpose of controlling the motion of the robotic fish is thereby achieved by changing the input of the CPG.

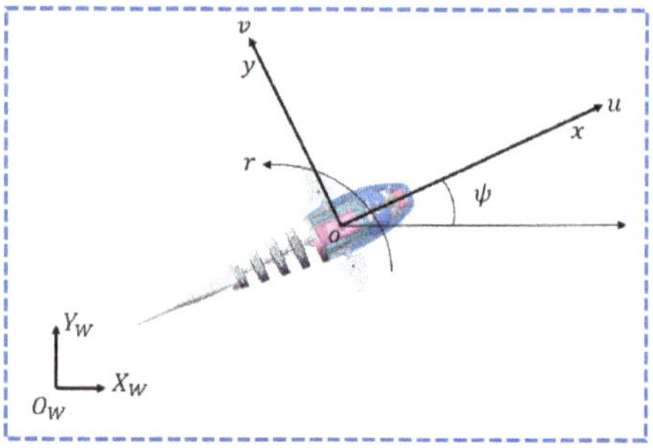

Figure 2. Kinematic diagram of robotic fish.

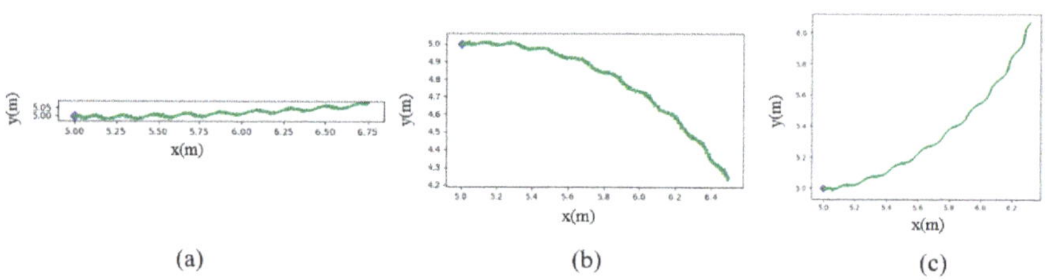

Figure 3. Motion curves of robotic fish under different CPG inputs. (**a**) Straight swimming: $M = 1.07,\ w = 6.28,\ b = 0,\ R = 1$; (**b**) turn right: $M = 1.07,\ w = 6.28,\ b = 0.1,\ R = 1$; and (**c**) turn left: $M = 1.07,\ w = 6.28,\ b = -0.1,\ R = 1$.

3.2. Outer-Loop Controller Design

To facilitate description, let us establish a coordinate system as shown in Figure 4. The desired path is defined as $P_d = (x_d(\varpi), y_d(\varpi))$, where ϖ is a coefficient related to the path. Taking a point on the desired path as the origin and its tangential direction as the X-axis, we establish the $frame_{SF}$ coordinate system. Based on the geometric relationships in Figure 4, the tracking error of the robotic fish can be expressed in the $frame_{SF}$ coordinate system as follows:

$$e = \begin{bmatrix} x_e \\ y_e \end{bmatrix} = \boldsymbol{R}^T(\psi_P)(P - P_d) \tag{3}$$

where $\boldsymbol{R}(\psi_P)$ represents the rotation matrix between the world coordinate system and the $frame_{SF}$ coordinate system, and it is defined as:

$$\boldsymbol{R}(\psi_P) = \begin{bmatrix} \cos(\psi_P) & -\sin(\psi_P) \\ \sin(\psi_P) & \cos(\psi_P) \end{bmatrix} \tag{4}$$

where ψ_P is the tangent angle or path angle at the current path point, which can be defined as follows. Assuming that $x_d(\varpi)$ and $y_d(\varpi)$ are twice differentiable with respect to ϖ

$$\psi_P(\varpi) = atan2(\dot{y}_d(\varpi),\ \dot{x}_d(\varpi)) \tag{5}$$

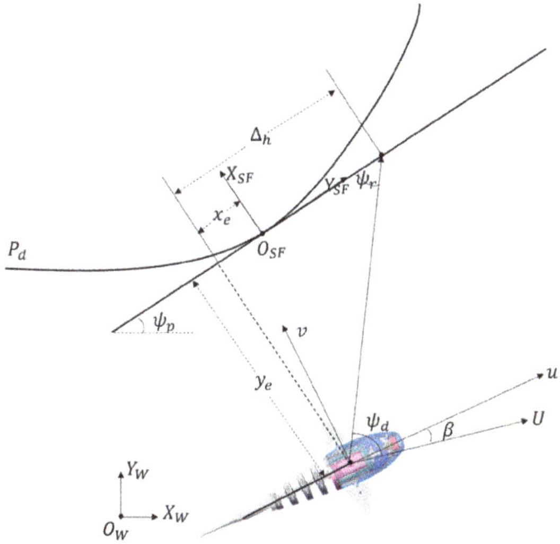

Figure 4. Line-of-sight guidance principle for curved path.

To achieve path following for the robotic fish, a certain guidance method is required to steer the fish towards the desired path and ensure it swims along that path. In the traditional LOS, the desired heading angle is defined as:

$$\begin{aligned}\psi_d &= \psi_P + \psi_r \\ &= \psi_P + \operatorname{atan}\left(-\frac{y_e}{\Delta_h}\right)\end{aligned} \tag{6}$$

By causing the controlled object to move according to the desired heading angle mentioned above, the controlled object can eventually converge to the desired path. However, the traditional line-of-sight guidance law has three drawbacks:

(1) coupling and mutual constraint exist between the convergence of along-track and cross-track errors;
(2) when the cross-track error is zero, the desired heading angle is outputted by the guidance subsystem equals the path angle;
(3) the constant and relatively large lookahead distance results in a slow convergence speed.

Due to drawback 1, the along-track tracking error of the robotic fish may not be zero throughout the entire motion process, and it can be comparable to the cross-track error [20]. When the desired path is a straight line, only a zero cross-track error is required for the robotic fish to follow the intended path. However, when the desired path is a curved path, both components in two directions must be zero simultaneously for successful path following. Therefore, the fish may have difficulty in accurately implementing path following. In fact, T. Fossen et al. have shown that, for a non-closed path, there exists an optimal path point $P(\varpi^*)$ where the along-track tracking error x_e is zero [19]. For a closed path, multiple optimal path points may exist at the same time, and in this case, the one closest to the previous optimal path point is selected. Therefore, the optimal path point within a single control cycle can be obtained using the following equation:

$$\begin{aligned} \varpi^* &= \underset{\varpi \geq 0}{\operatorname{argmin}} \left(\dot{\varpi} \right) \\ s.t. \;\; y - y_d(\varpi) &= -\frac{1}{\tan(\psi_P(\varpi))}(x - x_d(\varpi)) \end{aligned} \tag{7}$$

For drawback 2, when the cross-track error reaches zero, regardless of the magnitude of the along-track error, the desired yaw-angle output from the guidance subsystem will be strictly equal to the path angle. In this case, if the along-track error is non-zero, the angle will guide the robotic fish toward the desired path. However, when the along-track error is zero, indicating that the robotic fish is already on the desired path, it will cause the robotic fish to deviate from the desired path, thus affecting the path-following performance. In the traditional LOS approaches, the derivation process commonly disregards the influence of the sideslip angle. Despite its typically small magnitude, it becomes significant when the cross-track tracking error of the robotic fish nears zero and cannot be overlooked. Moreover, the sideslip angle can provide a certain heading angle compensation to mitigate the deviation phenomenon in the traditional law. Therefore, we proposed a side-slip angle observer to correct the output of the guidance system and improve the path-following performance of the robotic fish.

By differentiating Equation (3), the tracking-error kinematics can be obtained as follows:

$$\begin{cases} \dot{x}_e = -U_P + Ucos(\psi - \psi_P(\varpi) + \beta) + y_e\dot{\psi}_P \\ \dot{y}_e = Usin(\psi - \psi_P(\varpi) + \beta) - x_e\dot{\psi}_P \end{cases} \tag{8}$$

where U_P represents the velocity of the path point, U is the real-time synthesized motion velocity of the robotic fish; and β is the sideslip angle which is defined as the angle between U and u, i.e., atan(v/u). By calculating the optimal path point $P(\varpi^*)$ that minimizes x_e according to Equation (7) and assuming that sin$\beta \approx \beta$ and cos$\beta \approx 1$, based on the fact that beta is generally <10°. At the same time, it is assumed that the side-slip angle of the robotic fish changes slowly during the movement, i.e., there exists $\dot{\beta} = 0$. The equation can be reduced to:

$$\begin{aligned} \dot{y}_e &= Usin(\psi - \psi_P + \beta) \\ &= Usin(\psi - \psi_P)cos\beta + Ucos(\psi - \psi_P)sin\beta \\ &\approx Usin(\psi - \psi_P) + Ucos(\psi - \psi_P)\beta \end{aligned} \tag{9}$$

The side-slip angle observer can be designed as below:

$$\begin{cases} \dot{\hat{y}}_e = Usin(\psi - \psi_P) + Ucos(\psi - \psi_P)\,\hat{\beta} - \lambda sig^{0.5}\left(\tilde{y}_e\right) \\ \dot{\hat{\beta}} = -Ucos(\psi - \psi_P)\tilde{y}_e \end{cases} \tag{10}$$

where $\hat{y}_e$ is the estimated value of y_e, and $\tilde{y}_e = \hat{y}_e - y_e$ represents the estimation error of y_e. $\hat{\beta}$ is the estimated value of the side-slip angle β, and $sig^{0.5}(x) = |x|^{0.5}sign(x)$ denotes the square root sign function. With the action of this observer, the estimation errors $\tilde{y}_e$ and $\tilde{\beta}$ will converge to zero, i.e., $\hat{y}_e \to y_e$ and $\hat{\beta} \to \beta$.

Proof. Subtract Equation (9) from Equation (10), and we can get:

$$\dot{\tilde{y}}_e = Ucos(\psi - \psi_P)\tilde{\beta} - \lambda sig^{0.5}\left(\tilde{y}_e\right) \tag{11}$$

Choosing the Lyapunov function as $V_1 = \frac{1}{2}\left(\tilde{y}_e^2 + \overset{\sim}{\beta}^2\right)$ and taking the derivative of V_1:

$$\begin{aligned}\dot{V}_1 &= \tilde{y}_e\dot{\tilde{y}}_e + \tilde{\beta}\dot{\tilde{\beta}} \\ &= \tilde{y}_e\left(Ucos(\psi - \psi_P)\tilde{\beta} - \lambda sig^{0.5}\left(\tilde{y}_e\right)\right) - Ucos(\psi - \psi_P)\tilde{y}_e\tilde{\beta} \\ &= -\lambda\tilde{y}_e sig^{0.5}\left(\tilde{y}_e\right) \\ &= -\lambda\left|\tilde{y}_e\right|^{1.5} \\ &= -\lambda\left|\tilde{y}_e\right|^{1.5} - \lambda\left|\tilde{\beta}\right|^{1.5} + \lambda\left|\tilde{\beta}\right|^{1.5}\end{aligned} \tag{12}$$

With **Lemma 1** in [28], let $(a_1, a_2, a_3, \cdots, a_n) \in R^n$ and $0 < \rho < 2$. Then, the following inequality holds:

$$\left(a_1^2 + a_2^2 + a_3^2 \cdots + a_n^2\right)^\rho \le \left(a_1^\rho + a_2^\rho + a_3^\rho + \cdots + a_n^\rho\right)^2 \tag{13}$$

So that:

$$\begin{aligned}\dot{V}_1 &= -\lambda\left(\left|\tilde{y}_e\right|^{1.5} + \left|\tilde{\beta}\right|^{1.5}\right) + \lambda\left|\tilde{\beta}\right|^{1.5} \\ &\le -\lambda\left(\left|\tilde{y}_e\right|^2 + \left|\tilde{\beta}\right|^2\right)^{\frac{1.5}{2}} + \lambda\left|\tilde{\beta}\right|^{1.5} \\ &= -2^{\frac{1.5}{2}}\lambda V_1^{\frac{1.5}{2}} + \lambda\left|\tilde{\beta}\right|^{1.5}\end{aligned} \tag{14}$$

It is noted that there should exist a constant θ_0 satisfying $0 < \theta_0 < 1$, such that $2^{\frac{1.5}{2}}\lambda\theta_0 V^{\frac{1.5}{2}} = \lambda\theta_0\left(\tilde{y}_e^2 + \overset{\sim}{\beta}^2\right)^{\frac{1.5}{2}} = \lambda\left|\tilde{\beta}\right|^{1.5}$, which implies $\lambda\left|\tilde{\beta}\right|^{1.5} - 2^{\frac{1.5}{2}}\lambda\theta_0 V^{\frac{1.5}{2}} = 0$. Therefore, we can derive the following equation:

$$\begin{aligned}\dot{V}_1 &\le \gamma_\beta - 2^{\frac{1.5}{2}}\lambda\theta_0 V_1^{\frac{1.5}{2}} - 2^{\frac{1.5}{2}}\lambda(1 - \theta_0)V_1^{\frac{1.5}{2}} \\ &= -2^{\frac{1.5}{2}}\lambda(1 - \theta_0)V_1^{\frac{1.5}{2}}\end{aligned} \tag{15}$$

According to **Lemma 2** in [28], for the nonlinear autonomous system $\dot{x} = f(x), f(0) = 0$, if there exists a positive definite continuously differentiable scalar function $V(x)$, $\mathbb{R}^n \to \mathbb{R}$ satisfies the following conditions:

$$\dot{V}(x) \le -\alpha V^\rho(x) \tag{16}$$

where $\alpha > 0,\ 0 < \rho < 1$. Then, the system is finite time stable at the origin, i.e., there exists a time T:

$$T \le \frac{1}{\alpha(1 - \rho)}V^{1-\rho}(x) \tag{17}$$

such that $V(x(t)) = 0,\ \forall t > T$.

The estimation error $\left(\tilde{y}_e, \tilde{\beta}\right)$ will converge to $(0, 0)$ within a finite time, and the convergence time T_1 satisfies the following condition:

$$T_1 \le \frac{V_1(0)^{\frac{1.5}{2}}}{2^{\frac{1.5}{2}}\lambda(1 - \theta_0)\left(1 - \frac{1.5}{2}\right)}$$

Therefore, under the action of the observer, we can approximate $\hat{y}_e$ to y_e and $\hat{\beta}$ to β. Based on the estimated value of the sideslip angle obtained from the observer, the yaw control law of the guidance subsystem is given by:

$$\begin{aligned} \psi_d &= \psi_r + \psi_p \\ &= \operatorname{atan}\left(-\frac{y_e}{\Delta_h} - \hat{\beta}\right) + \psi_p \end{aligned} \tag{18}$$

Under the influence of this desired yaw angle, the robotic fish's lateral error can converge to zero. □

Proof. Assuming that the robotic fish can perfectly track the desired yaw angle, when the desired yaw angle is substituted into the observer, the following can be obtained:

$$\begin{aligned} \dot{y}_e &= -\frac{U(y_e+\hat{\beta}\Delta_h)}{\sqrt{\Delta_h^2+(y_e+\hat{\beta}\Delta_h)^2}} + \frac{U\beta\Delta_h}{\sqrt{\Delta_h^2+(y_e+\hat{\beta}\Delta_h)^2}} \\ &= -\frac{Uy_e}{\sqrt{\Delta_h^2+(y_e+\hat{\beta}\Delta_h)^2}} \end{aligned} \tag{19}$$

Taking the Lyapunov function as $V_2 = \frac{1}{2}y_e^2$ and differentiating it, it can be derived that:

$$\begin{aligned} \dot{V}_2 &= y_e\dot{y}_e \\ &= -\frac{Uy_e^2}{\sqrt{\Delta_h^2+(y_e+\Delta_h\hat{\beta})^2}} \\ &\le -kV_2 \\ &\le 0 \end{aligned} \tag{20}$$

where $k = U/\sqrt{\Delta_h^2 + (y_e + \Delta_h\hat{\beta})^2}$. According to the Lyapunov stability theory, it can be concluded that y_e will eventually converge to zero.

To address drawback 3, the selection of the lookahead distance plays a crucial role. Thus, an adaptive LOS strategy is proposed. When encountering a large lateral tracking error y_e, indicating that the fish is significantly deviating from the desired path, it is advantageous for the fish to swiftly approach the desired path. In this case, opting for a smaller lookahead distance would result in the desired yaw angle being nearly perpendicular to the path angle, thereby enabling faster convergence with the desired path. On the other hand, conversely, when facing a small lateral tracking error, it is preferable to utilize a larger lookahead distance in order to minimize overshooting during convergence. In this scenario, emphasis should be placed on aligning the desired yaw angle with the path angle. It is important to note that, while a smaller lookahead distance facilitates convergence speedily, an excessive focus on convergence can lead to oscillations. Therefore, employing a time-varying formula for calculating lookahead distance can mitigate these issues:

$$\Delta_h = (\Delta_{h_{\max}} - \Delta_{h\min})\left(-\tanh^2(ky_e) + 1\right) + \Delta_{h\min} \tag{21}$$

where Δ_{hmax} represents the upper limit of the time-varying lookahead distance, while $\Delta_{h\min}$ represents the lower limit. The parameter k serves as a positive design parameter that controls the sensitivity of the lookahead distance to the lateral tracking error. By modifying the value of k, the sensitivity of the lookahead distance to variations in the lateral tracking error can be regulated, which facilitates the trade-off of convergence speed and oscillations during the tracking process. The flowchart of ALOS is shown in Figure 5.

Figure 6 illustrates the overall framework of the FTALOS control method for path tracking.

$$\begin{cases} fh = fhan(x_1(k) - v(k), x_2(k), r, h) \\ x_1(k+1) = x_1(k) + hx_2(k) \\ x_2(k+1) = x_2(k) + hfhan \end{cases} \tag{22}$$

where $v(k)$ is the real-time position of the robotic fish, $x_1(k)$ is the tracking signal of $v(k)$, $x_2(k)$ is the tracking signal of $\dot{v}(k)$, and $fhan$ is the optimal control synthesis function, which is defined as [28]:

$$\begin{cases} d = rh \\ d_0 = dh \\ a_0 = hx_2 \\ y = x_1 + a_0 \\ a_1 = \sqrt{d(d+8|y|)} \\ a_2 = a_0 + sign(y)(a_1 - d)/2 \\ a = (a_0 + y)fsg(y,d) + a_2(1 - fsg(y,d)) \\ fhan = -r(\frac{a}{d})fsg(a,d) - rsign(a)(1 - fsg(a,d)) \\ fsg(x,d) = (sign(x+d) - sign(x-d))/2 \end{cases} \tag{23}$$

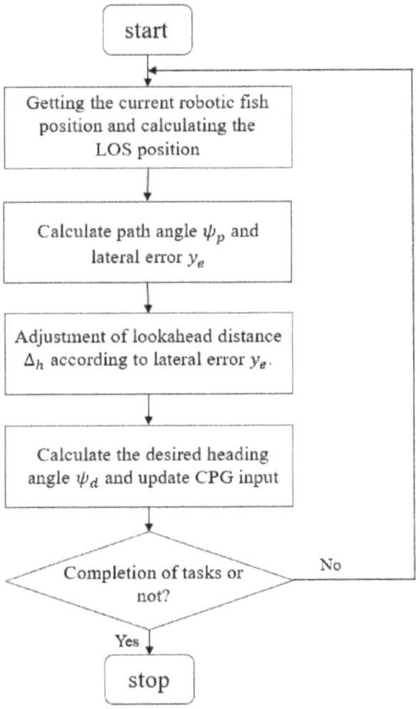

Figure 5. Flowchart of the working of ALOS.

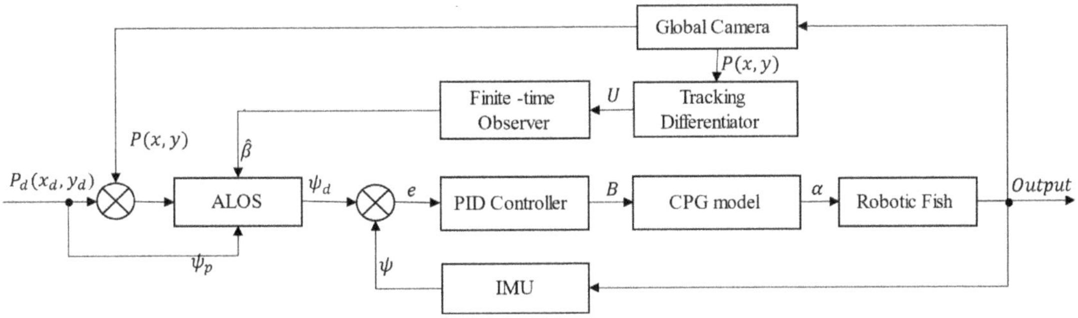

Figure 6. Control diagram of path-following control for the wire-driven robotic fish. □

3.3. Inner-Loop Controller Design

In the previous section, the primary function of the outer-loop controller is to simplify the path-tracking control problem by transforming it into a lateral tracking-error stabilization control problem. This is further enhanced through the implementation of an improved LOS, which subsequently converts it into a yaw-tracking control problem. Therefore, an effective yaw-tracking controller is necessary, involving the utilization of LOS as the outer-loop controller for deriving the desired yaw-angle signal. Subsequently, the inner-loop yaw-tracking controller generates the corresponding motor control signal, thereby driving the robotic fish to accurately follow the intended path.

The compliant and flexible design of the tail in the proposed biomimetic robotic fish structure in Section 2 enhances energy efficiency while achieving natural fish-like swimming. However, accurately establishing the dynamic model of the robotic fish presents challenges, making the application of model-based control methods impractical. Conversely, PID controllers have been widely employed in various control tasks due to their simplicity, robustness, and ability to achieve satisfactory control performance without requiring a precise model of the controlled system. Therefore, we utilize a PID controller as the inner-loop yaw-tracking controller, with its specific control output defined as follows:

$$\begin{aligned} B(k) &= K_p e(k) + K_i T_s \sum_{i=1}^{k} e(i) + K_d \frac{e(k) - e(k-1)}{T_s} \\ &= K_P e(k) + K_I \sum_{i=1}^{k} e(i) + K_D (e(k) - e(k-1)) \end{aligned} \tag{24}$$

where B is bias, one of the inputs of the CPG model. $e(i) = \psi(i) - \psi_d(i)$ represents the tracking error at the i-th time instant, where $\psi(i)$ is the current yaw angle and $\psi_d(i)$ is the desired yaw angle. The parameters K_p, K_i, and K_d are the proportional, integral, and derivative gains of the controller, respectively. T_s represents the sampling period.

To improve the robustness of the controller, an incremental PID approach is employed. By taking the difference of Equation (24), we obtain:

$$\begin{aligned} \Delta B = \; & K_P(e(k) - e(k-1)) + K_I e(k) \\ & + K_D(e(k) - 2e(k-1) + e(k-2)) \end{aligned} \tag{25}$$

The inclusion of the integral term introduces the potential for integral windup. As a result, it is essential to evaluate whether the control signal has reached saturation when calculating the adjustment to the control signal. If saturation is identified, the integral component is disregarded, and only the proportional and derivative terms are retained. This approach, commonly known as an anti-windup mechanism, effectively mitigates the negative impacts of integral saturation.

It is important to note that the MPU6050 obtains the current orientation angles through velocity integration, where the Z-axis angle represents a relative angle rather than an absolute angle. Upon power-up, the Z-axis angle is initialized to 0°, indicating that the fish's head direction at power-on is considered the 0° direction. Typically, the range of variation for the Z-axis angle is $[-180°, 180°]$. Therefore, significant angle jumps may occur when the fish's yaw direction is opposite to the initial direction at power-up, such as a jump from 180° to $-180°$ or from $-180°$ to 180°. In order to mitigate abrupt changes in the yaw-tracking error and prevent erroneous control commands due to angle discontinuities, it is essential to preprocess the yaw angle obtained from the fish's attitude sensor before updating the control signal. The specific preprocessing steps are as follows:

$$e = \begin{cases} e - 2\pi & , \text{if } e \geq \pi \\ e & , \text{if } -\pi < e < \pi \\ e + 2\pi & , \text{if } e \leq -\pi \end{cases} \tag{26}$$

Note that the Z-axis angle sampled by the MPU6050 is not the actual yaw angle but the direction in which the fish head is pointing because the fish head will oscillate during the swimming process. In order to get the actual yaw angle, we need to approximate the yaw angle by first-order filtering:

$$\psi_{act}(k) = \alpha\psi_{act}(k-1) + (1-\alpha)\psi_{IMU}(k) \tag{27}$$

where α is the filtering factor, $\psi_{act}(k)$ is the yaw angle corresponding to the current sampling cycle, $\psi_{act}(k-1)$ is the yaw angle of the previous cycle, and $\psi_{IMU}(k)$ is the pointing of the head of the fish captured by the MPU6050 in the current cycle, with a sampling period of 0.02 s.

4. Experiment

The size of the experimental pool is 2.5 m × 1.6 m, and the water depth is 0.6 m. The experimental parameter settings are shown in Table 2.

Table 2. Parameters setting for experiment.

dh_{max}	dh_{min}	k	r	K_P	K_I
150	30	60	20	0.6	0.0001
K_D	λ	M	B	ω	R
0.0003	1.5	10	-	1.5	1

4.1. Yaw Control

In Section 3, the path-tracking problem is theoretically derived to be transformed into a yaw-tracking control problem. The internal control system is tasked with guiding the robotic fish to follow the real-time desired yaw-angle output by the outer guidance subsystem, thereby achieving effective path-following control. Consequently, it is imperative to conduct tests on the efficacy of the internal control system in order to validate the feasibility of the proposed control scheme. During the path-following process, the desired yaw angle provided by the guidance subsystem continuously changes, it becomes essential to assess and test both constant and dynamically changing yaw angles for their trackability by our proposed controller. As a result, the yaw-tracking control experiment is divided into two parts, namely tracking of static yaw angles and tracking of dynamic yaw angles.

In the static yaw-angle tracking experiment, we conducted tracking for angles of 30°, 45°, and 60°. The trajectory of the robotic fish during the experiment is illustrated in Figure 7. A careful examination of the angle diagram in the bottom right corner of the figure reveals a close alignment between the fish's movement direction and the desired angles. To provide a more intuitive representation of the relationship between actual and desired values, Figure 8 displays the real-time yaw angle of the fish compared to the desired yaw angle. Remarkably, the implementation of the proposed incremental PID yaw tracking controller results in effective tracking performance for static yaw angles in robotic fish.

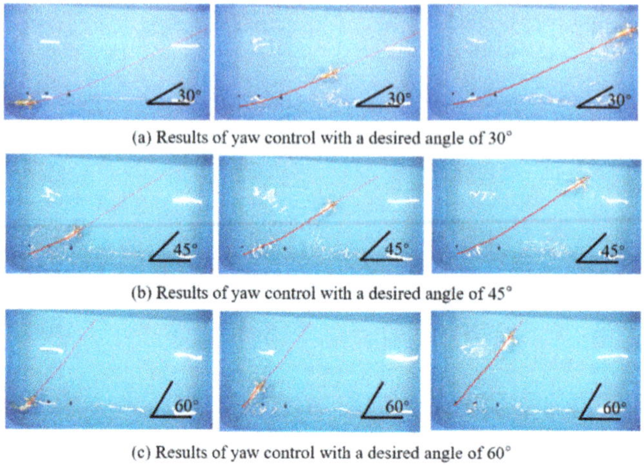

Figure 7. Results of directional tracking experiment.

In the experiment of dynamic yaw-angle tracking, we designated the desired yaw angle as a linear time-varying variable, with incremental changes set at $1°/0.2$ s, $4°/0.2$ s, and $6°/0.2$ s, respectively. The comparison between the actual yaw angle of the robotic fish and the time-varying desired yaw angle during the experiment is illustrated in Figure 9. It is evident that the proposed PID incremental controller demonstrates commendable tracking performance for dynamic yaw angles with varying angular growth rates.

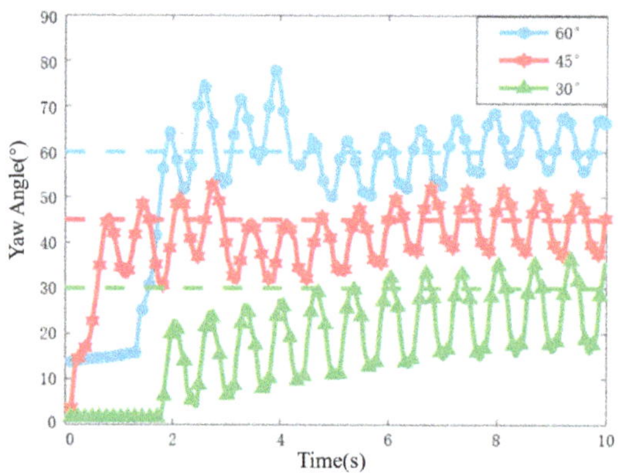

Figure 8. Variation of actual yaw angle with different desired angles.

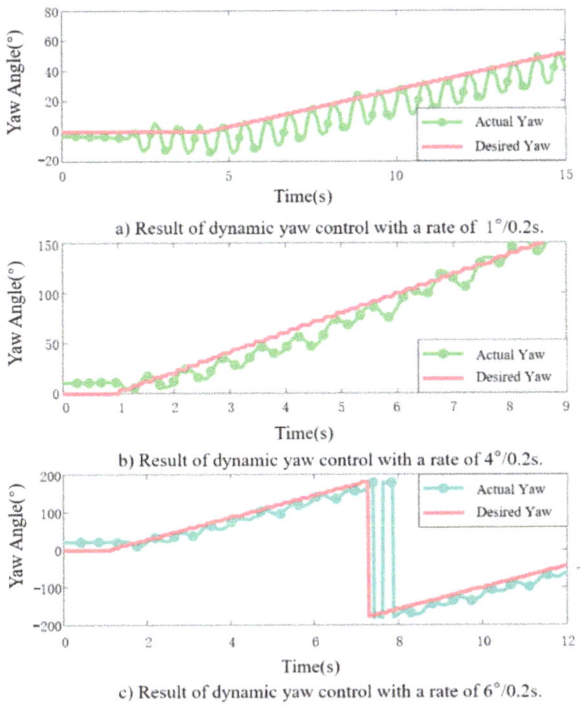

Figure 9. Variation of actual yaw angle for desired angles with different rates of change.

Due to the inherent limitations of the fish's bias, there is a slight delay in tracking the dynamic yaw angle with an angular growth rate of $6^\circ/0.2$ s, which represents the maximum limit that the fish can effectively track. However, during the path-tracking stage, the desired yaw-angle changes typically do not exceed $3^\circ/0.2$ s. Consequently, adopting the incremental PID controller as the inner-loop controller for path following presents a practical and viable solution.

4.2. Path-Following Control

The control block diagram for the path following is shown in Figure 6, where the real-time position of the robotic fish serves as feedback. However, traditional positioning modules are not suitable due to the physical dimensions of the fish and constraints imposed by the operating environment. Therefore, this study utilizes real-time experimental images captured using a global camera, with image-processing techniques employed to extract the fish's position. Considering both the image-processing stage and the characteristics of the robotic fish, increasing the control frequency does not significantly improve performance. Due to limitations in the fish's actuators, instantaneous mode transitions cannot be achieved. As a result, the control period is set to 200 ms in this study, taking into consideration the requirements of the image-processing stage and the fish's actuation capabilities. Unlike trajectory tracking, path-following control is not constrained by time but rather focuses on ensuring that the robotic fish can move along a predetermined path. Therefore, to simplify the control implementation, in this study, the parameters M, ω, and R in the CPG high-level control commands (M, B, ω, R) are set as constants for each experiment. The fish only adjusts the bias of the caudal fin (B) during the tracking process to achieve turning. Specifically, the beating frequency (ω) is set to 1.5 Hz, the time ratio (R) is set to 1, and the oscillation amplitude (M) is set to 10.

The initial position of the robotic fish is set to $(250, 150)$, and the target path is set as $x_P(\varpi) = 1000 + 300cos(2\pi\varpi)$, $y_P(\varpi) = 500 + 300\sin(2\pi\varpi)$ with $\varpi(0) = 0$.

Figure 10 depicts a series of snapshots demonstrating the path-following process guided by the proposed strategy (Video S1). The anticipated path is denoted by a yellow dotted line, while the actual swimming trajectory of the robotic fish during the experimental trial is represented by a red solid line. It can be observed from Figure 10 that the robotic fish enters the tracking phase at T = 14.33 s. At this point, the controller initiates the turning motion because there is a significant difference between the heading angle of the robotic fish and the path angle. However, due to the inherent control lag, the robotic fish deviates from the desired path between 14.33 s and 20.53 s. Subsequently, under the guidance of our proposed strategy, at T = 20.53 s, the robotic fish successfully returns to the proximity of its intended path and maintains stability with minimal deviations from it thereafter.

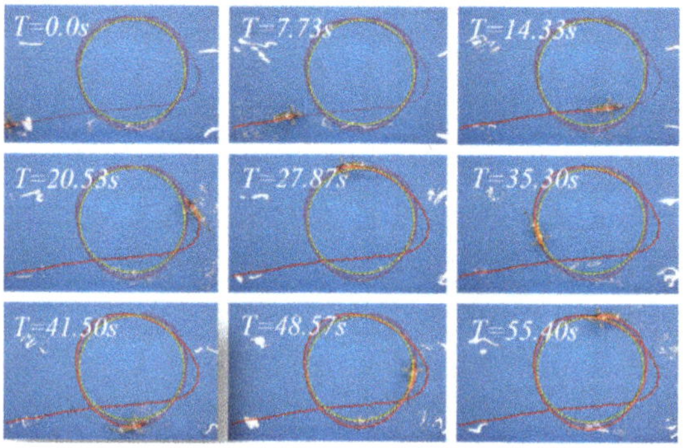

Figure 10. Snapshot sequence for circle path following. The yellow dashed line represents the expected trajectory, while the red implementation represents the actual trajectory of the robotic fish.

A comparison was made between the traditional fixed lookahead distance LOS, enclosed circular lookahead distance (CLOS), and the proposed ALOS in this study. Figure 11 shows the comparison of the path-following experiments using different strategies, depicting the actual swimming trajectories of the robotic fish under each navigation method. It can be seen that the traditional LOS is slow to converge and prone to oscillation in the process of path following, which is caused by the inherent limitations of the traditional LOS, i.e., drawback 3 of the traditional LOS, as mentioned in this paper. Too large a lookahead distance will result in slow convergence, while too small a lookahead distance will result in oscillation in the vicinity of the trajectory, making it difficult to obtain the appropriate lookahead distance. Similarly, CLOS has significant overshooting but performs a little better than LOS at sharp corners. In contrast, the ALOS proposed in this paper significantly outperforms the previous two in both sharp corners and the amount of overshooting and tracks the desired path more smoothly, which is due to the fact that the ALOS can adaptively adjust the lookahead distance. But, the path-tracking accuracy is still insufficient, due to the interference from the sideslip angle of the robotic fish. However, the FTALOS strategy proposed in this paper not only adaptively adjusts the lookahead distance but also takes into account the interference of the sideslip angle of the robotic fish, estimates the side-slip angle, and obtains a more realistic line-of-sight angle, and the final experimental results are satisfactory and significantly better than the first three.

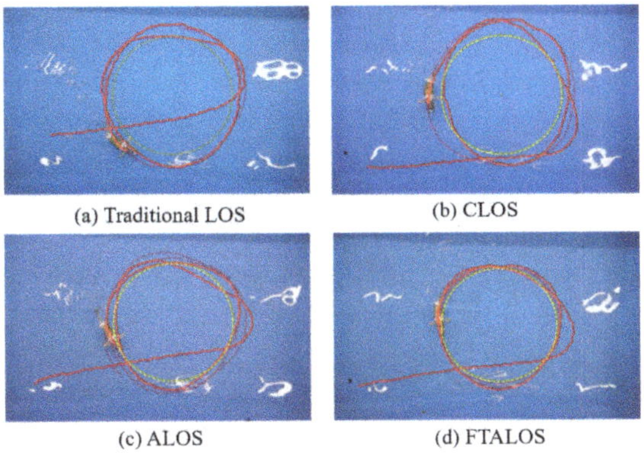

Figure 11. Comparison of the actual trajectories and desired paths under different guidance strategies. The yellow dashed line represents the expected trajectory, while the red implementation represents the actual trajectory of the robotic fish.

Figure 12a depicts the real-time variation of the yaw angle during the experimental process, demonstrating that the FTALOS continuously updates the desired yaw angle. The robotic fish achieves accurate tracking of this signal under the influence of the incremental yaw-tracking PID controller, thereby accomplishing the path-tracking task. Figure 12b presents the tail-fin deflection signal in the CPG high-level control command throughout the experiment. When combined with Figure 10, it becomes apparent that, during the initial stage, slight turns are executed by the robotic fish to reach the starting tracking point, resulting in a small positive deflection value. However, at T = 14.33 s, a substantial turn is undergone by the robotic fish, leading to a rapid transition of the deflection value to a larger negative magnitude. At T = 20.53 s, the robotic fish successfully returns to the desired path and, thereafter, swims along the desired trajectory, gradually stabilizing the subsequent deflection value, which consistently hovers around −10.

To quantitatively assess the advantages of the proposed FTALOS, we measured the experimental tracking errors, as presented in Table 3 and Figure 13. The results demonstrate that the FTALOS exhibits superior tracking performance across all three error dimensions, including RMSE, MSE, and MAE. Specifically, it achieves the lowest RMSE value of 0.0544, the lowest MSE value of 0.0425, and the lowest MAE value of 0.0030. Meanwhile, Figure 13 clearly shows that FTALOS has the smallest average tracking error and standard deviation. This indicates that the FTALOS outperforms both the CLOS and traditional LOS methods. The ALOS guidance strategy, incorporated in the FTALOS, enables faster convergence to the desired path compared to the traditional LOS, which exhibits slower convergence speed. Additionally, the inclusion of FTO enhances the performance of the ALOS, further improving its tracking capabilities.

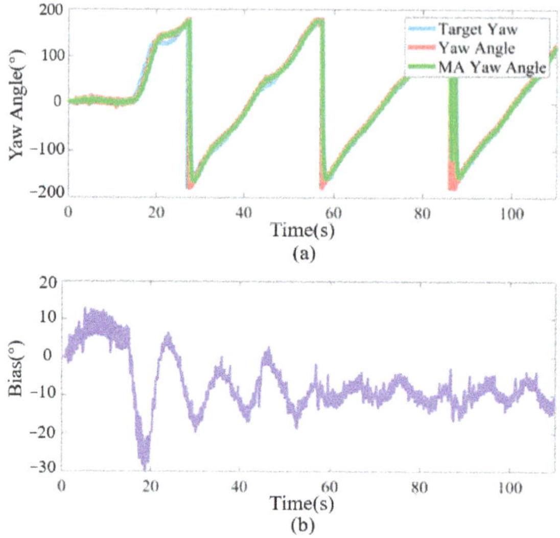

Figure 12. Experiment results. (**a**) Comparison of target yaw angle and actual yaw angle; (**b**) real-time bias of CPG high-level command, it represents the swinging center of the caudal fin of robotic fish.

Table 3. Quantitative analysis of circle path-following error for experiments.

Items	Quantitative Analysis(m)		
	RMSE.	**MAE**	**MSE**
Traditional LOS	0.0986	0.0880	0.0097
CLOS	0.0980	0.0831	0.0096
ALOS	0.0706	0.0550	0.0050
FTALOS	**0.0544**	**0.0425**	**0.0030**

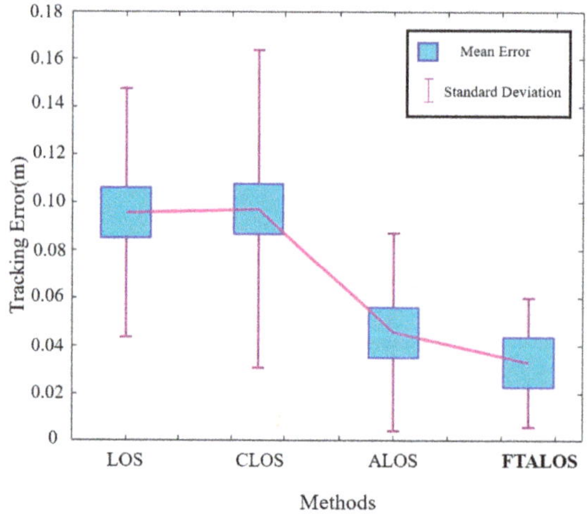

Figure 13. Tracking-error comparison of different control methods.

However, our proposed FTALOS path-tracking method has certain limitations as it relies on the assumption that the path is expected to be second-order differentiable. Consequently, the tracking results may not meet expectations for paths with folded lines.

5. Conclusions

In this paper, an FTALOS guidance subsystem is proposed to achieve a planar path following a bio-inspired robotic fish equipped with a compliant and flexible tail fin driven by a cable mechanism. Initially, the kinematic model of the underwater bio-inspired robotic fish is introduced based on coordinate transformations. Subsequently, an ALOS guidance strategy is proposed, which incorporates compensation for the sideslip angle. Moreover, an incremental PID controller for yaw-tracking control is introduced and experimentally validated to evaluate its ability to track the yaw angle. The integration of the guidance subsystem and the PID heading controller is synergistically implemented, and the efficacy of the proposed tracking strategy is verified through experiments. The experimental results demonstrate that the robotic fish guided by the proposed FTALOS exhibits minimal overshoot and achieves a lower steady-state error compared to the traditional LOS and CLOS.

In the future, we will develop advanced mechanical structures to enhance the swimming efficiency and experimental reproducibility of robotic fish. Moreover, our focus will remain on refining control algorithms, including model-free control based on reinforcement learning, to enable robotic fish to autonomously learn and improve controller robustness. Furthermore, we will conduct outdoor experiments with random perturbations to track more complex paths.

Supplementary Materials: The following supporting information can be downloaded at: https://www.mdpi.com/article/10.3390/biomimetics9090556/s1, Video S1: demo video.

Author Contributions: All authors were involved in the conceptualization and discourse surrounding the paper. Y.Z. was responsible for the conceptualization of the paper's overarching framework and supervision. Y.L. conducted a detailed review and revision of the paper. Y.M. and W.S. were tasked with theoretical derivation, experimental design, and paper writing. Z.H. conducted a detailed review and revision of the paper. All authors have read and agreed to the published version of the manuscript.

Funding: This work was supported in part by GDNRC[2023]33, the National Natural Science Foundation of China under Grant 62103152, 62473161, and the Guangzhou applied Basic Research project 2024A04J9942.

Institutional Review Board Statement: Not applicable.

Data Availability Statement: No new data were created or analyzed in this study. Data sharing is not applicable to this article.

Conflicts of Interest: The authors declare that they have no known competing financial interests or personal relationships that could have appeared to influence the work reported in this paper.

References

1. Chen, G.; Li, Y.; Chen, J.; Du, R.; Zhong, Y. Investigation on Yaw Stability of Bionic Propulsion in Flow Field. In Proceedings of the 2021 IEEE International Conference on Robotics and Biomimetics (ROBIO), Sanya, China, 27–31 December 2021; pp. 726–731.
2. Sfakiotakis, M.; Lane, D.M.; Davies, J.B.C. Review of fish swimming modes for aquatic locomotion. *IEEE J. Ocean. Eng.* **1999**, *24*, 237–252. [CrossRef]
3. Chen, B.; Jiang, H. Swimming performance of a tensegrity robotic fish. *Soft Robot.* **2019**, *6*, 520–531. [CrossRef] [PubMed]
4. Chen, Z.; Shatara, S.; Tan, X. Modeling of biomimetic robotic fish propelled by an ionic polymer–metal composite caudal fin. *IEEE/ASME Trans. Mechatron.* **2009**, *15*, 448–459. [CrossRef]
5. Triantafyllou, M.S.; Triantafyllou, G.S. An efficient swimming machine. *Sci. Am.* **1995**, *272*, 64–70. [CrossRef]
6. Du, S.; Wu, Z.; Wang, J.; Qi, S.; Yu, J. Design and control of a two-motor-actuated tuna-inspired robot system. *IEEE Trans. Syst. Man Cybern. Syst.* **2019**, *51*, 4670–4680. [CrossRef]
7. Dong, H.; Wu, Z.; Chen, D.; Tan, M.; Yu, J. Development of a whale-shark-inspired gliding robotic fish with high maneuverability. *IEEE/ASME Trans. Mechatron.* **2020**, *25*, 2824–2834. [CrossRef]

8. Liao, X.; Zhou, C.; Wang, J.; Fan, J.; Zhang, Z. A wire-driven elastic robotic fish and its design and cpg-based control. *J. Intell. Robot. Syst.* **2023**, *107*, 4. [CrossRef]
9. Zhong, Y.; Li, Z.; Du, R. A novel robot fish with wire-driven active body and compliant tail. *IEEE/ASME Trans. Mechatron.* **2017**, *22*, 1633–1643. [CrossRef]
10. Yan, S.; Wu, Z.; Wang, J.; Huang, Y.; Tan, M.; Yu, J. Real-world learning control for autonomous exploration of a biomimetic robotic shark. *IEEE Trans. Ind. Electron.* **2022**, *70*, 3966–3974. [CrossRef]
11. Chen, J.; Yin, B.; Wang, C.; Xie, F.; Du, R.; Zhong, Y. Bioinspired closed-loop CPG-based control of a robot fish for obstacle avoidance and direction tracking. *J. Bionic Eng.* **2021**, *18*, 171–183. [CrossRef]
12. Chen, J.; Chen, Y.; Yu, B.; Du, R.; Zhong, Y.; Xie, F. Obstacle avoidance control strategy of a biomimetic wire-driven robot fish. In Proceedings of the 2019 IEEE International Conference on Robotics and Biomimetics (ROBIO), Dali, China, 6–8 December 2019; pp. 326–331.
13. Zhong, Y.; Hong, Z.; Li, Y.; Yu, J. A General Kinematic Model of Fish Locomotion Enables Robot Fish to Master Multiple Swimming Motions. *IEEE Trans. Robot.* **2024**, *40*, 750–763. [CrossRef]
14. Zhong, Y.; Chen, Y.; Wang, C.; Wang, Q.; Yang, J. Research on target tracking for robotic fish Based on low-cost scarce sensing information fusion. *IEEE Robot. Autom. Lett.* **2022**, *7*, 6044–6051. [CrossRef]
15. Wang, J.; Wu, Z.; Yan, S.; Tan, M.; Yu, J. Real-time path planning and following of a gliding robotic dolphin within a hierarchical framework. *IEEE Trans. Veh. Technol.* **2021**, *70*, 3243–3255. [CrossRef]
16. Chen, Y.; Qiao, J.; Liu, J.; Zhao, R.; An, D.; Wei, Y. Three-dimensional path following control system for net cage inspection using bionic robotic fish. *Inf. Process. Agric.* **2022**, *9*, 100–111. [CrossRef]
17. Yu, J.; Tan, M.; Wang, S.; Chen, E. Development of a biomimetic robotic fish and its control algorithm. *IEEE Trans. Syst. Man Cybern. Part B (Cybern.)* **2004**, *34*, 1798–1810. [CrossRef]
18. Liu, J.; Liu, Z.; Wu, Z.; Yu, J. Three-dimensional path following control of an underactuated robotic dolphin using deep reinforcement learning. In Proceedings of the 2020 IEEE International Conference on Real-Time Computing and Robotics (RCAR), Asahikawa, Japan, 28–29 September 2020; pp. 315–320.
19. Fossen, T.I.; Pettersen, K.Y.; Galeazzi, R. Line-of-sight path following for dubins paths with adaptive sideslip compensation of drift forces. *IEEE Trans. Control Syst. Technol.* **2014**, *23*, 820–827. [CrossRef]
20. Qu, Y.; Cai, L.; Xu, H. Curved path following for unmanned surface vehicles with heading amendment. *IEEE Trans. Syst. Man Cybern. Syst.* **2019**, *51*, 4183–4192. [CrossRef]
21. Dai, S.; Wu, Z.; Wang, J.; Tan, M.; Yu, J. Barrier-based adaptive line-of-sight 3-D path-following system for a multijoint robotic fish with sideslip compensation. *IEEE Trans. Cybern.* **2022**, *53*, 4204–4217. [CrossRef]
22. Rome, L.C.; Swank, D.; Corda, D. How fish power swimming. *Science* **1993**, *261*, 340–343. [CrossRef]
23. Van Leeuwen, J.; Lankheet, M.; Akster, H.; Osse, J. Function of red axial muscles of carp (*Cyprinus carpio*): Recruitment and normalized power output during swimming in different modes. *J. Zool.* **1990**, *220*, 123–145. [CrossRef]
24. Ijspeert, A.J.; Crespi, A.; Ryczko, D.; Cabelguen, J.-M. From swimming to walking with a salamander robot driven by a spinal cord model. *Science* **2007**, *315*, 1416–1420. [CrossRef] [PubMed]
25. Ijspeert, A.J.; Crespi, A. Online trajectory generation in an amphibious snake robot using a lamprey-like central pattern generator model. In Proceedings of the 2007 IEEE International Conference on Robotics and Automation, Rome, Italy, 10–14 April 2007; pp. 262–268.
26. Manduca, G.; Santaera, G.; Miraglia, M.; Jansen Van Vuuren, G.; Dario, P.; Stefanini, C.; Romano, D. A Bioinspired Control Strategy Ensures Maneuverability and Adaptability for Dynamic Environments in an Underactuated Robotic Fish. *J. Intell. Robot. Syst.* **2024**, *110*, 69. [CrossRef]
27. Korkmaz, D.; Ozmen Koca, G.; Li, G.; Bal, C.; Ay, M.; Akpolat, Z.H. Locomotion control of a biomimetic robotic fish based on closed loop sensory feedback CPG model. *J. Mar. Eng. Technol.* **2021**, *20*, 125–137. [CrossRef]
28. Han, J. From PID to active disturbance rejection control. *IEEE Trans. Ind. Electron.* **2009**, *56*, 900–906. [CrossRef]

biomimetics MDPI

Review

Learning from Octopuses: Cutting-Edge Developments and Future Directions

Jinjie Duan [1,2], Yuning Lei [1,2], Jie Fang [1,2], Qi Qi [1,2], Zhiming Zhan [3,*] and Yuxiang Wu [2,*]

1 School of Optoelectronic Materials and Technology, Jianghan University, Wuhan 430056, China; djjstudy@sina.com (J.D.); 13638665980@163.com (Y.L.); jf32607@163.com (J.F.); qiqi07230324@163.com (Q.Q.)
2 Institute of Intelligent Sport and Proactive Health, Department of Health and Physical Education, Jianghan University, Wuhan 430056, China
3 School of Artificial Intelligence, Jianghan University, Wuhan 430056, China
* Correspondence: zhanzhm@jhun.edu.cn (Z.Z.); yxwu@jhun.edu.cn (Y.W.)

Abstract: This paper reviews the research progress of bionic soft robot technology learned from octopuses. The number of related research papers increased from 760 in 2021 to 1170 in 2024 (Google Scholar query), with a growth rate of 53.95% in the past five years. These studies mainly explore how humans can learn from the physiological characteristics of octopuses for sensor design, actuator development, processor architecture optimization, and intelligent optimization algorithms. The tentacle structure and nervous system of octopus have high flexibility and distributed control capabilities, which is an important reference for the design of soft robots. In terms of sensor technology, flexible strain sensors and suction cup sensors inspired by octopuses achieve accurate environmental perception and interaction. Actuator design uses octopus muscle fibers and movement patterns to develop various driving methods, including pneumatic, hydraulic and electric systems, which greatly improves the robot's motion performance. In addition, the distributed nervous system of octopuses inspires multi-processor architecture and intelligent optimization algorithms. This paper also introduces the concept of expected functional safety for the first time to explore the safe design of soft robots in failure or unknown situations. Currently, there are more and more bionic soft robot technologies that draw on octopuses, and their application areas are constantly expanding. In the future, with further research on the physiological characteristics of octopuses and the integration of artificial intelligence and materials science, octopus soft robots are expected to show greater potential in adapting to complex environments, human–computer interaction, and medical applications.

Keywords: octopus; sensor; actuator; processor

1. Introduction

In recent years, the field of robotics has undergone a revolutionary transformation, with soft robotics emerging as a pioneering technology. Soft robots demonstrate remarkable potential across a variety of applications due to their unique characteristics, including flexibility and adaptability. Their durability, deformability, and ability to perform complex tasks in unstructured environments have attracted considerable attention from researchers [1–3].

Soft robotics is an interdisciplinary field that integrates concepts from biology, engineering, materials science, and computer science. By examining natural organisms—particularly soft-bodied creatures, such as octopuses, worms, and jellyfish—scientists have drawn significant inspiration for innovative robot designs [4,5]. These organisms exhibit exceptional movement, manipulation, and adaptability, providing a wealth of ideas for the development of advanced soft robotic systems [6,7].

Marine animals have evolved to thrive in fluid environments, developing highly adaptable movement and body structures over time [8]. The wave-like swimming of fish, the flexible tentacle movements of octopuses, and the pulsatile propulsion of jellyfish are all remarkable examples of nature's ingenuity in achieving efficient locomotion within complex environments [9]. The octopus, a quintessential invertebrate, is particularly notable for its highly flexible tentacles that possess multi-degree-of-freedom movement capabilities. Composed of both muscular and gelatinous tissues, the tentacles of octopuses can perform a range of complex actions, including grasping, coiling, and propulsion. This unique structure and functionality serve as valuable inspiration for the design of soft robots [10,11].

Interestingly, the neural architecture of octopuses is decentralized; approximately 40% of their neurons reside in the brain, while the remaining 60% are distributed throughout their tentacles [12]. This distributed processing significantly enhances the octopus's responsiveness and ability to execute multiple tasks simultaneously, offering insights for the development of distributed processing systems in robotics, as well as strategies for task allocation between primary and auxiliary processors [13–15]. The development of soft robots inspired by octopuses primarily manifests in the implementation of flexible strain sensors, the design of sensor arrays on suction cups, and the innovative creation of bionic sensor materials [16,17]. Soft robots can be classified based on their actuators' power sources into three main categories: pneumatic actuators, electric actuators, and hydraulic actuators [18–20]. Additionally, control algorithms, such as proportional–integral-derivative (PID) and linear quadratic regulator (LQR), significantly enhance the motion accuracy of these actuators when integrated with traditional systems. Research into the neural network systems of octopuses has led to the development of distributed control methods, which improve the overall control efficiency of robotic systems. This work also provides crucial insights for the exploration of asynchronous parallel optimization algorithms [21–23]. The significance of these studies is particularly pronounced in the context of the era characterized by big data and advanced computational power.

In recent years, significant advancements have emerged from bionic research focused on octopuses, particularly in the areas of perception, control, actuator design, and processing systems. Further investigation into the physiological characteristics of octopuses holds substantial importance for the ongoing development of soft robots. This review systematically presents the bionic inspiration derived from octopuses and outlines the related research progress in soft robotics, specifically in the domains of sensing, actuation, and distributed processing (Figure 1). Furthermore, this paper introduces the concept of expected functional safety into the realm of soft robot research for the first time [24].

The primary function of sensors is to facilitate accurate data interaction between robots and their environments, while actuators are responsible for enabling movement and task execution. Processors play a crucial role in data processing and movement control [25]. Accordingly, this review begins with a succinct exploration of the physiological characteristics of octopuses, specifically examining the functions of their sensory organs, motor systems, and nervous systems. Subsequently, the discussion extends to the development of multi-processor systems modeled after the octopus nervous system. It also addresses advancements in flexible robotic sensing technologies and multi-sensor fusion algorithms inspired by the sensory organs of octopuses. Additionally, the evolution of various movement modalities and control technologies in flexible robots, influenced by the motor capabilities of octopuses, is introduced. The coordination between the main processor and co-processor is also examined.

Finally, the review concludes by outlining prospective future research directions in the field of bionics that can be applied to the development of flexible robots.

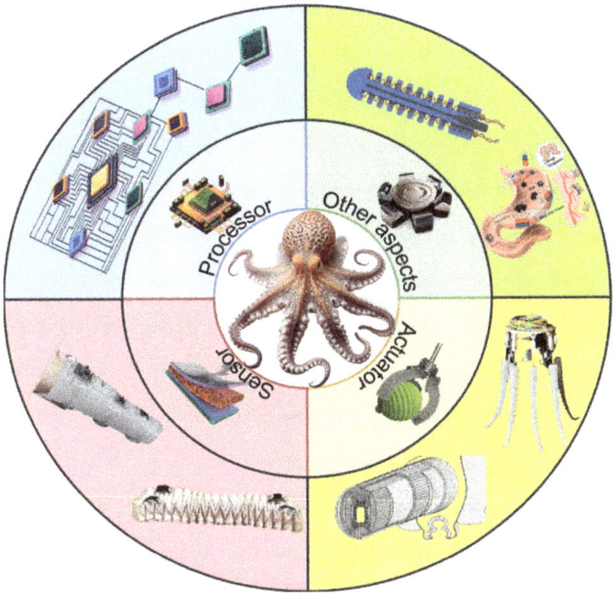

Figure 1. Schematic diagram of the relationship between the octopus's body structure and the bionic soft robot.

2. Octopus and Soft Robot Sensor Technology

Perception technology serves as the foundation for soft robots to achieve autonomous operation and intelligent behavior, and it is a crucial factor for their successful deployment in complex environments. The sensing technology of soft robots is central to their ability to perceive and interact with their surroundings. To enable these robots to respond intelligently to external stimuli, it is essential to integrate soft sensing components and flexible strain sensors into their structural design, allowing them to perceive a variety of environmental signals as feedback [26,27].

Currently, the availability of suitable strain sensors is vital for the perception and autonomy of soft robots; however, the variable shape and dynamic actuation of these robots present challenges for sensor manufacturing and long-term durability [6,7,28,29]. To address these challenges, a deterministic crack propagation mechanism has been proposed, characterized by a programmed crack array utilizing a micro-wrinkle strategy. A computational design of strain sensors has been developed to enhance both sensor modeling and stability. As illustrated in Figure 2a, environmentally stable single-walled carbon nanotubes (SWNTs) are employed to fabricate piezoresistive strain sensors. This design features a two-stage process that integrates the function of "programmed crack arrays in micro-wrinkles", resulting in what is referred to as the PCAM sensor.

To validate the effectiveness of this mechanism, an origami robot was designed, as depicted in Figure 2d. Innovatively, this study also incorporated deep learning technology by employing an artificial neural network (ANN) to predict the motion trajectory of the origami robot based on data collected from the strain sensors [30,31]. The test results are presented in Figure 2e. This technology holds potential for applications in automated navigation and terrain mapping for robotic systems in future endeavors [26].

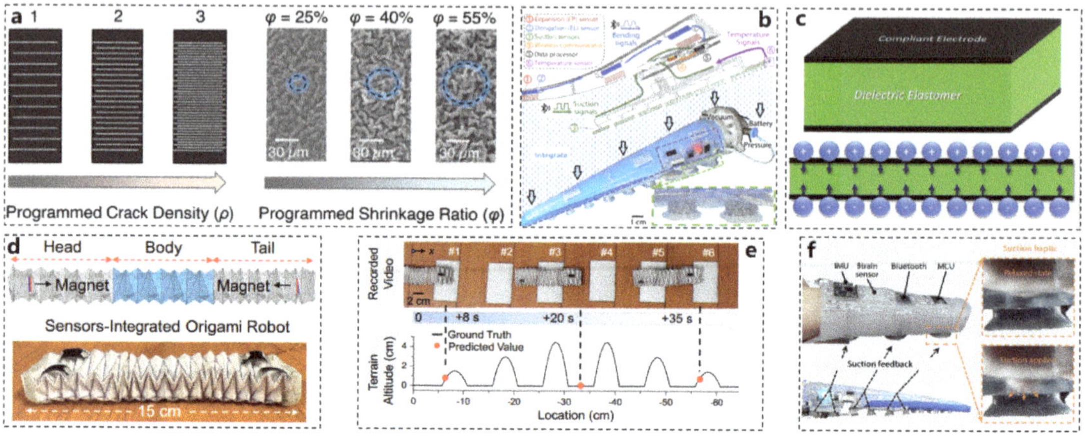

Figure 2. New sensor materials and their applications. (**a**) SWNT material characterization diagram, computationally guided PCAM sensor design during origami robot design, examples of different ρ and φ values. (**b**) E-SOAM integrates bending sensing (through strain sensors) and suction, embeds metal electronic components into the terminal holder, processes all signals at the terminal, and then transmits them wirelessly. (**c**) Schematic diagram of the charge distribution of piezoelectric materials that can be measured and grasped [32]. (**d**) Schematic diagram of the sensor-integrated origami robot. (**e**) ANN predicts the motion trajectory of the origami robot. (**f**) The three internal suction cups on the ventral inner surface of the E-SOAM haptic device transmit the sensory feedback of the end gripper suction to the human finger. The suction force applied to the finger to establish the tactile interaction between the end gripper and the finger is described in detail.

The perception of soft continuum robots presents a significant challenge in the field of robotics. The inherent high degree of freedom and redundancy in soft continuum structures complicates the simultaneous attainment of speed and accuracy in numerical approximate solutions to their nonlinear partial differential equation models [33,34]. Environmental perception in soft continuum robots faces several issues, including difficulties in aligning the sensor circuitry with the Young's modulus of the robot's body and challenges in designing a functional distribution of structures [5,26,35,36].

Observing how octopuses capture prey reveals an effective strategy: their slender tentacles utilize a unique mode of "bending wave transmission" to approach targets. This process begins with bending at the base, propagating along the arm toward the prey [37,38] Once a suction cup attaches to the target, the highly sensitive nerves within the tentacles and suckers quickly detect and secure the object. Inspired by this remarkable predatory behavior, the electronic integrated soft octopus arm (E-SOAM), illustrated in Figure 2b, is designed to enable extensive sensing and interaction capabilities. The E-SOAM features a wearable finger glove equipped with three internal suction cups on its ventral surface. When the suction cup attaches to an object, it generates a vacuum that pulls the skin of the glove, effectively translating the robot's sensory feedback into a tangible human sensation. Consequently, the operator can discern the softness of the object through the tactile feedback provided by the terminal fixture and the wearable glove [39]. The E-SOAM offers a valuable design reference for enhancing the perception and interaction capabilities of soft continuum robots with their environments. By leveraging the bending transmission motion mode characteristic of octopus tentacles, along with wearable flexible tactile devices, we propose a sensor-integrated octopus arm robot prototype. This prototype successfully realizes an operation mode akin to that of octopuses, encompassing sensing, motion, and environmental interaction functionalities.

In terms of materials, the design of the bionic octopus sensor incorporates the functionality of a sensor into a conventional soft robot sucker, drawing inspiration from the octopus sucker. When a voltage is applied across the sucker material, the capacitance increases proportionally. Notably, when the sucker makes contact with an object, the capacitance experiences a significant increase, eliminating the need to lift the object to confirm a successful grasp [40–42]. The material structure of the sensor is illustrated in Figure 2c.

The bionic octopus sensor utilizes a CNT/Ecoflex conductive elastomer as its primary material, enabling the collection of the normal vector of the three-dimensional (3D) spatial force. Additionally, a laser-induced graphene sensing film is integrated into the Ecoflex to serve as the tactile wrist of the sensor, allowing for the discernment of both the magnitude and direction of the tangential component of the force. This integration endows the sensor with perceptual capabilities. Various shapes and scanning electron microscopy (SEM) images of the sensor are presented in Figure 3a [43].

Common sensors operating underwater face significant challenges in complex environments due to issues such as signal attenuation, biological fouling, and interference from water flow [44,45]. Drawing inspiration from the tactile system of the octopus sucker, a novel sucker tactile system has been developed that employs triboelectric tactile receptors to simulate the cephalopod-specific chemical sensor (CR) mechanism. The application of superhydrophobic treatment enhances the microstructure of the sensor's surface, resulting in substantial improvements in output voltage, sensitivity, and response time. The constructed underwater material identification system (UMIS), based on the OI-TENG (Octopus-Inspired Triboelectric Nanogenerator), is illustrated in Figure 3b [46].

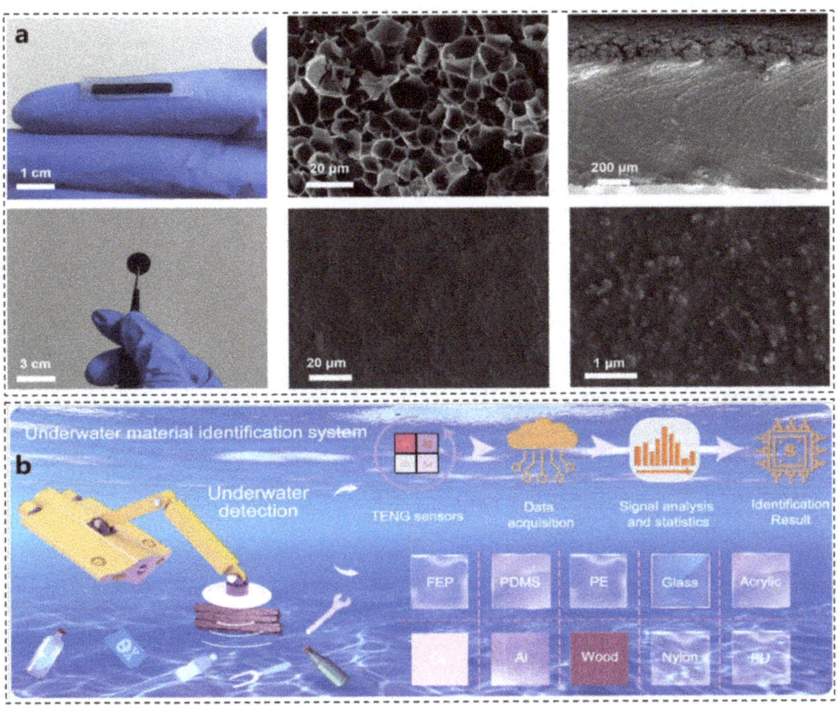

Figure 3. Sensor material design and detection performance optimization. (**a**) CNT/Ecoflex material diagram and SEM characterization [43]. (**b**) OI-TENG underwater material recognition system [46].

In addition to the continuous expansion of sensor types and technical applications, significant progress has been made in the development of sensor post-processing algorithms for soft robot perception technology. To achieve a comprehensive understanding of their environment, soft robots must integrate multiple sensors that can detect various parameters, including force, temperature, and shape [47]. This multimodal sensing capability enables soft robots to better comprehend and adapt to their operational environments [26,47,48]. Among the various techniques employed in multi-sensor fusion technology, the Kalman filter and its variants hold a prominent position. The Kalman filter, along with its extended versions—such as the extended Kalman filter (EKF) and the unscented Kalman filter (UKF)—are widely utilized for multi-sensor data fusion [49]. These methods are particularly effective in addressing sensor noise and uncertainty in both linear and nonlinear systems. In recent years, deep learning technologies, including convolutional neural networks (CNNs) and recurrent neural networks (RNNs), have gained traction in the realm of multi-sensor fusion. These advanced techniques can automatically extract and learn features from sensor data, enhancing the overall performance of perception systems [50–52].

The rapid development of multi-sensor fusion technology has made it a revolutionary enabler for soft robotic systems, especially for robust environmental interaction through adaptive sensor-actuator coupling. As shown in Table 1, a comparative analysis of Kalman filters, CNNs, and RNNs reveals the advantages, disadvantages, and application scenarios of several common sensor processing methods. This synergistic integration of model-based estimation and data-driven learning paradigms is driving the development of embodied intelligence in soft robots, enabling autonomous operation with sub-millimeter precision in unstructured environments.

Table 1. Comparison of the advantages and disadvantages of Kalman filters, CNNs, and RNNs in the sensor fusion of soft robots.

Method	Advantages	Disadvantages	Applicable Scenarios
Kalman	Real-time performance Efficient calculation Multidimensional data fusion	Dependent linear models Noise sensitivity	Soft tentacle pose estimation Dynamic environment position
CNN	Spatial feature extraction High-dimensional data processing End-to-end learning	Highly data-dependent Limited real-time performance	Object recognition (grasping) Tactile texture perception
RNN/LSTM	Temporal modeling capabilities Dynamic adaptation Multi-modal fusion	High training complexity Computational delay	Continuous motion planning Abnormal state detection

3. Octopus and Soft Robotic Actuators

The actuator is a critical component of a soft robot, enabling movement and functionality. Table 2 shown the differences between octopus-inspired soft robot actuators and other biomimetic soft robot actuators. Given the inherent characteristics of soft robots, actuators must possess flexibility, deformability, and the ability to adapt to complex environments. Common modes of actuation in soft robots are primarily powered by various energy sources, including gas, liquid, electricity, heat, magnetism, and chemical reactions [6,53–55].

The octopus serves as an ideal animal model for studying slender, flexible appendages. As illustrated in Figure 4a, the muscle system of the octopus arm consists of a dense three-dimensional array of muscle fibers and connective tissue, which extends along the entire length of the arm and encases the axial nerve cord [55–57]. Similar to the tentacles of the octopus, the movement of a soft robot is primarily facilitated by soft actuators. These actuators are crucial for enabling the movement of the soft robot, making their design and functionality vital to the overall composition of the robotic system. Depending on the specific application scenarios, the design, manufacturing, and control of soft actuators may vary significantly.

Table 2. Comparison between octopus-inspired soft robot actuators and other biomimetic soft robot actuators.

Bionic Objects	Design Features	Drive Mode	Application Scenarios	Technical Challenges and Limitations
Octopus	(1) Highly flexible (2) Dynamic grasping (3) Stable attachment	Pneumatic/ hydraulic drive SMA	Flexible gripping Underwater detection Medical endoscope	Nonlinear deformation control is complex
Elephant Trunk	(1) Flexible and rigid, can carry heavy objects (2) Layered muscle structure	Pneumatic/ hydraulic drive	Industrial handling Rescue robots	Insufficient dynamic stability under high loads
Earthworm	(1) Segmented structure (2) Low energy consumption	Pneumatic Electroactive polymers	Underground exploration Pipeline inspection	Slower movement speed
Fish	(1) Streamlined body (2) Low noise	Motor Electroactive polymers	Marine ecological monitoring Underwater military reconnaissance	Limited steering flexibility

Pneumatic actuators are among the most prevalent types of actuators used in soft robots, utilizing compressed air to facilitate movement. By adjusting the air pressure and flow, pneumatic actuators can expand or contract, generating a range of movements. These actuators offer high flexibility and rapid response times, enabling complex deformation and movement patterns. They are widely employed in tasks such as grasping, walking, and crawling. However, the implementation of pneumatic actuators necessitates complex pneumatic systems and control devices, which may be constrained by the availability of air sources and the configuration of pipelines [20,58–60].

Inspired by the muscle structure of the octopus arm, researchers have designed a soft robotic arm constructed from soft silicone, as depicted in Figure 4d. This robotic arm comprises three slender segments arranged in series, with each segment containing multiple pneumatic chambers embedded within its silicone body. The solenoid valves illustrated in Figure 4e are employed to regulate the flow of compressed air into these pneumatic chambers. The control logic for the system is presented in Figure 4b. By selectively inflating the corresponding chambers, the arm can achieve bending, extension, and twisting movements, thereby facilitating the deformation of the current segment of the robotic arm. The overall control framework is depicted in Figure 4c.

By integrating segment deformation and simulating the function of the antagonistic muscles found in octopuses, this robotic arm can bend in multiple directions, adjust its length, and even twist around its central axis, thereby achieving movement capabilities akin to those of an octopus arm [60–62].

Hydraulic actuators utilize liquid pressure to facilitate movement and are particularly well-suited for applications that require substantial force. By controlling the flow of liquid through hydraulic pumps and valves, these actuators can generate significant thrust and torque, providing both high output force and precise control. They are ideal for tasks that demand high loads and meticulous regulation. However, hydraulic systems tend to be complex, incur high maintenance costs, and carry a risk of leakage [18,63,64].

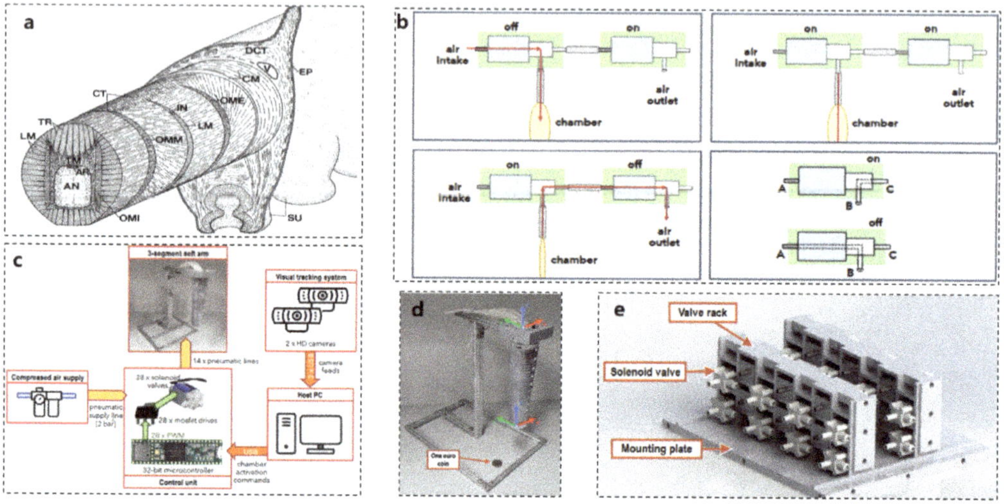

Figure 4. Pneumatic actuator working example reference [60]. (**a**) Octopus vulgaris arm muscle system [57]. (**b**) Pneumatic actuator control logic. (**c**) Diagram of the soft arm pneumatic actuation and control system, including an air source, 28 controllable solenoid valves, a 32-bit microcontroller, a visual tracking system, a host computer, and a three-segment octopus bionic soft arm. (**d**) An actual picture of the three-segment octopus bionic soft arm, comparing its size with a coin. (**e**) Pneumatic actuator control component solenoid valve.

Inspired by the morphology of octopus movement, novel actuators can be designed based on the principles of fluid drive. The actuator model developed for this purpose is illustrated in Figure 5c. This design employs silicone material for the drive chamber, with the driving fluid supplied to the robot through a flexible pipe that channels it to each arm via internal conduits within the robot's body. As depicted in Figure 5d, the arm features a conical shape, and the drive chamber is constructed from silicone reinforced with multiple layers of polyester wire. The polyester wire is arranged in a spiral configuration along the conical clamping wall, ensuring that the actuator can bend and stretch effectively.

Testing has demonstrated that the bionic actuator shown in Figure 5e can achieve movement with minimal drive pressure when the drive is controllable. This actuator is capable of forward movement, rotation, and twisting around its main axis [60,65].

Electric actuators typically achieve motion tasks through direct motor control or methods such as electromagnetism and electrothermal effects. Among these, motor control provides high-precision angle and position regulation, enabling the actuator to perform complex grasping and manipulation tasks. The motor drive system is characterized by its rapid response, allowing for quick starts and stops. Additionally, motors exhibit high-energy conversion efficiency and can deliver substantial output torque within a compact volume. Their ease of integration with other sensors facilitates the creation of comprehensive intelligent control systems. Given the maturity of motor control technology, dynamic modeling is relatively straightforward, allowing for improved design of the entire system [66–70].

Figure 5a illustrates a novel bionic robotic swimming platform inspired by octopus locomotion. This platform achieves efficient swimming using only two standard motors. It combines asymmetric passive deformation arms with umbrella-shaped rapid return structures, effectively mimicking the arm movements and stroke timing of octopuses to enhance swimming efficiency. This actuator design significantly reduces the complexity of the drive system while maintaining excellent swimming performance, offering valuable insights for the development of low-cost underwater vehicles. The motion posture of the modified bionic soft robot is depicted in Figure 5b [71].

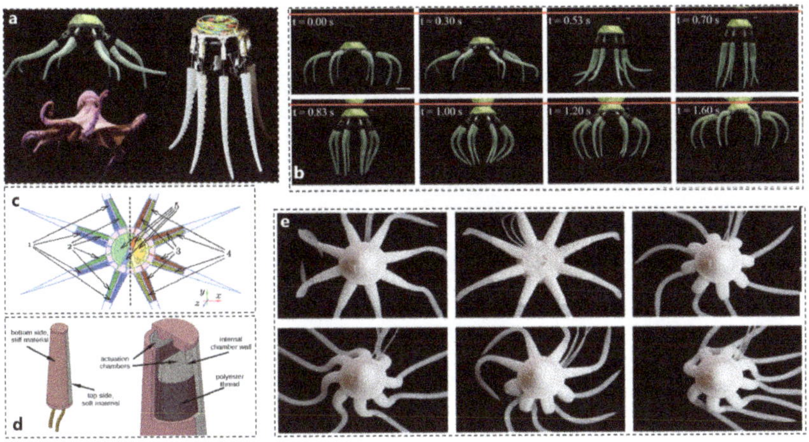

Figure 5. References to other power source actuator working examples. (**a**) Electric actuator complete machine example [71]. (**b**) Electric actuator motion posture [71]. (**c**) Hydraulic actuator modeling [65]. (**d**) Hydraulic actuator tentacle structure design [65]. (**e**) Hydraulic actuator motion posture [65].

In addition to the novel actuators previously discussed, the control algorithm governing the actuator is also of paramount importance. The flexible deformation of the robotic arm structure complicates accurate control. Figure 6b illustrates a proportional–integral (PI) control algorithm based on self-inductance resistance feedback from shape memory alloys (SMAs), which enhances the control accuracy of the bionic robotic arm. The soft robotic arm designed using SMA in conjunction with the PI control algorithm can be flexibly bent and precisely controlled. The overall system control block diagram is presented in Figure 6a.

In this design, SMA wire is selected as the actuator, while silicone material serves as the foundation for the flexible robotic arm, resulting in a bionic flexible arm. The SMA wire acts as an actuator that simulates the longitudinal muscle fibers found in octopus arms, embedded parallel to the central axis of the robot arm, as shown in Figure 6c. Three groups of SMA wires are symmetrically arranged around the axis of the flexible robot arm interface. By constructing a constitutive model for the SMA, the changes in parameters such as stress, strain, and martensite percentage due to temperature variations during the SMA phase transition can be effectively described.

The SMA is driven by pulse-width modulation (PWM) for electric heating. By adjusting the PWM duty cycle, the temperature of the SMA can be controlled, thereby regulating the movement of the flexible robotic arm. This logical relationship allows for the construction of a thermodynamic model for the SMA. Together, these two models elucidate the relationship between variations in the PWM pulse signal's duty cycle and the resulting deformation of the SMA. It is well-known that flexible actuators exhibit significant inertia and hysteresis, leading to slow changes in motion error during operation. Consequently, the differentiating component of the PID algorithm is less effective. Therefore, a PI controller is employed to accurately manage the movement of the bionic flexible robot arm. Control angle measurements were conducted using the device depicted in Figure 6d. Experimental results confirmed that varying driving voltages produced different bending angles and response speeds of the robot arm. Furthermore, it was observed that higher voltages corresponded to faster angle changes, as illustrated in Figure 6e [72–76].

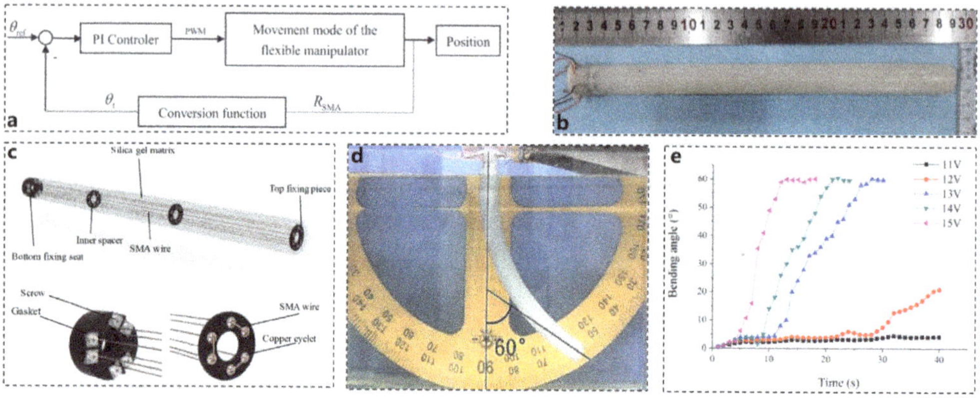

Figure 6. Actuator control algorithm and actuator materials [76]. (**a**) PI control logic suitable for SMA. (**b**) Actual picture of actuator designed with SMA. (**c**) Schematic diagram of actuator power based on SMA material. (**d**) SMA + PI control experimental device diagram. (**e**) Experimental data of SMA motion performance under different voltages.

By studying the movement and associated organ characteristics of the octopus, we have significantly broadened our perspectives on actuator design. This represents a crucial milestone in the optimization of soft robot design and the development of enhanced functionalities.

4. Octopus and Processor Architecture and Intelligent Optimization Algorithms

Octopuses are regarded as the most intelligent of all invertebrates. Research has shown that only 40% of the neurons in octopuses are located in the brain, while the remaining 60% are distributed throughout their eight tentacles. The distribution of these structural functions is illustrated in Figure 7a. A significant portion of the octopus's nervous system resides within its arms. Each arm contains a nerve cord, known as the arm or axial nerve cord, which runs parallel to the suckers. Along the ventral side of each arm, 200 to 300 suckers are staggered, and the nerve cord comprises a dense, continuous network of neuropil that extends from the surrounding layer of monopolar neuronal cell bodies [77–80].

The nerve fibers at the base of each sucker are enlarged, forming structures commonly referred to as ganglia, while the segments between these ganglia are termed interganglionic regions. Notably, the octopus brain does not issue commands for every minor movement of the arms; instead, many decisions are made autonomously by the arms themselves. This distributed arrangement of neurons enables the octopus's arms to solve problems independently. For instance, when an octopus explores a cave and discovers edible items, its arm can simultaneously open shellfish. This distributed control architecture facilitates effective and computationally efficient control of the arms [79,81].

The structure of the octopus's nervous system and the manner in which the brain communicates with the arms can inform the design of system architectures. A multi-processor distributed processing system allocates control between a main processor and a coprocessor, allowing the coprocessor to share workload tasks and thereby reducing the burden on the main processor. This distributed control approach is often referred to as hybrid control, which describes a hierarchically organized system where a discrete control mode is selected at a higher level, while different continuous controllers operate at a lower level for each mode. The control strategy employed by octopuses exhibits several characteristics that align with hybrid control principles. The evolution of the distributed processing architecture inspired by the octopus nervous system is illustrated in Figure 7b [22,82–85].

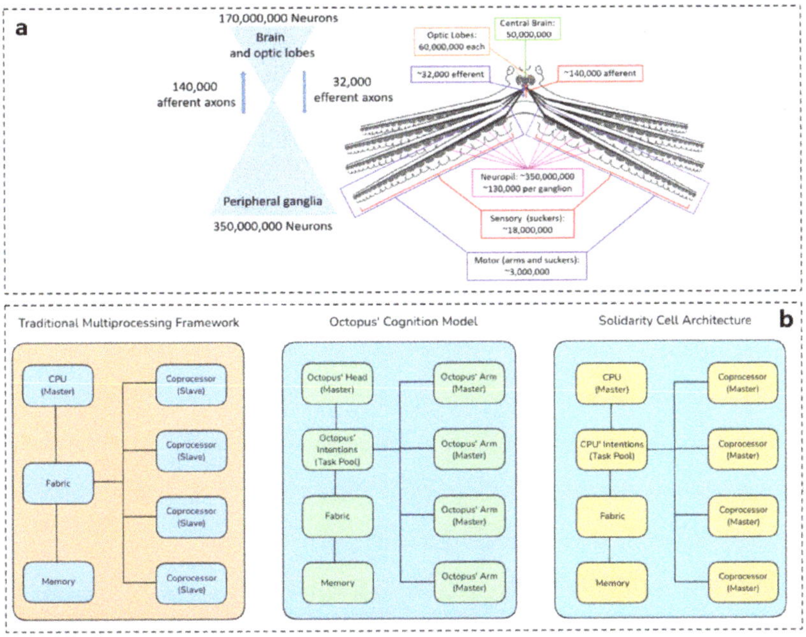

Figure 7. Octopus neural network explaining distributed processing. (**a**) Schematic diagram of octopus neural network and neuron distribution [83]. (**b**) Evolution diagram of distributed processing architecture inspired by octopus [82].

In addition, an intelligent optimization algorithm inspired by the characteristics of the octopus neural network has been developed. The Octopus-Inspired Optimization Algorithm (OIO) is an advanced optimization technique that leverages the unique features of the octopus's neural system. It employs a multi-level hierarchical strategy to achieve global search and local optimization through asynchronous parallelism. This effective combination of search methodologies significantly enhances both search efficiency and adaptability.

The multi-level structure of the OIO consists of four levels: tentacles, suction cups, individuals, and groups, simulating the octopus's nervous system to facilitate information transfer and interaction. As illustrated in Figure 8a, the suction cups are responsible for the perception and processing of local information. At the beginning of each iteration, information is collected through an evaluation function and relayed to the tentacles. The tentacles then adjust the parameters and update the performance indicators based on the feedback received from the suction cups. As shown in Figure 8b, the tentacles are tasked with distributed search and autonomous decision making. They integrate information from the suction cups, evaluate the environment, and adjust their movements accordingly, providing feedback to the individual and influencing subsequent movement instructions. Figure 8c depicts the role of individuals as macro decision makers, who make decisions by synthesizing information from the tentacles and environmental characteristics. They also implement a reward and punishment mechanism to adjust the group optimization algorithm, preventing prolonged stagnation in inefficient areas.

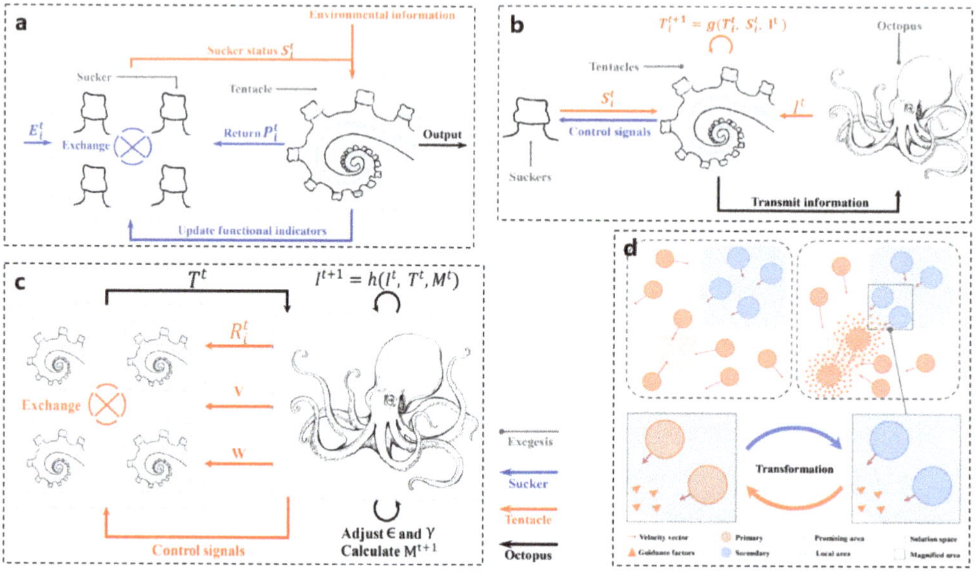

Figure 8. Intelligent optimization algorithms inspired by octopuses [22]. (**a**) Suction cups and perception. (**b**) Tentacles and distributed processing. (**c**) Octopuses and individual decision making. (**d**) Octopus swarms and group optimization algorithms.

Figure 8d illustrates how the group achieves collaborative optimization by aggregating information to comprehensively describe the solution process, enabling all individuals to work in concert. The OIO utilizes an asynchronous parallel mechanism to allocate computing resources efficiently, thereby improving execution efficiency. Different tentacles and suction cups can search and process information in parallel, significantly reducing the algorithm run time. Moreover, the OIO possesses the capability for adaptive parameter control, allowing it to adjust search step sizes, exploration–exploitation balance factors, and other parameters based on real-time feedback. This adaptability enhances the robustness and convergence speed of the algorithm. A regional regeneration mechanism is employed to prevent the tentacles from remaining in local optimal areas for extended periods, thereby enhancing global search capabilities. The OIO not only emphasizes individual intelligence simulation but also highlights the emergence of group intelligence. The exchange of information between individuals improves the algorithm's adaptability. In terms of design, the OIO rationally implements parallel and asynchronous iteration strategies, fully utilizing multi-core processor resources to avoid inefficiencies associated with synchronous updates, thus further enhancing execution efficiency. The adaptive and dynamic adjustment mechanisms enable the algorithm to modify strategies in real time according to varying optimization problems, improving its versatility. Additionally, the OIO employs various mechanisms, such as tentacle extension and contraction and regional regeneration, to avoid local optima and continuously explore new areas in search of global optimal solutions. This flexible design, combined with diverse mechanisms, equips the OIO with superior search capabilities and optimization performance when addressing complex optimization problems, offering novel ideas and methodologies for applications in related fields [22,86].

The octopus's high-performance distributed control layer is notably distinct compared to that of other organisms. Its intricate distributed local control system enables the octopus to manage thousands of actuated degrees of freedom, specifically in its suction cups and arm musculature [80]. While it is currently unfeasible to create a robot with as many local control degrees of freedom as an octopus possesses, the architectural principles of the octopus can be effectively applied to robots with a reduced number of local control degrees of freedom.

In the era of big data, studying the nervous system of octopuses holds great significance. For example, distributed processing can allocate data across multiple nodes for parallel computing, enabling large-scale data integration. As the volume of data increases, the number of computing nodes can also be expanded, thereby enhancing the processing capacity of the entire system. In the field of deep learning, these insights have important implications for the development of edge computing, federated learning, and multimodal data processing [87–89]. In the future, a better understanding of octopus-distributed control could help humans develop innovative methods for robot planning and control, as well as provide new inspiration for computational optimization. Distributed processing can integrate data from various sensors in real time. Through parallel computing, perception and control tasks are distributed across multiple computing nodes, allowing robots to process complex computational tasks more quickly and enhancing the overall intelligence and flexibility of the system [90,91].

5. Octopus Inspires Other Aspects of Robotics

In addition to inspiring the design of robotic components, octopuses also influence the development of various application fields. As shown in Figure 9b, the numerous suction cups on octopus tentacles can expel air through muscle contractions, generating a pressure difference. When the muscle tension is relaxed, the suction cups firmly adhere to objects. This phenomenon has inspired numerous research directions, including the integration of electrophysiological signal acquisition equipment with measurement targets, the development of wearable hydrogels, wound-healing methods, and high-performance robotic arm suction cups. Given the strong adhesion capabilities of octopus suction cups, incorporating this structure into the probes of electrophysiological signal acquisition equipment can ensure a secure connection to measurement targets, such as rough skin, thereby enabling the continuous monitoring of vital signs in living organisms [92,93].

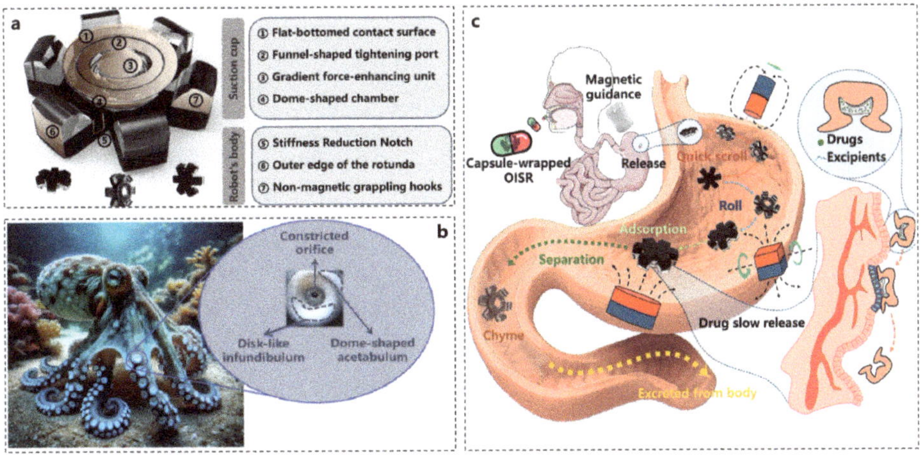

Figure 9. Octopus-inspired drug-releasing robot [94]. (**a**) Bionic structure. (**b**) Schematic diagram of the characteristics of the octopus suction cup. (**c**) Schematic diagram of the working of the drug-releasing robot.

Figure 9a illustrates a drug-release soft robot designed based on the structural characteristics of octopus suction cups, capable of slow drug release in the body for therapeutic purposes. The overall design of the robot is magnetic; under the influence of an external magnetic field, as depicted in Figure 9c, the soft robot can be positioned at lesion sites in the stomach and securely attach via instantaneous suction generated by the magnetic gradient force. The drug contained within the suction cup is gradually released at the lesion site, achieving targeted and sustained drug delivery [94].

In addition, the development of Safety of the Intended Functionality (SOTIF) continues in the industry. The concept of SOTIF originated in the automotive industry and is defined by the ISO21448 standard [95]. It aims to address the risk of harm caused by functional limitations or reasonably foreseeable misuse under non-system failures. Its core goal is to minimize the proportion of known and unknown hazardous scenarios and improve the boundaries of the system in complex environments through design optimization and verification. We believe that this concept is also applicable to bionic robots. This paper introduces the concept of SOTIF into the study of soft robots for the first time. Research on SOTIF primarily focuses on countermeasures in the event of soft robot failures or situations outside the scope of system modeling, ensuring that the soft robot actuator does not cause harm to itself or the external environment. Similar to how an octopus protects itself in unknown situations through means such as camouflage and ink spraying, the design of predictive functional safety can guarantee the safe operation of the soft robot's actuator during the control process. In response to these challenges, some researchers have proposed feedback strategies to ensure the safe operation of soft actuators when controlling soft robots. The supervisory controller monitors the actuator's state and dynamically adjusts the control input to prevent conditions that could lead to physical damage. As shown in Figure 10b, relevant researchers utilized shape memory alloys as actuators along with thermal sensors. The temperature of the soft limb robot is maintained in a stable state to prevent overheating during contact scenarios, such as environmental restrictions, human interaction, or unreasonable control inputs. Figure 10a presents experimental results demonstrating that when the robot is touched, the system automatically lowers the temperature to prevent burns. The experiment confirms that the expected functional safety monitor is effective, demonstrating the stability and verifiable safety of the supervisory controller under specific conditions [96].

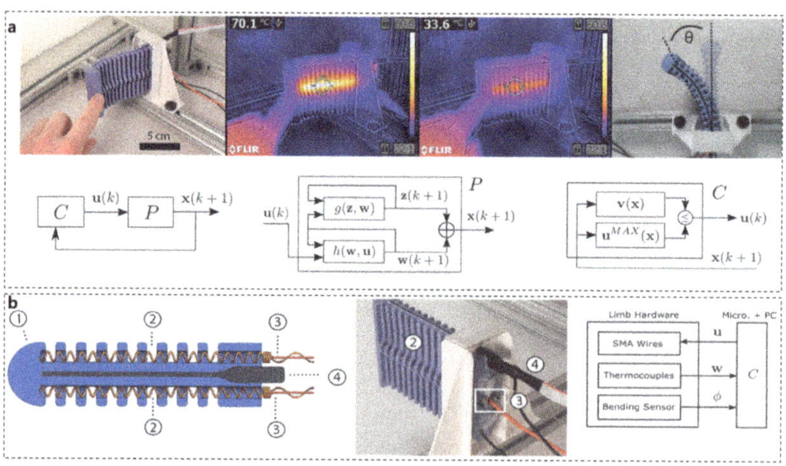

Figure 10. SOTIF functional design example [96]. (**a**) Test solution design and demonstration. (**b**) SOTIF functional design and system block diagram.

6. Conclusions

In the realm of soft robotics, the octopus serves as an excellent teacher. From it, we can derive insights into sensor design and fusion, innovative actuator design, optimization of processing systems, and application ideas across various fields. This paper discusses the application of advanced materials inspired by octopuses in the design of soft robot sensors, including new combinations of classic sensor materials to achieve novel sensor functions and improved sensor performance driven by octopus designs.

In the area of actuator design, we have presented the latest research progress from perspectives including pneumatic, hydraulic, electric, and energy-derived actuation, along with the development of new actuator materials, drive structures, and control algorithms that enhance control accuracy, inspired by the movement and locomotion of octopuses.

Regarding processing systems, we explored the design and optimization of distributed systems inspired by octopuses and introduced intelligent optimization algorithms based on octopus behavior. Finally, we integrated the concept of expected functional safety into soft robot design and outlined its practical applications.

Human exploration of the ocean is deepening, biotechnology is advancing rapidly, and research on mollusks, such as octopuses, is becoming increasingly sophisticated. Through bionic research on octopuses, combined with the continuous progress in artificial intelligence and materials science, breakthrough developments are expected in many fields [4,97]. The tentacles of octopuses exhibit exceptional freedom of movement and flexibility, while their unique nervous system and learning capabilities provide a rich source of inspiration for artificial intelligence and robotics. In the realm of intelligent soft robots, future bionic octopus robots will be able to simulate octopus movements more accurately, optimize grasping paths and strengths using artificial intelligence algorithms, and enhance their adaptability in complex environments [98,99].

In the realms of human–computer interaction and medical applications, octopus-like flexible robotic arms are expected to see widespread use in surgical procedures and rehabilitation treatments [17,76,100]. Their softness and high degree of freedom enable more precise operation of surgical instruments, thereby reducing trauma to patients. Additionally, when combined with artificial intelligence and flexible materials, smart wearable devices resembling octopus tentacles may be developed in the future for rehabilitation training or assisted movement [101].

In the context of integrating artificial intelligence and neuroscience, the nervous system and learning capabilities of the octopus offer a new research direction for artificial intelligence. By simulating the octopus's neural network, more efficient decision-making algorithms can be developed in the future, enabling robots to autonomously learn and adapt to complex environments [22,102–108]. Additionally, further research into the learning and memory mechanisms of octopuses may lead to breakthroughs in artificial intelligence within the fields of cognitive and emotional computing. Regarding multifunctional integrated bionic systems, robots that can monitor environmental changes in real time and autonomously adjust their behavior will be developed by combining the octopus's perceptual abilities with the responsiveness of smart materials [32,101–106].

Author Contributions: Conceptualization, J.D. and Y.W.; methodology, Y.W.; investigation, Y.L. and J.F.; data curation, Q.Q.; writing—original draft preparation, J.D. and Z.Z.; writing—review and editing, Y.W. and Z.Z.; supervision, Y.W.; funding acquisition, Y.W. All authors have read and agreed to the published version of the manuscript.

Funding: This research was funded by the National Natural Science Foundation of China (82071970) and the Science Fund for Distinguished Young Scholars of Hubei Province (2023AFA109).

Institutional Review Board Statement: Not applicable.

Informed Consent Statement: Not applicable.

Data Availability Statement: Not applicable.

Conflicts of Interest: The authors declare no conflicts of interest.

References

1. Zhang, Y.; Li, P.; Quan, J.; Li, L.; Zhang, G.; Zhou, D. Progress, Challenges, and Prospects of Soft Robotics for Space Applications. *Adv. Intell. Syst.* **2023**, *5*, 2200071. [CrossRef]
2. Hartmann, F.; Baumgartner, M.; Kaltenbrunner, M. Becoming sustainable, the new frontier in soft robotics. *Adv. Mater.* **2021**, *33*, 2004413. [PubMed]
3. Fang, J.; Zhuang, Y.; Liu, K.; Chen, Z.; Liu, Z.; Kong, T.; Xu, J.; Qi, C. A Shift from Efficiency to Adaptability: Recent Progress in Biomimetic Interactive Soft Robotics in Wet Environments. *Adv. Sci.* **2022**, *9*, 2104347. [CrossRef]
4. Chen, Y.; Zhang, Y.; Li, H.; Shen, J.; Zhang, F.; He, J.; Lin, J.; Wang, B.; Niu, S.; Han, Z.; et al. Bioinspired hydrogel actuator for soft robotics: Opportunity and challenges. *Nano Today* **2023**, *49*, 101764. [CrossRef]
5. Yang, Y.; He, Z.; Jiao, P.; Ren, H. Bioinspired soft robotics: How do we learn from creatures? *IEEE Rev. Biomed. Eng.* **2022**, *17*, 153–165.
6. Sarker, A.; Ul Islam, T.; Islam, M.R. A Review on Recent Trends of Bioinspired Soft Robotics: Actuators, Control Methods, Materials Selection, Sensors, Challenges, and Future Prospects. *Adv. Intell. Syst.* **2025**, *7*, 2400414.
7. Dou, W.; Zhong, G.; Cao, J.; Shi, Z.; Peng, B.; Jiang, L. Soft Robotic Manipulators: Designs, Actuation, Stiffness Tuning, and Sensing. *Adv. Mater. Technol.* **2021**, *6*, 2100018. [CrossRef]
8. Somero, G.N. Solutions: How adaptive changes in cellular fluids enable marine life to cope with abiotic stressors. *Mar. Life Sci. Technol.* **2022**, *4*, 389–413.
9. Luo, Y. A Numerical Study of Fin and Jet Propulsions Involving Fluid-Structure Interactions. Ph.D. Thesis, University of Strathclyde, Glasgow, UK, 2021.
10. Li, G.; Wong, T.W.; Shih, B.; Guo, C.; Wang, L.; Liu, J.; Wang, T.; Liu, X.; Yan, J.; Wu, B.; et al. Bioinspired soft robots for deep-sea exploration. *Nat. Commun.* **2023**, *14*, 7097. [CrossRef]
11. Hasan, K.; Ahmad, S.; Liaf, A.F.; Karimi, M.; Ahmed, T.; Shawon, M. Oceanic Challenges to Technological Solutions: A Review of Autonomous Underwater Vehicle Path Technologies in Biomimicry, Control, Navigation and Sensing. *IEEE Access* **2024**, *12*, 46202–46231.
12. Mather, J. Octopus consciousness: The role of perceptual richness. *NeuroSci* **2021**, *2*, 276–290. [CrossRef]
13. Li, Z.; Chao, X.; Hameed, I.; Li, J.; Zhao, W.; Jing, X. Biomimetic omnidirectional multi-tail underwater robot. *Mech. Syst. Signal Process.* **2022**, *173*, 109056. [CrossRef]
14. Tong, R.; Feng, Y.; Wang, J.; Wu, Z.; Tan, M.; Yu, J. A Survey on Reinforcement Learning Methods in Bionic Underwater Robots. *Biomimetics* **2023**, *8*, 168. [CrossRef] [PubMed]
15. Hao, T.; Xiao, H.; Ji, M.; Liu, Y.; Liu, S. Integrated and Intelligent Soft Robots. *IEEE Access* **2023**, *11*, 99862–99877. [CrossRef]
16. Giordano, G.; Carlotti, M.; Mazzolai, B. A perspective on cephalopods mimicry and bioinspired technologies toward proprioceptive autonomous soft robots. *Adv. Mater. Technol.* **2021**, *6*, 2100437.
17. Zhang, D.; Xu, J.; Liu, X.; Zhang, Q.; Cong, Q.; Chen, T.; Liu, C. Advanced Bionic Attachment Equipment Inspired by the Attachment Performance of Aquatic Organisms: A Review. *Biomimetics* **2023**, *8*, 85. [CrossRef]
18. Shi, H.; Tan, K.; Zhang, B.; Liu, W. Review on Research Progress of Hydraulic Powered Soft Actuators. *Energies* **2022**, *15*, 9048. [CrossRef]
19. Pagoli, A.; Chapelle, F.; Corrales-Ramon, J.A.; Mezouar, Y.; Lapusta, Y. Review of soft fluidic actuators: Classification and materials modeling analysis. *Smart Mater. Struct.* **2021**, *31*, 013001.
20. Su, H.; Hou, X.; Zhang, X.; Qi, W.; Cai, S.; Xiong, X.; Guo, J. Pneumatic Soft Robots: Challenges and Benefits. *Actuators* **2022**, *11*, 92. [CrossRef]
21. Liu, Z.; Peng, K.; Han, L.; Guan, S. Modeling and Control of Robotic Manipulators Based on Artificial Neural Networks: A Review. *Iran. J. Sci. Technol. Trans. Mech. Eng.* **2023**, *47*, 1307–1347. [CrossRef]
22. Wang, X.; Xu, L.; Wang, Y.; Dong, Y.; Li, X.; Deng, J.; He, R. Octopus Inspired Optimization Algorithm: Multi-Level Structures and Parallel Computing Strategies. *arXiv* **2024**, arXiv:2410.07968.
23. Zhao, W.; Zhang, Y.; Lim, K.M.; Yang, L.; Wang, N.; Peng, L. Research on control strategy of pneumatic soft bionic robot based on improved CPG. *PLoS ONE* **2024**, *19*, e0306320. [CrossRef]
24. Dharmdas, A.; Patil, A.Y.; Baig, A.; Hosmani, O.Z.; Mathad, S.N.; Patil, M.B.; Kumar, R.; Kotturshettar, B.B.; Fattah, I.M.R. An Experimental and Simulation Study of the Active Camber Morphing Concept on Airfoils Using Bio-Inspired Structures. *Biomimetics* **2023**, *8*, 251. [CrossRef] [PubMed]

25. Javaid, M.; Haleem, A.; Singh, R.P.; Rab, S.; Suman, R. Significance of sensors for industry 4.0: Roles, capabilities, and applications. *Sens. Int.* **2021**, *2*, 100110. [CrossRef]
26. Hegde, C.; Su, J.; Tan, J.M.R.; He, K.; Chen, X.; Magdassi, S. Sensing in soft robotics. *ACS Nano* **2023**, *17*, 15277–15307. [CrossRef]
27. Yang, H.; Ding, S.; Wang, J.; Sun, S.; Swaminathan, R.; Ng, S.W.L.; Pan, X.; Ho, G.W. Computational design of ultra-robust strain sensors for soft robot perception and autonomy. *Nat. Commun.* **2024**, *15*, 1636. [CrossRef]
28. Chen, Z. A review on robotic fish enabled by ionic polymer–metal composite artificial muscles. *Robot. Biomim.* **2017**, *4*, 24. [CrossRef]
29. Zhou, S.; Li, Y.; Wang, Q.; Lyu, Z. Integrated Actuation and Sensing: Toward Intelligent Soft Robots. *Cyborg Bionic Syst.* **2025**, *6*, 0105. [CrossRef]
30. Hoang, A.T.; Nižetić, S.; Ong, H.C.; Tarelko, W.; Pham, V.V.; Le, T.H.; Chau, M.Q.; Nguyen, X.P. A review on application of artificial neural network (ANN) for performance and emission characteristics of diesel engine fueled with biodiesel-based fuels. *Sustain. Energy Technol. Assess.* **2021**, *47*, 101416. [CrossRef]
31. Kurani, A.; Doshi, P.; Vakharia, A.; Shah, M. A Comprehensive Comparative Study of Artificial Neural Network (ANN) and Support Vector Machines (SVM) on Stock Forecasting. *Ann. Data Sci.* **2023**, *10*, 183–208. [CrossRef]
32. Jamali, A.; Mishra, D.B.; Goldschmidtboeing, F.; Woias, P. Soft octopus-inspired suction cups using dielectric elastomer actuators with sensing capabilities. *Bioinspir. Biomim.* **2024**, *19*, 036009. [CrossRef]
33. Della Santina, C.; Duriez, C.; Rus, D. Model-based control of soft robots: A survey of the state of the art and open challenges. *IEEE Control Syst. Mag.* **2023**, *43*, 30–65.
34. Mengaldo, G.; Renda, F.; Brunton, S.L.; Bächer, M.; Calisti, M.; Duriez, C.; Chirikjian, G.S.; Laschi, C. A concise guide to modelling the physics of embodied intelligence in soft robotics. *Nat. Rev. Phys.* **2022**, *4*, 595–610. [CrossRef]
35. Chen, Z.; Renda, F.; Le Gall, A.; Mocellin, L.; Bernabei, M.; Dangel, T.; Ciuti, G.; Cianchetti, M.; Stefanini, C. Data-Driven Methods Applied to Soft Robot Modeling and Control: A Review. *IEEE Trans. Autom. Sci. Eng.* **2024**, *22*, 2241–2256. [CrossRef]
36. Rus, D.; Tolley, M.T. Design, fabrication and control of soft robots. *Nature* **2015**, *521*, 467–475. [CrossRef]
37. Good, J.; Jeffs, A.; Spreitzenbarth, S. The Effect of Prey Size and Prey Density on Paralarval Feeding Success of the Common Sydney Octopus, Octopus tetricus. *Aquac. Fish Fish.* **2024**, *4*, e70028.
38. Awa, N.; Dan, S.; Nagatsuka, K.; Sekiguchi, Y.; Shimba, A.; Anaguchi, Y.; Kamei, Y.; Hamasaki, K. Ontogeny of predatory capacity and prey choice during early life of the holobenthic octopus Amphioctopus fangsiao (d'Orbigny, 1841): Switching prey-choice strategy. *Mar. Biol.* **2024**, *171*, 206. [CrossRef]
39. Xie, Z.; Yuan, F.; Liu, J.; Tian, L.; Chen, B.; Fu, Z.; Mao, S.; Jin, T.; Wang, Y.; He, X.; et al. Octopus-inspired sensorized soft arm for environmental interaction. *Sci. Robot.* **2023**, *8*, eadh7852.
40. Wang, Z.; Sun, G.; Fan, X.; Xiao, P.; Zhu, L. Biomimetic Octopus Suction Cup with Attachment Force Self-Sensing Capability for Cardiac Adhesion. *Soft Robot.* **2024**, *11*, 1043–1054.
41. Wu, M.; Afridi, W.H.; Wu, J.; Afridi, R.H.; Wang, K.; Zheng, X.; Wang, C.; Xie, G. Octopus-Inspired Underwater Soft Robotic Gripper with Crawling and Swimming Capabilities. *Research* **2024**, *7*, 0456. [CrossRef]
42. Sun, Y.; Tang, L.; Wang, L.; Guo, Q.; Lv, Y.; Yang, L. Liquid-Amplified Electrostatically Driven Octopus Sucker-Inspired Suction Cup. *Adv. Intell. Syst.* **2025**, *7*, 2400279. [CrossRef]
43. Ruan, D.; Chen, G.; Luo, X.; Cheng, L.; Wu, H.; Liu, A. Bionic octopus-like flexible three-dimensional force sensor for meticulous handwriting recognition in human-computer interactions. *Nano Energy* **2024**, *123*, 109357. [CrossRef]
44. Teng, S.; Liu, A.; Ye, X.; Wang, J.; Fu, J.; Wu, Z.; Chen, B.; Liu, C.; Zhou, H.; Zeng, Y.; et al. Review of intelligent detection and health assessment of underwater structures. *Eng. Struct.* **2024**, *308*, 117958. [CrossRef]
45. Mohsan, S.A.H.; Mazinani, A.; Othman, N.Q.H.; Amjad, H. Towards the internet of underwater things: A comprehensive survey. *Earth Sci. Inform.* **2022**, *15*, 735–764. [CrossRef]
46. Hao, Y.; Sun, Y.; Wen, J.; Gao, X.; Wang, Y.; Zhu, Z.; Wang, Z.L.; Chen, B. Octopus-inspired multichannel tactile sensor for enhanced underwater material identification. *Chem. Eng. J.* **2025**, *507*, 160604. [CrossRef]
47. Qiu, Y.; Ashok, A.; Nguyen, C.C.; Yamauchi, Y.; Do, T.N.; Phan, H. Integrated Sensors for Soft Medical Robotics. *Small* **2024**, *20*, e2308805. [CrossRef]
48. Wang, T.; Jin, T.; Lin, W.; Lin, Y.; Liu, H.; Yue, T.; Tian, Y.; Li, L.; Zhang, Q.; Lee, C. Multimodal sensors enabled autonomous soft robotic system with self-adaptive manipulation. *ACS Nano* **2024**, *18*, 9980–9996.
49. Wang, Y.; Xie, C.; Liu, Y.; Zhu, J.; Qin, J. A Multi-Sensor Fusion Underwater Localization Method Based on Unscented Kalman Filter on Manifolds. *Sensors* **2024**, *24*, 6299. [CrossRef]
50. Qiu, S.; Zhao, H.; Jiang, N.; Wang, Z.; Liu, L.; An, Y.; Zhao, H.; Miao, X.; Liu, R.; Fortino, G. Multi-sensor information fusion based on machine learning for real applications in human activity recognition: State-of-the-art and research challenges. *Inf. Fusion* **2022**, *80*, 241–265. [CrossRef]
51. Khusheef, A.S.; Shahbazi, M.; Hashemi, R. Deep Learning-Based Multi-Sensor Fusion for Process Monitoring: Application to Fused Deposition Modeling. *Arab. J. Sci. Eng.* **2024**, *49*, 10501–10522. [CrossRef]

52. Kim, Y.; Zhao, X. Magnetic soft materials and robots. *Chem. Rev.* **2022**, *122*, 5317–5364. [CrossRef]
53. Wang, J.; Chortos, A. Control strategies for soft robot systems. *Adv. Intell. Syst.* **2022**, *4*, 2100165. [CrossRef]
54. Ahmed, F.; Waqas, M.; Shaikh, B.; Khan, U.; Soomro, A.M.; Kumar, S.; Ashraf, H.; Memon, F.H.; Choi, K.H. Multi-material Bio-inspired Soft Octopus Robot for Underwater Synchronous Swimming. *J. Bionic Eng.* **2022**, *19*, 1229–1241. [CrossRef]
55. Di Clemente, A. Neural Control and Biomechanics of the Octopus Arm Muscular Hydrostat. Ph.D. Thesis, University of Genoa, Genova, Italy, 2022.
56. Mathews, R.K. EMG Analysis of Octopus Arms' Muscles. Ph.D. Thesis, Arizona State University, Tempe, AZ, USA, 2022.
57. Kier, W.; Stella, M. The arrangement and function of octopus arm musculature and connective tissue. *J. Morphol.* **2007**, *268*, 831–843. [CrossRef]
58. Xie, Z.; Domel, A.G.; An, N.; Green, C.; Gong, Z.; Wang, T.; Knubben, E.M.; Weaver, J.C.; Bertoldi, K.; Wen, L. Octopus Arm-Inspired Tapered Soft Actuators with Suckers for Improved Grasping. *Soft Robot.* **2020**, *7*, 639–648. [CrossRef]
59. Xavier, M.S.; Tawk, C.D.; Zolfagharian, A.; Pinskier, J.; Howard, D.; Young, T.; Lai, J.; Harrison, S.M.; Yong, Y.K.; Bodaghi, M.; et al. Soft pneumatic actuators: A review of design, fabrication, modeling, sensing, control and applications. *IEEE Access* **2022**, *10*, 59442–59485. [CrossRef]
60. Papadakis, E.; Tsakiris, D.P.; Sfakiotakis, M. An Octopus-Inspired Soft Pneumatic Robotic Arm. *Biomimetics* **2024**, *9*, 773. [CrossRef]
61. Flash, T.; Zullo, L. Biomechanics, motor control and dynamic models of the soft limbs of the octopus and other cephalopods. *J. Exp. Biol.* **2023**, *226* (Suppl. S1), jeb245295. [CrossRef]
62. Hochner, B.; Zullo, L.; Shomrat, T.; Levy, G.; Nesher, N. Embodied mechanisms of motor control in the octopus. *Curr. Biol.* **2023**, *33*, R1119–R1125. [CrossRef]
63. Kotak, P.; Maxson, S.; Weerakkody, T.; Cichella, V.; Lamuta, C. Octopus-Inspired Muscular Hydrostats Powered by Twisted and Coiled Artificial Muscles. *Soft Robot.* **2024**, *11*, 432–443. [CrossRef] [PubMed]
64. Li, Y.; Li, R.; Yang, J.; Yu, X.; Xu, J. Review of Recent Advances in the Drive Method of Hydraulic Control Valve. *Processes* **2023**, *11*, 2537. [CrossRef]
65. Fras, J.; Noh, Y.; Macias, M.; Wurdemann, H.; Althoefer, K. Bio-Inspired Octopus Robot Based on Novel Soft Fluidic Actuator. In Proceedings of the 2018 IEEE International Conference on Robotics and Automation (ICRA), Brisbane, QLD, Australia, 21–25 May 2018; pp. 1583–1588. [CrossRef]
66. Ma, Z.; Sameoto, D. A review of electrically driven soft actuators for soft robotics. *Micromachines* **2022**, *13*, 1881. [CrossRef] [PubMed]
67. Albukhari, A.; Mescheder, U. Inchworm Motors and Beyond: A Review on Cooperative Electrostatic Actuator Systems. *Actuators* **2023**, *12*, 163. [CrossRef]
68. Jahanshahi, H.; Zhu, Z.H. Review of Machine Learning in Robotic Grasping Control in Space Application. *Acta Astronaut.* **2024**, *220*, 37–61.
69. Febrer-Nafría, M.; Nasr, A.; Ezati, M.; Brown, P.; Font-Llagunes, J.M.; McPhee, J. Predictive multibody dynamic simulation of human neuromusculoskeletal systems: A review. *Multibody Syst. Dyn.* **2023**, *58*, 299–339. [CrossRef]
70. Chu, L.; Li, Q.; Gu, F.; DU, X.; He, Y.; Deng, Y. Design, modeling, and control of morphing aircraft: A review. *Chin. J. Aeronaut.* **2022**, *35*, 220–246. [CrossRef]
71. Zhang, B.; Zhang, Y.; Li, Y.; Xuan, S.; Ng, H.W.; Liufu, Y.; Tang, Z.; Laschi, C. Octopus-Swimming-Like Robot with Soft Asymmetric Arms. *arXiv* **2024**, arXiv:2410.11764.
72. Shang, D.; Li, X.; Yin, M.; Li, F. Dynamic modeling and fuzzy adaptive control strategy for space flexible robotic arm considering joint flexibility based on improved sliding mode controller. *Adv. Space Res.* **2022**, *70*, 3520–3539.
73. Zhu, J.; Cherubini, A.; Dune, C.; Navarro-Alarcon, D.; Alambeigi, F.; Berenson, D.; Ficuciello, F.; Harada, K.; Kober, J.; Li, X.; et al. Challenges and outlook in robotic manipulation of deformable objects. *IEEE Robot. Autom. Mag.* **2022**, *29*, 67–77.
74. Khan, A.M.; Kim, Y.; Shin, B.; Moghadam, M.H.; Mansour, N.A. Mansour Modeling and control analysis of an arc-shaped SMA actuator using PID, sliding and integral sliding mode controllers. *Sens. Actuators A Phys.* **2022**, *340*, 113523. [CrossRef]
75. Li, Y.; Ni, L.; Wang, G.; Aphale, S.S.; Zhang, L. Q-Learning-Based Dumbo Octopus Algorithm for Parameter Tuning of Fractional-Order PID Controller for AVR Systems. *Mathematics* **2024**, *12*, 3098. [CrossRef]
76. Ye, T.; Wang, Y.; Xu, S.; Wang, Y.; Li, J. Modeling and motion control of an octopus-like flexible manipulator actuated by shape memory alloy wires. *J. Intell. Mater. Syst. Struct.* **2021**, *33*, 3–16. [CrossRef]
77. Mather, J.A. Ethics and invertebrates: The problem is us. *Animals* **2023**, *13*, 2827. [CrossRef]
78. Hoffmann, C.H. Intelligence in Light of Perspectivalism: Lessons from Octopus Intelligence and Artificial Intelligence. *J. Hum. Earth Future* **2022**, *3*, 288–298.
79. Carls-Diamante, S. Where is it like to be an octopus? *Front. Syst. Neurosci.* **2022**, *16*, 840022. [CrossRef]
80. Sivitilli, D.M. Reverse Engineering the Octopus Arm. Ph.D. Thesis, University of Washington, Washington, DC, USA, 2023.
81. Kuuspalu, A.; Cody, S.; Hale, M.E. Multiple nerve cords connect the arms of octopuses, providing alternative paths for inter-arm signaling. *Curr. Biol.* **2022**, *32*, 5415–5421.e3. [CrossRef]

82. Íñiguez, A. The Octopus as a Model for Artificial Intelligence-A Multi-Agent Robotic Case Study. In Proceedings of the 9th International Conference on Agents and Artificial Intelligence: ICAART, Porto, Portugal, 24–26 February 2017; pp. 439–444. [CrossRef]
83. Sivitilli, D.M.; Smith, J.R.; Gire, D.H. Lessons for robotics from the control architecture of the octopus. *Front. Robot. AI* **2022**, *9*, 862391.
84. Wang, T.; Halder, U.; Gribkova, E.; Gillette, R.; Gazzola, M.; Mehta, P.G. Neural Models and Algorithms for Sensorimotor Control of an Octopus Arm. *arXiv* **2024**, arXiv:2402.01074.
85. Dai, Y.; Zhang, S.; Cheng, W.; Li, P. Neural Network-Based Shape Analysis and Control of Continuum Objects. *Biomimetics* **2024**, *9*, 772. [CrossRef]
86. Aras, S.; Gedikli, E.; Kahraman, H.T. A novel stochastic fractal search algorithm with fitness-distance balance for global numerical optimization. *Swarm Evol. Comput.* **2021**, *61*, 100821. [CrossRef]
87. Bajaj, A.; Sangwan, O.P.; Abraham, A. Improved novel bat algorithm for test case prioritization and minimization. *Soft Comput.* **2022**, *26*, 12393–12419.
88. Usman, S.; Mehmood, R.; Katib, I.; Albeshri, A. Data Locality in High Performance Computing, Big Data, and Converged Systems: An Analysis of the Cutting Edge and a Future System Architecture. *Electronics* **2023**, *12*, 53. [CrossRef]
89. Yin, F.; Shi, F. A comparative survey of big data computing and HPC: From a parallel programming model to a cluster architecture. *Int. J. Parallel Program.* **2022**, *50*, 27–64.
90. Andronie, M.; Lăzăroiu, G.; Karabolevski, O.L.; Ștefănescu, R.; Hurloiu, I.; Dijmărescu, A.; Dijmărescu, I. Remote Big Data Management Tools, Sensing and Computing Technologies, and Visual Perception and Environment Mapping Algorithms in the Internet of Robotic Things. *Electronics* **2023**, *12*, 22. [CrossRef]
91. Rao, A.S.; Radanovic, M.; Liu, Y.; Hu, S.; Fang, Y.; Khoshelham, K.; Palaniswami, M.; Ngo, T. Real-time monitoring of construction sites: Sensors, methods, and applications. *Autom. Constr.* **2022**, *136*, 104099. [CrossRef]
92. Li, W.; Zhou, R.; Ouyang, Y.; Guan, Q.; Shen, Y.; Saiz, E.; Li, M.; Hou, X. Harnessing Biomimicry for Controlled Adhesion on Material Surfaces. *Small* **2024**, *20*, e2401859. [CrossRef]
93. Ullah, H.; Wahab, M.A.; Will, G.; Karim, M.R.; Pan, T.; Gao, M.; Lai, D.; Lin, Y.; Miraz, M.H. Recent Advances in Stretchable and Wearable Capacitive Electrophysiological Sensors for Long-Term Health Monitoring. *Biosensors* **2022**, *12*, 630. [CrossRef]
94. Tong, D.; Zhao, Y.; Wu, Z.; Chen, Y.; Xu, X.; Chen, Q.; Fan, X.; Yang, Z. Octopus-Inspired Soft Robot for Slow Drug Release. *Biomimetics* **2024**, *9*, 340. [CrossRef]
95. *ISO 21448:2022*; Road Vehicles—Safety of the Intended Functionality, International Standard. International Organization for Standardization: Geneva, Switzerland, 2022.
96. Sabelhaus, A.P.; Patterson, Z.J.; Wertz, A.T.; Majidi, C. Safe Supervisory Control of Soft Robot Actuators. *Soft Robot.* **2024**, *11*, 561–572.
97. Fattepur, G.; Patil, A.Y.; Kumar, P.; Kumar, A.; Hegde, C.; Siddhalingeshwar, I.G.; Kumar, R.; Khan, T.M.Y. Bio-inspired designs: Leveraging biological brilliance in mechanical engineering—An overview. *3 Biotech* **2024**, *14*, 312. [CrossRef]
98. Laschi, C.; Mazzolai, B.; Mattoli, V.; Cianchetti, M.; Dario, P. Design and Development of a Soft Actuator for a Robot Inspired by the Octopus Arm. In *Experimental Robotics*; Springer Tracts in Advanced Robotics; Khatib, O., Kumar, V., Pappas, G.J., Eds.; Springer: Berlin/Heidelberg, Germany, 2009; Volume 54. [CrossRef]
99. Gong, H.; Li, Z.; Meng, F.; Tan, B.; Hou, S. Octopus Predation-Inspired Underwater Robot Capable of Adsorption through Opening and Closing Claws. *Appl. Sci.* **2024**, *14*, 2250. [CrossRef]
100. Chin, K.; Hellebrekers, T.; Majidi, C. Machine Learning for Soft Robotic Sensing and Control. *Adv. Intell. Syst.* **2020**, *2*, 1900171. [CrossRef]
101. Liu, H.; Teng, X.; Qiao, Z.; Yang, W.; Zou, B. Magnetically Driven Quadruped Soft Robot with Multimodal Motion for Targeted Drug Delivery. *Biomimetics* **2024**, *9*, 559. [CrossRef] [PubMed]
102. Mo, Y.; Su, W.; Hong, Z.; Li, Y.; Zhong, Y. Finite-Time Line-of-Sight Guidance-Based Path-Following Control for a Wire-Driven Robot Fish. *Biomimetics* **2024**, *9*, 556. [CrossRef]
103. Wu, M.; Zheng, X.; Liu, R.; Hou, N.; Afridi, W.H.; Afridi, R.H.; Guo, X.; Wu, J.; Wang, C.; Xie, G. Glowing Sucker Octopus (*Stauroteuthis syrtensis*)-Inspired Soft Robotic Gripper for Underwater Self-Adaptive Grasping and Sensing. *Adv. Sci.* **2022**, *9*, 2104382. [CrossRef]
104. Ponte, G.; Chiandetti, C.; Edelman, D.B.; Imperadore, P.; Pieroni, E.M.; Fiorito, G. Cephalopod Behavior: From Neural Plasticity to Consciousness. *Front. Syst. Neurosci.* **2022**, *15*, 787139. [CrossRef]
105. Kudithipudi, D.; Aguilar-Simon, M.; Babb, J.; Bazhenov, M.; Blackiston, D.; Bongard, J.; Brna, A.P.; Chakravarthi Raja, S.; Cheney, N.; Clune, J.; et al. Biological underpinnings for lifelong learning machines. *Nat. Mach. Intell.* **2022**, *4*, 196–210. [CrossRef]
106. Wu, H.; Yang, G.; Zhu, K.; Liu, S.; Guo, W.; Jiang, Z.; Li, Z. Materials, Devices, and Systems of On-Skin Electrodes for Electrophysiological Monitoring and Human–Machine Interfaces. *Adv. Sci.* **2021**, *8*, 2001938. [CrossRef]

107. Jin, J.; Zhang, C.; Zhao, J.; Yu, M.; Lei, M.; Jin, C.; Yin, R.; Zhao, W. An adaptive bionic sensor: Enhancing ankle joint tracking with high sensitivity and superior cushioning performance. *Chem. Eng. J.* **2024**, *500*, 157332. [CrossRef]
108. Shamilyan, O.; Kabin, I.; Dyka, Z.; Kuba, M.; Langendoerfer, P. Octopuses: Biological facts and technical solutions. In Proceedings of the 2021 10th Mediterranean Conference on Embedded Computing (MECO), Budva, Montenegro, 7–11 June 2021; IEEE: Piscataway, NJ, USA, 2021; pp. 1–7. [CrossRef]

MDPI AG
Grosspeteranlage 5
4052 Basel
Switzerland
Tel.: +41 61 683 77 34

Biomimetics Editorial Office
E-mail: biomimetics@mdpi.com
www.mdpi.com/journal/biomimetics

www.ingramcontent.com/pod-product-compliance
Lightning Source LLC
Chambersburg PA
CBHW041947260926
51850CB00018B/1875